W0235015

DETECTION
AND IDENTIFICATION
OF ORGANIC
COMPOUNDS

DETECTION
AND IDENTIFICATION
OF ORGANIC
COMPOUNDS

by

MIROSLAV VEČEŘA

Professor of Organic Chemistry, Institute of Chemical Technology, Pardubice

and

JIŘÍ GASPARIČ

*Associate Professor of Analytical Chemistry, Institute of Chemical Technology
and Senior Research Fellow, Institute of Organic Syntheses, Pardubice-Rybitví*

PLENUM PRESS, NEW YORK • LONDON

SNTL — PUBLISHERS OF TECHNICAL LITERATURE, PRAGUE

Distributed throughout the world with the exception
of the Socialist countries by

PLENUM PRESS
Plenum Publishing Corporation
227 West 17th Street
New York, New York 10011
Distributed in the United Kingdom by
Plenum Publishing Company Limited
Davis House
8 Scrubs Lane, Harlesden,
London NW106SE, England

ISBN-13: 978-1-4684-1835-4   e-ISBN-13: 978-1-4684-1833-0
DOI: 10.1007/978-1-4684-1833-0

Library of Congress Catalog Card Number 72-110605

© 1971 Miroslav Večeřa and Jiří Gasparič
Softcover reprint of the hardcover 1st edition 1971
Translated by Želimír Procházka
English edition first published in 1971 simultaneously by
Plenum Publishing Corporation and
SNTL — Publishers of Technical Literature, Prague

CONTENTS

6

Nitrous Acid — 21. Hydroxamate Test — 22. Picric Acid — 23.
Ninhydrin — 24. Liebermann Reaction I — 25. Liebermann Reaction II — 26. Formaldehyde and Sulfuric Acid — 27. Ceric
Ammonium Nitrate — 28. Sodium Iodide — 29. Fluorescein
Chloride — 30. 3,5-Dinitrobenzoic Acid Anhydride + SnCl$_2$ —
31. Iodoform Reaction — 32. Stannous Chloride and Hydrochloric Acid — 33. Iodine and Sulfanilic Acid — 34. Ferric
Hydroxide — 35. Nitrole Reaction — References

PREFACE TO THE AMERICAN EDITION

The American edition of our monograph is not a mere translation of the Czech edition, which appeared some five years ago. We have had to respect the fact that even such a short period has sufficed for progress in this field, and that the field of application of methods of organic analysis has widened. We have therefore revised a number of chapters in Part 1, the general part of the monograph—mainly those devoted to chromatographic methods, which have been extended and complemented by methods of thin-layer chromatography and electrophoresis. The chapters on the theory of color reactions and on analytical literature have also been extended; the chapter on spectral methods has been extended by including the use of proton magnetic resonance in organic analysis, and the list of references has been enlarged by adding books of importance for organic analysis.

In Part 2, the part dealing specifically with various elements and chemical groups, we have extended the chapters on solubility and on acids and bases. The methods for the detection and identification of given classes of compounds have also been supplemented by references to recent papers.

The Czech edition was adapted to the working conditions of Czech chemists, mainly with respect to apparatus and equipment. In the American edition we have not included this specific information, since the English-speaking worker has at his disposal a number of commercial publications on new apparatus. Hopefully, the present monograph will be able to serve as a practical handbook of modern chemical analytical methods in organic laboratories in the USA.

PREFACE TO THE CZECH EDITION

The rapid world-wide evolution of industrial organic chemistry requires a continually more intense and deeper study of organic reactions. An essential part of this research is organic analysis, and ever greater demands are being made on it. The increasing number of papers in this field proves that this is so.

A series of specialized monographs on qualitative organic analysis has already been published. The majority of these monographs were written as textbooks for universities and colleges and their extent and structure are given by the requirements and aims of the given school. In addition to this, special monographs appear which treat exhaustively and with much detail specific areas of organic analysis.

In preparing our monograph, we were guided by the knowledge that in practice chemists in organic or biochemical laboratories, both in research institutes and in industrial enterprises, are often faced with problems of the detection and identification of organic compounds which they are not sufficiently trained or experienced to solve. We therefore decided to write a handbook of moderate size and to incorporate in it the experience we have gained in this field, acquired by the performance of practical analyses as well as by the elaboration of original identification methods published in more than 50 papers.

We have endeavored, in contrast to university textbooks, to work up the material primarily from the point of view of practical application, i.e., to show the possibilities of organic analysis and to help laboratory workers in the choice of suitable methods and procedures. Organic analysis has no systematic method analogous to the hydrogen sulfide method of inorganic analysis, and it has, therefore, to combine several methods; a satisfactory result depends on the ability of the analyst to utilize these procedures and to interpret their results correctly.

In the general part of the book the extent and use of various methods are presented; references are made to special publications for methodical details. The main emphasis is laid on the importance of these methods for the identification of organic compounds and the results obtained by their use, although these methods are usually treated independently and not in connection with classical organic analysis (chromatographic and spectral methods).

In the special part of the book procedures are given for the practical detection and identification of given classes of organic compounds by means of color reactions, the preparation of derivatives, paper chromatography, and gas chromatography, often with reference to other methods as well. We have kept in mind the fact that modern practice requires to an ever-increasing extent work with micro quantities of substances, and we therefore also give instructions for work on a microanalytical scale, using simple procedures which should be possible to carry out in every laboratory. All procedures for the preparation of derivatives have been checked in our laboratories, which has enabled us to give actual yields and melting points found for given single derivatives. We did not wish to rely merely on general procedures, because we know from our own experience that they must often be adapted for given compounds. This is why in this part we give procedures for the identification of specific substances, the yields of which should facilitate the choice of a suitable method for the compounds studied. We have also used the majority of color reactions in our own laboratories, which has enabled us to incorporate our own observations. Moreover, we give certain special color reactions or references to such reactions in order to give the chemist a choice and to save time on literary search. We have devoted great attention to the choice of methods of paper chromatography, because the larger part of our experience is with this method and we cannot imagine the identification of organic compounds without it.

In order to prevent a too schematic approach, we give in typical examples explanations and analyses based on the present state of theoretical chemistry. These examples also demonstrate that complicated problems of organic chemistry cannot be solved by simply following brief directions, but demand deeper thought.

Exhaustive tables of physical constants — for example, melting points of derivatives — should be an integral supplement to a handbook on organic analysis. However, the size of this book does not allow us to insert these tables, and lists of melting points of derivatives are only occasionally given as examples. We plan to publish these tables in a separate volume.

Our aim has been for this book to aid the largest possible number of research chemists in solving problems of organic analyses, and also to be of use to laboratory workers with technical educations in their routine work.

Miroslav Večeřa   Jiří Gasparič

PART 1

CHAPTER I

AIMS AND METHODS
OF ORGANIC QUALITATIVE ANALYSIS

The basic aims of organic qualitative analysis are the detection and the identification of organic compounds. As other terms for both ideas sometimes appear in the literature, we shall first give their definitions.

By detection we understand a qualitative test based usually on a color or precipitation reaction, on the liberation of a gas, or on behavior during heating, dissolution, etc. This test is usually characteristic of a whole class of compounds; in certain cases it can be specific for individual compounds.

Identification of an organic substance can be carried out by a number of methods used in organic analysis. They differ from detection methods by the fact that measurable data are obtained which can be compared with literature data (concerning the given substance) or with the results of one's own measurements of an authentic sample. Identification can be made, for example, by the comparison of physical constants, by preparing a derivative, or by chromatographic and spectral methods.

The difference between detection and identification will best be explained by the following two examples:

1. If a solution of the tested substance forms a precipitate on the addition of a 2,4-dinitrophenylhydrazine solution in 2 N HCl, this represents a detection of the presence of a carbonyl compound. If the precipitate is isolated and purified by crystallization and its melting point determined, identification has been made, because on comparison with the published data or on measuring the mixture melting point with an authentic sample we can determine the identity of the original substance.

2. The detection of the presence of phenols in a sample can be carried out by coupling them with a diazonium salt in alkaline medium, producing a characteristically colored azo dye. If a diazonium salt without a sulfo or carboxyl group in the molecule is used, the dye formed can be extracted from the reaction mixture with an organic solvent and then identified by paper or thin-layer chromatography.

The identity of two substances can be confirmed even if their structure is not known. The structure determination of an organic substance is a problem which goes beyond organic analysis, and which has to be solved in combination with organic and physical chemistry. In order to complete the elucidation of the basic ideas and aims of organic analysis, the definition of "determination" should also be given. Under the idea of determination we understand exclusively a quantitative analysis or measurement — for example, the determination of a certain substance, in its crude or pure product or in a mixture, determination of small amounts of impurities, determination of the amount of a functional group in a substance, melting-point determination, molecular-weight determination, etc. As far as methods are concerned, we speak about gravimetric, titrimetric, colorimetric, polarographic, or similar determination. In the chemical literature we often meet such ideas as, for example, qualitative determination, quantitative determination, or qualitative detection. From the point of view of the definitions given above these ideas are unnecessary and sometimes even illogical.

The nature of problems to be solved by an analyst is often such that the basic ideas mentioned are sometimes superposed or appear in various modifications. What is required from organic analysis may vary widely, and if we take into account the enormous variety of organic compounds in general, it will become evident that a true picture of the whole range of problems can only be gained through actual practice.

Let us try to outline at least a few problems with which an analyst may be faced in practice in research institutes or industrial laboratories in the organic chemistry field:

a) A test is to be carried out or the identity of a known compound confirmed. Examples: initial control of the identity of raw materials, determination or confirmation of the identity of a known main or by-product of a known reaction.

b) A substance unknown to the analyst, but described in the literature, is to be identified. Examples: identification of unlabeled raw material in the store room, identification of an isolated unknown by-product from a reaction or from production, identification of some industrial product of a competitive firm, the composition of which is not specified, etc.

c) The structure of an unknown substance, as yet undescribed in the literature, has to be determined. Examples: structure determination of an individual compound isolated from natural material, or of an industrial product the composition of which is not known and is not given even in the patents literature.

d) The presence of one or several known compounds has to be proved in a known mixture. Examples: during the initial control it should be confirmed that all components of the product were really employed; the proof should be carried out that all known substances formed in a reaction are really present in the product; a special example consists in the proof – even negative – that an expected or known impurity is present in excess in a known substance.

e) A qualitative analysis of a mixture of substances, unknown to the analyst, but described in the literature, is to be carried out. Examples: when following the course of a new reaction a mixture of compounds is formed which has to be identified; individual components of an industrial product of a competitive company have to be determined, the product being an unknown mixture of substances described in the literature; the necessity might arise of identifying an unknown impurity in an organic substance which causes the substance to be unsuitable for further technological processing; in a common organic-technological procedure unexpectedly low yields are obtained, or a fire or explosion or undesirable course of reaction takes place, and it is necessary to determine the cause by analyzing the residues, etc.

f) An unknown mixture of as yet undescribed compounds needs to be analyzed. Examples: determination of compounds in natural material, isolation and structure determination of products of unknown reactions, or analysis of industrial products of unknown composition of competitive firms.

In addition to the examples of problems quoted above which may face analysts in the field of organic chemistry and technology, organic analysis offers valuable service in other special fields – for example, in biochemistry, forensic chemistry, air-pollution studies, in food technology, and in mineral oil chemistry. In these fields the analyst adapts the method to suit the analyzed medium – for example, the biochemical material or food.

In recent years the demands on the control and recognition of organic reactions and production processes have steadily increased, as have those concerning the quality of products (trace impurities).

The call for analytical methods and their results increases incessantly. Practice requires closer collaboration between organic chemists and technologists with the analysts on the one hand, and a continuous perfection and development of analytical methods on the other. It very often happens that for certain problems it is very difficult to decide where the work of the synthesist or the technologist ends and where the task of the analyst begins. In other instances it is essential that the analyst becomes well acquainted, from the chemical, technological, and economic points of view, with the problem or the field for which the analyses are carried out. Only then can he understand the analytical results obtained.

To fulfill all these requirements, the analyst has a series of methods at his disposal in organic chemistry, and success in his work usually depends on the correctness of their choice and combination. In addition to the classical chemical methods of organic qualitative analysis, i.e., color reactions and the conversion of organic compounds to such derivatives which can be characterized on the basis of their melting points and other physical constants, other methods that can be applied include those of quantitative organic analysis (determination of elements and functional groups), further basic physical and physicochemical methods, spectral methods (such as emission spectrophotometry, absorption spectrophotometry in the ultraviolet, visible, and infrared regions, and Raman spectra), X-ray diffraction method, chromatographic methods (paper chromatography or column chromatography, partition or adsorption chromatography, thin-layer chromatography, gas-liquid chromatography), electromigration methods (electrophoresis), mass spectrometry and nuclear magnetic resonance. Each of these methods is per se an extensive and independent area of knowledge, i.e., branch of science, and it would be impossible to require the analyst to know them all. It is, however, almost essential for the analyst to master at least one or several of the methods mentioned in addition to those of classical organic analysis, and have a good idea of the possibilities of use of the remaining ones; only then will he be able, if necessary, to choose the proper method for the solution of his problems.

Our experience shows that when solving difficult problems the organic analyst cannot do without the help of other experts. Special fields usually require apparatus and equipment, the price, accessibility, and maintenance of which are beyond the possibilities of an ordinary chemical laboratory and would require a perfectly equipped physicochemical department.

The choice of the method suitable for the solution of a certain problem is significantly influenced by the scale on which the analyst can or must work in a given case. A limited amount of sample, a very low concentration of the substance to be detected, or the presence of impurities in the sample exclude the application of common macro-methods, but compels the analyst to work on a semimicro, micro, or ultramicro scale. The limits of these individual scales cannot be determined unambiguously and they vary from method to method; therefore, we shall indicate them in each particular method. We consider it necessary to point out that throughout the world the development of analytical chemistry is characterized by a trend toward microscale work and even lower, because this usually leads to an increase in the speed of analytical work, even though training in work on such a scale seems rather tiresome. Nowadays this transition is facilitated by the perfection of analytical, especially physical, apparatus, and also the automation or mechanization of numerous methods.

The question of equipment for an analytical laboratory bears upon the preceding facts. It is absolutely valueless to give a list of items necessary for an analytical laboratory, because this would depend on the type of the work performed and on the problems to be solved, on the economic possibilities of the laboratory, and also on the personal preferences of the analyst. We know from our own experience that, for example, ascending and descending development in paper chromatography give very similar results; nonetheless, when solving the same problem some analytical workers give preference to one method over the other, either because they have only one particular kind of equipment at their disposal, or simply because they first learned one particular modification which they then have come personally to prefer.

Numerous monographs, advertising literature, and catalogs offer the analyst various possible devices and equipment for basic analytical operations. According to our experience, it is important that the analyst follow these possibilities systematically and keep them in mind; though we recommend carrying out the basic operations as simply as possible: heating in a sealed glass tube or ampoule is often as serviceable as refluxing in a complicated micro apparatus. To separate two substances by paper chromatography, no expensive commercial apparatus is necessary: the same result can often be obtained on a strip of filter paper in a test tube. These views cannot be generalized, however; we wish only to show that the analyst's personal skill and his ability to adjust conditions often play a predominant role.

In conclusion, we can state that the aims of organic analysts are numerous, and the solutions of the problems they face are made possible by a whole set of analytical procedures. Systematic study of analytical methods and continuous attention to the world-wide evolution in the field represent a very effective armament for the analyst in organic chemistry, who does not have a well-elaborated analytical procedure at his disposal (as, for example, the hydrogen sulfide method of inorganic analysis) and therefore has to deal operatively and intelligently with individual cases as they arise.

CHAPTER II

APPROACH TO ANALYSIS

In the previous chapter we mentioned that the analyst in organic chemistry does not have a systematic procedure at his disposal for the solution of his problems. There are cases, of course, when he is able to elaborate a basic systematic procedure for a series of analyses, but in the case of one anomalous sample such a procedure might already become useless.

In spite of this, it is necessary to ·outline certain directives for the solution of problems of qualitative organic analysis and to show the way these problems should be approached.

Generally, for all cases of organic analysis the following five principles apply:

1. The analyst must be told exactly the aim and extent of the task presented to him. This is very important, because the analyst has to choose the method according to the requirements of the analytical result. We have often been witnesses to misunderstandings in this respect and we consider it necessary to elucidate this question. The analyst may be asked to do a complicated job, but after discussion between the analyst and the customer it might turn out that the problem could have been solved in a much simpler way. In other cases just the opposite may occur. It makes a great difference whether a mixture is to be analyzed for the main component or for a minor component or even trace amounts; the kind of substances in whose presence it is to be determined also matters.

2. The analyst must always be maximally informed on the problem to be solved and on the sample to be analyzed. This facilitates his work and ensures a speedier solution of the problem. In practice, the approach to this problem is different than in a teaching situation, where for educational reasons the student has to reach his results by a strictly experimental method. To the analyst, the mere knowledge of the origin of the sample may be of great value, as may a knowledge of its use, the conditions under which it was obtained, etc., although these details may seem unimportant to the customer.

3. For analytical work, proper systematic work is necessary. It is of

no use to carry out a great number of haphazardly chosen tests and thus consume an irreplaceable sample. An experienced analyst works with deliberation, the tests which he wants to carry out with the sample are critically chosen, and the results obtained are carefully interpreted so that he may obtain a definite result by the evaluation of a minimum number of experiments. This caution is sometimes dictated by the small amount of the sample, limiting the number of experiments. However, it is in the analyst's own interest to work in the same manner even when a sufficient amount of the sample is available; he should consider each sample as a preparation or training for his further activity. Practice shows that when in apparently simple cases the preliminary experiments are carried out unsystematically, the analyst soon becomes disoriented and has to return to systematic experiments, which means a loss of time and material.

4. A correct, careful, and purposeful recording of results in laboratory notebooks is part of systematic work. In all instances it is preferable to record the results of experiments in great detail than to rely on the memory and oral reports. Imperfect records may lead to the necessity of repeating certain tests and hence to time losses and an unnecessary consumption of the sample.

5. In all stages of the analysis the analyst must be in close familiarity with the literature, not only analytical, but, if necessary, also theoretical and technological, in order to be able to evaluate and interpret the obtained results correctly.

Let us now show the proper approach to the identification of unknown substances or their mixtures.

After having obtained a maximum of information along the lines shown in the preceding five points, the analyst may proceed to the actual work with the sample. In the first place the following preliminary tests should be carried out, if necessary:

1. Attention is given to the external characters of the sample.

a) *Consistency of the sample.* In addition to the main observation, whether the sample is a liquid, solid, or gas, one should also observe how viscous or syrupy the sample is, the form of crystals, etc. In the case of liquid samples a precipitate or crystals sticking to the sides of the container may be an indication that the sample is a mixture, or that an impurity or a decomposition product, etc., may be present in it. In the case of emulsions or semisolid and semiliquid samples the mere appearance may be a guide. Attention should also be paid to the homogeneity of the sample.

b) *Color of the sample.* As the majority of organic substances are colorless or white, the color of the sample always has a definite meaning for the analyst. The hue and the intensity of the color may be characteristic of certain groups of substances or they may represent an evidence of impurities or of the fact that the sample has been artificially dyed. For example, a yellowish color may indicate that aromatic nitro compounds, azoxy compounds, or certain nitroso compounds may be present in the sample; a yellow color is characteristic of α-diketones, quinones, nitrophenols, nitrated aromatic amines, and osazones; an orange to orange-red color is indicative of azo compounds, nitroaminophenols, hydroxyanthraquinones, isatin, nitrophenylhydrazines, and corresponding hydrazones; a red, violet, or brown color may belong to azo compounds and aminoderivatives of anthraquinone. Nitroso compounds are green. Nitrobenzene and homologs are pale green if dissolved or melted, nitrosodialkylanilines are intensely green even in the solid state. A blue color is typical of hydroxyaminoanthraquinone derivatives and also of certain azo dyes, terpenes, and 1-cyano-1-nitrosocyclohexane. Quinhydrones are also dark-colored. A black color of the sample may indicate the presence of charcoal, soot, or black dyes, which, of course, very often represent a mixture of dyes.

For the relation between the structure and the color of organic substances see p. 79. Pale gray, violet, or brown, or strongly brown colors, caused by air oxidation, are characteristic of aromatic amines, polyphenols, and hydrazines, and need not mean that the substance contains other impurities than the oxidation products.

c) *Odor of the sample.* In view of the fact that a great number of organic substances possess a characteristic odor, a test by olfaction has become the first operation carried out by the analyst with the sample. However, in doing this a certain caution is indicated, because vapors of a number of organic substances may be dangerous. An odor test may disclose the presence of certain substances occurring in such minute amounts that they cannot be detected chemically. They can be either trace impurities or, on the contrary, substances added purposely to the product to conceal its own odor. In certain fields, for example, in testing perfumes, the analysis by olfaction is well established and it is carried out by well-trained experts, who use filter-paper strips dipped into the tested mixture and follow the gradual evaporation of single components.

Olfactory impressions are of course very individual and depend on the experience of the worker. An odor can be characteristic of a whole class of substances in one case, and in other cases only of a certain limited group of substances (the smokey odor of medium alcohols). Sometimes two completely different compounds possess an almost identical odor (for example, the camphorlike odor of hexachloroethane), and very often the odor changes even in a homologous series (lower, medium, and higher aliphatic alcohols or acids).

d) *Taste of the sample.* Although many substances have quite typical tastes and in spite of the fact that many "experts" are used to "analyzing" in such a way, we do not recommend this method, because a great number of organic substances are toxic and the analyst has much more suitable methods at his disposal. However, we do not wish to deny the importance of tests by taste, especially in certain fields, as, for example in food technology.

e) *Wrapping of the sample.* Certain conclusions on the properties of the sample can be made on the basis of the way it is packed; for example, attention may be drawn to the hygroscopic character of the sample or to its photosensitivity, volatility, or toxicity.

2. The test by heating or burning has several meanings. In the case of completely unknown samples, the use of which is not known, it is absolutely necessary to carry out a test by heating and burning on a platinum sheet (spatula). First it must be determined whether the sample is explosive or whether it decomposes at a dangerous rate. Such properties can be displayed by certain apparently innocuous materials (for example, insidious means of attack), and it is therefore important to eliminate in advance any possible danger to the worker. The ignition test can indicate certain properties of the sample on the basis of the way it burns, which is also characteristic of different types of substances. The residue on the platinum sheet (spatula) after ignition may be indicative of the presence of inorganic components. If a drop of water and a droplet of indicator are added to the residue, an alkaline reaction proves the presence of alkali elements. If the sample neither carbonizes nor burns, it might not be organic in character, and a test on the presence of carbon in it is indicated.

3. Approximately at this stage of the analysis it is necessary to make sure that the sample is individual, i.e., to decide whether the sample is a single compound or a mixture. In the case of liquid samples the analyst should determine its boiling point (see p. 40) and then, by evaporating it on a small dish over a water or oil bath, whether or not the sample leaves a solid dry residue. After this the purity of the sample can be checked by gas chromatography, best at least on two suitably chosen stationary phases (see p. 72). If this method cannot be used, and if the amount of the sample permits it, the character of the distillation curve should be determined. We should not, however, be misled by a constant boiling point to consider a constant-boiling azeotropic mixture as a pure substance. The mixture can be recognized by distilling at two different pressures and by measuring the refractive index of the distillate fractions. Solid substances should be crystallized and their melting point determined both before crystallization and after, or after chromatographic purification. If the

melting point does not change after two subsequent crystallizations, the substance can be considered as pure. If the melting point changes on crystallization, or if the substance changes visibly when put in contact with a solvent (for example, a colored substance is extracted and the residue remains colorless), the substance is evidently a mixture. For a general discussion on the purification of substances see p. 25. When the analyst has thus ensured that the sample is a pure chemical substance, and when he has purified it, it may be considered ready for analysis. If such is the case, the following experiments should be carried out:

a) Determination of basic physical constants (p. 36).

b) Solubility test (p. 105).

c) Detection of specific elements (C, H, N, S, and halogens, and if necessary, also P, B, Si, F) and analysis by emission spectrometry (p. 95).

d) Elemental analysis. For the identification of common substances elemental analysis may seem a luxury, but in the case of more complicated substances it is necessary.

e) Group and classification tests (p. 104).

f) Preparation of derivatives (p. 53).

g) Determination of functional groups either in the original substances or in its derivatives.

During all these tests enumerated the literature should be consulted as much as possible, because it is in these phases of the work we determine if the substance being analyzed is identical with a known substance, or whether we are analyzing an as yet undescribed compound.

CHAPTER III

PREPARATION OF CHEMICAL
SUBSTANCES FOR ANALYSIS

Some modern methods of organic analysis permit the analysis of substances which are not quite pure, or even true mixtures. This is sometimes necessary because of the nature of the sample or its quantity. Certain methods, for example, chromatography, permit simultaneous separation and identification of components. Nevertheless, the first aim of the analyst when identifying a substance remains the preparation of individual substances, using basic separation and purification operations; only then he can proceed with carrying out the analysis proper. The separation and purification methods are met by the analyst in subsequent preparation of derivatives. Sometimes the separation and purification of the sample are more difficult and time-consuming than the analysis itself. Therefore, the mastering and the knowledge of basic separation and purification methods are prerequisites without which the analyst's work in an organic laboratory cannot be conceived. Various individual operations are well described in the literature in connection with the techniques of organic chemistry, because, especially in the micropreparative work recommended so much in this field, organic chemistry and analysis use the same techniques for their work. It is beyond the scope of this monograph to describe the details of given operations, as these are accessible in various basic works (1, 2, 3).

When separating and purifying organic substances use is generally made of the fact that on transition from one phase to the other single fractions become enriched in one, purified substance to the detriment of other components of the mixture. For example, distillation is based on the transition of a liquid phase to a gaseous one and vice versa; sublimation, on the transition from solid to gaseous phase and vice versa; and crystallization, on the transition of a solid phase to a liquid one. The analyst is faced by such phenomena according to the nature of his samples and working conditions. For example, someone studying terpenes will master and use a different technique than someone working in the field of dye chemistry. It is therefore important for the analyst to get gradually acquainted with all methods, to know about them and to know how to use them.

References

1. Keil, B., Herout, V., and Protiva, M.; Laboratoriumstechnik der organischen Chemie; Akademie Verlag, Berlin, 1961.
2. Cheronis, N. D.: Micro and Semimicromethods. Vol. 6: Weissberger (Editor), Techniques of Organic Chemistry. Interscience, New York 1954.
3. Hecht, F., and Zacherl, M. K. (Editors): Handbuch der mikrochemischen Methoden. Springer, Vienna 1954.

1. Crystallization

Crystallization is one of the most efficient purification operations for solid organic substances. The analyst uses it both for the purification of the sample before the analysis, and for the preparation of derivatives necessary for the final determination of basic constants.

Most important for crystallization are a correct choice of solvent, its volume, and the size of the vessel used. A suitable solvent must be found empirically with a few milligrams of substance, in micro test tubes (approx. dimensions: $0.4 - 0.6 \times 5 - 6$ cm), and on addition of a few drops of the solvent. This is added either with Pasteur pipettes or with evenly-cut, narrow glass tubes (with ends treated in a flame) $15 - 20$ cm long and $2 - 3$ mm in diameter, which must always be available, dry and clean, in sufficient quantity in the analyst's laboratory table. We recommend carrying out this test even when the literature indicates a suitable solvent for crystallization in the procedure for the preparation of the derivative. It is important to ascertain the amount of solvent necessary for crystallization. It would be very difficult to enumerate all utilizable solvents. They are legion, and a correct choice depends on the nature of the crystallized substance and also on the nature of the impurities. The analyst looks for a solvent in which the crystallized substance is well soluble at elevated temperature and poorly soluble at low temperatures, while the impurities are well soluble even in the cold; the solvent should not react with the crystallized substance, the substance should not decompose at the boiling point, and the solvent must be completely removable from the purified substance on drying. Crystallization is also required to give very good yields, especially if the available amount of the substance is low. In certain cases, where analytical purification is also carried out by other methods, we prefer a lower yield of a truly pure substance over a higher yield of still impure substance.

When looking for a suitable solvent one should proceed according to basic experience, according to which a definitely nonpolar substance (for example, a hydrocarbon) is best soluble in nonpolar solvents (for example, hexane or benzene), while substances containing polar substituents are better soluble in polar solvents (for example, water). The first attempts are better carried out with volatile solvents, because if the substance is negligibly insoluble or too soluble, the solvent can be more easily evaporated before carrying out further tests with other solvents. If no single solvent is suitable, a mixture of two solvents should be tried — one in which the substance is easily soluble and the other in which the substance is poorly soluble or almost insoluble. A mixture of both solvents is used either directly (care should be taken to see that the mixture does not get enriched by one, higher-boiling component on prolonged boiling), or else the substance is dissolved in that component in which it is better soluble, and the second component (solvent) is added gradually to the warm solution until it becomes turbid. The turbidity is eliminated either by heating the mixture, or by the addition of a drop of the first solvent. After cooling, the substance crystallizes out. This method of crystallization must not be confused with the precipitation of a dissolved substance by the addition of another solvent in which it is insoluble. This method is used mainly for isolation purposes, and for purification only when the impurities remain soluble.

After finding or checking a suitable solvent the subsequent procedure will depend on the amount of the crystallized substance and the amount of solvent necessary for this operation. For amounts from 0.1 to 1.0 and more grams, 5 − 20 ml Erlenmayer or Phillips flasks are used; crystallization from organic solvents is never carried out from beakers, crystallization dishes, or round-bottomed flasks. We crystallize from these only if they already contained some evaporation or reaction residue, but when the substance is dissolved we filter the solution into an Erlenmayer flask for crystallization. For smaller amounts (milligrams) small test tubes are most suitable. It is convenient if they have small rods sealed to them (as holders) for better manipulation.

Crystallization from Erlenmayer flasks is usually carried out over an electric heater with perfectly protected spirals. When dealing with volatile and inflammable solvents it is sometimes advisable to heat the heater fully and to switch it off immediately before use, or to use electrically heated water baths. To the crystallized substance in the flask a smaller amount than necessary for a complete dissolution is added and the mixture is heated to boiling point. However, before starting to do this a capillary sealed on its upper end is inserted into the solution to ensure smooth boiling and to prevent overheating and bumping, and loss of the valuable sample. The dissolution is enhanced by trituration of larger particles with a glass rod, and the necessary amount of the solvent is gradually added. The dissolution is carefully controlled, because the substance itself may often be already dissolved and only impurities remain undissolved. On the addition of more solvent the substance might get "drowned," i.e., the solution too diluted. If too much substance remains undissolved, the hot solution is decanted and more solvent is added to the residue. If the substance does not dissolve further, this fraction is not added to the first one. If the substance dissolves slowly, it is better to boil it under reflux.

If the solution (after dissolution of the substance) is colored or if it contains impurities, it may be useful to use certain adsorbents, i.e., the solutions in polar solvents are briefly boiled with a small amount of charcoal, those in nonpolar solvents are shaken (without boiling) with alumina (the powder remaining after sieving alumina for chromatography may be used).

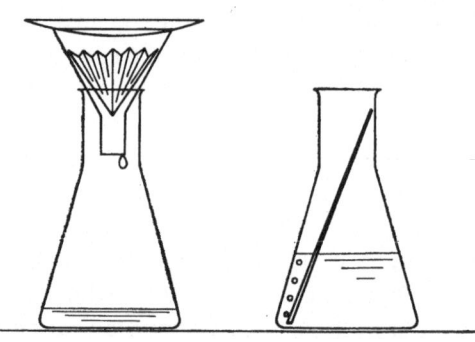

Adsorbents are added after complete dissolution of the sample and to partly cooled solution, in order to prevent the solution boiling over.

After this operation filtration follows. It is always carried out on the first and the last crystallization. If the amount of solution is larger than a few milliliters, a simple glass funnel may be used provided with a folded filter paper. The stems of the

Fig. 1. Crystallization on an electric heater.

funnels for crystallization should be as short as possible, to prevent crystallization or solidification of the solution in them. The filter and the funnel must be wetted with a few drops of the solvent, taking care that the drops reach the bottom of the flask. The flask is then placed on the burner (Fig. 1). The vapors of the solvent fill the space in the flask and reach the funnel, heating both, so that the filtered solution cannot solidify, i.e., crystallize out immediately on the filter and in the stem and the flask. If crystallization takes place in the neck of the funnel in spite of this, it usually suffices to bring the filtrate in the flask to boiling, and the climbing vapors which condense on the neck of the funnel dissolve the crystals in the neck and on the funnel.

When solutions in very volatile solvents are filtered through paper filters, losses may occur on the edges of the filter paper. Therefore, sintered-glass filters are useful for filtration in such cases.

After filtration the flask is allowed to stand undisturbed to cool. A slow crystallization is preferred, because larger crystals are more easily filtered and observed under the microscope during the melting-point determination. Sometimes, of course, the whole operation can be quite an ordeal, especially if the substance does not precipitate or if it precipitates in the form of an oil. If it will not precipitate, the crystallization may be induced by scratching the sides of the flask or the test tube with a glass rod or a metallic spatula; if volatile solvents are used for crystallization, it may help to take a drop of the solution on the spatula and to let it evaporate, and then to seed the bulk of the solution with the dry residue on the spatula for crystallization. On repeated crystallizations of poorly crystallizing substances it is convenient to keep a few crystals from preceding crystallization for the seeding of the next one. However, if the substance has been "drowned" in the solvent, the excess solvent must be evaporated by distillation.

The cause of precipitation of a substance in the form of an oil may be that it separated from the solution at a temperature above its melting point. If such is the case, a solvent has to be chosen whose boiling point is lower, or the crystallization should not be carried out from a saturated solution. In other instances a substance can separate out as an oil if it is not sufficiently pure. In many cases prolonged standing in a refrigerator or scratching with a glass rod or spatula may help the oil to solidify. The scratching is carried out best by taking a droplet of the oil out and scratching it in the neck of the flask. This often brings the substance to solidification, and the oil in the flask is then seeded with the solid crystals. In cases where the nature of impurities is known, it is advisable to pour the supernatant solvent off and to treat the remaining oil with a suitable solvent or reagent, to remove the impurities. For example, when benzoyl or acetyl derivatives are prepared it is convenient to separate the oily product from the free acid by mixing it with a solution of sodium hydrogen carbonate, upon which the oil usually solidifies. In other cases column chromatography is more suitable, possibly in combination with crystallization.

As soon as the crystallization itself is complete the separated substance has to be isolated. This is done by filtering it off under suction on a small Büchner funnel with a sintered-glass filter or on a filtering tube (Fig. 2). The condition for a successful isolation of the sample is a correct choice of the size of the filtering device. Losses may occur both when using a too large or too small a funnel, from which the product bubbles over. A hard suction is enhanced by compressing the substance on the funnel with a glass rod with a flat end, which can be easily prepared by melting the end in a flame and pressing it gently on an even asbestos or metal plate. The crystals are then washed with a small amount of a suitable solvent.

After suction and washing of the crystals, they must be dried. If they are not hygroscopic and if a sufficiently volatile solvent has been used, it is sufficient to suck the air through the crystals on the funnel; however, the funnel should be covered with a small Petri dish. To dry small amounts of substances in filtering tubes, it is very convenient and efficient to fit the tube with a rubber stopper (noneroding rubber) to a glass tube connected with a calcium chloride tube by a small piece of rubber tubing. In this case the substance is dried in a stream of air free of moisture (Fig. 2).

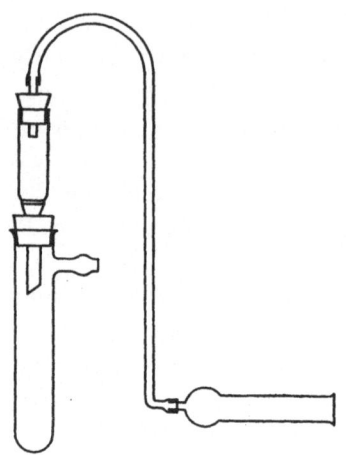

For micropreparative work, special devices are sometimes necessary. Particularly in this field (devices for crystallization and their manipulation), it is best if each analyst masters the technique which suits him best for the substances he is working with.

The effect of crystallization is controlled mainly by melting-point determinations; paper chromatography and thin-layer chromatography and functional-group determinations are also suitable. In all instances the substance should be crystallized until a constant melting point or a constant content of a functional group is attained, or until the substance is chromatographically pure.

Fig. 2. Scheme of a filtration set with a filtration tube and the connected calcium chloride tube, for drying the filtered crystals in a stream of dried air.

In conclusion, the analyst should have ready in his laboratory all devices and utensils mentioned here, in sufficient quantity, of various sizes, and absolutely clean, in order to be able to use them immediately. Filtration tubes of various sizes can be prepared by a glassblower, but he must be careful to seal the sintered-glass filter well to the tube around its whole circumference.

2. Distillation

This method of purification is used for liquid or liquefied substances. If some substances can be both distilled and crystallized, distillation is carried out first. By simple distillation, i.e., by transformation of a liquid substance into vapor and the condensation of the latter in a separate part of the apparatus, volatile substances can be separated from nonvolatile ones, or different volatile substances with pronounced differences in their boiling points can be separated. A mixture of volatile substances is separated by fractional distillation, during which the distillate is collected in a number of fractions. It is carried out on efficient distillation columns, and is characterized by the close contact of the escaping vapors with the refluxed liquid, which provides for the attainment of distillation equilibrium and so causes a maximum enrichment of the distillate by the most volatile component. For high-boiling substances or substances decomposing at their boiling point, distillation at a reduced pressure is indicated. For substances with

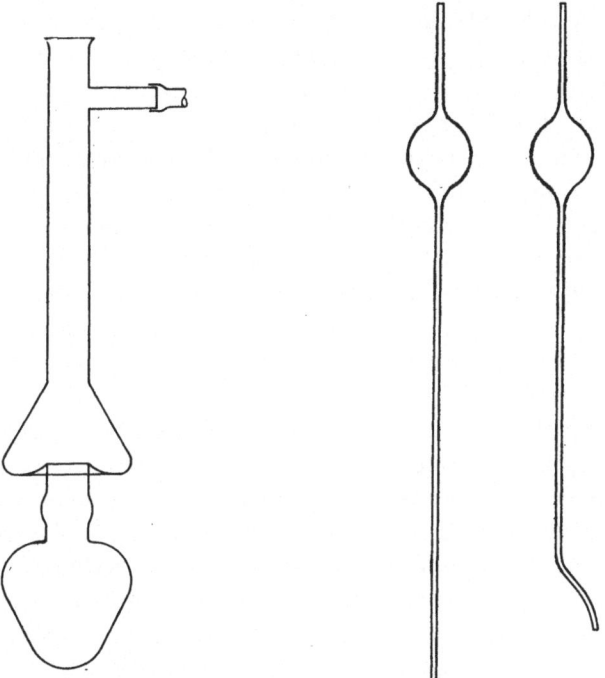

Fig. 3a. Hickmann distillation flask. Fig. 3b. Bubble pipettes.

an extremely low vapor pressure, molecular distillation can be used. In analytical chemistry the chemist uses distillation when purifying the sample for analysis or when separating mixtures.

Here mention will be made of the distillation of small amounts of substances for which so-called Hickmann flasks are used, which can be given various shapes. The simplest type is represented in Fig. 3a. They are advantageous because they do not cause losses due to the scattering of the distilled liquid and its vapors on superfluous surfaces. For the distillation of more volatile substances they can be provided with condensers; for the distillation of larger amounts of liquids they are provided with a side tube to lead the distillate from the collar; for vacuum distillation another tube is sealed to the flask used for the insertion of the capillary, and if the separation efficiency has to be increased, a deflagration part or a microcolumn with a suitable filling can be inserted between the distillation flask and the delivery tube of the collar. To prevent bumping, glass wool is used, introduced into the flask through a glass tube of convenient size and reaching to the bottom of the flask. The glass wool plug is pushed into the tube by means of a glass rod. The distilled substance is introduced into the flask with a pipette ("bubble pipette" prepared from a piece of a glass tube by blowing it out; see Fig. 3b). At vacuum distillation a glass capillary is used instead of the glass wool. The liquid is introduced into the flask through the side tube. The flask is heated in glycerol, oil, or paraffin bath. The distillate in the collar is taken out with the already-mentioned pipette sealed at its upper end. The blown-out part of the pipette is first gently warmed over a flame (otherwise, the substance might decompose at too high a temperature) and then quickly inserted with its bent end into the collar of the flask. On cooling of the pipette (or, rather, of the air in it, which contracts) the distillate is sucked in. When the liquid is completely aspirated the other end of the pipette can also be sealed, the pipette provided with a label, and the material stored.

Here we shall not give details on distillation because this operation is already very well described in a series of monographs [see (1) p. 26] and in the advertising literature; fractional distillation is automated and represents a field which an analyst not dealing regularly with liquids need not necessarily master; it is best if one member of the working group devotes himself to this technique more thoroughly. As regards a current common distillation or even a simple fractional distillation, the necessary equipment is commercially available, both complete sets and single parts of the distillation apparatus. Certain firms also produce special apparatus for microscale operations. It is therefore in the analyst's interest to study catalogues of such firms, because they are, to an appreciable extent, a reflection of present development.

In cases where the substances cannot be separated by distillation, preparative gas chromatography (p. 75) must be applied, especially if small amounts of mixtures of liquids (several milliliters) have to be separated which cannot be separated by fractional column distillation.

3. Sublimation

Sublimation is a very elegant method of purification of solid substances. These are transformed at temperature lower than their melting points to vapors, which are then condensed again to give solid matter. Sublimation can be carried out at normal or at a reduced pressure, or in a stream of an inert gas.

The temperature necessary for sublimation is best ensured with a Kofler block. A minute amount of the substance, 0.5–1 mg, is placed on a microscope slide, a glass ring is applied onto it (surrounding the substance) and covered with a microscope cover glass. The block is then heated and the process is observed with a microscope. As soon as a sublimate

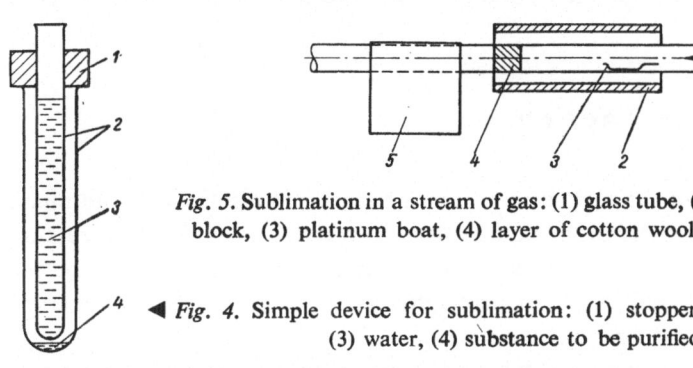

Fig. 5. Sublimation in a stream of gas: (1) glass tube, (2) electric heating block, (3) platinum boat, (4) layer of cotton wool, (5) wet paper.

◀ *Fig. 4.* Simple device for sublimation: (1) stopper, (2) test tubes, (3) water, (4) substance to be purified.

appears on the lower side of the cover glass the temperature is read. Sublimation can be carried out in a simpler manner by using two test tubes of different diameters. The smaller one is filled with water and inserted into the larger, outer one containing the sample. Both are heated slowly in a bath (Fig. 4). Larger amounts of substances can be purified by sublimation from a tube of 8–10 mm in diameter. The substance is placed into a simple boat and a glass wool plug is inserted into the tube to prevent mechanical movement of the substance into the sublimation space. The substance is heated with a suitable heating block (electrical or gas) to the required tempera-

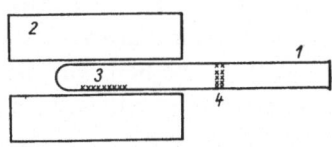

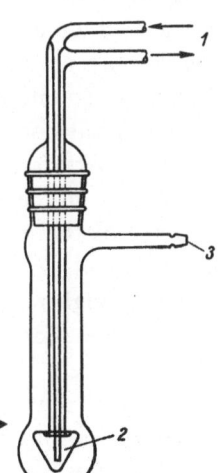

Fig. 6. A simple device for microsublimation: (1) glass tube, (2) electric heating block, (3) substance to be purified (may also be in a boat), (4) sublimate.

Fig. 7. Apparatus for microsublimation: (1) inlet and outlet of ▶ the cooling water, (2) place for condensation of sublimate, (3) connection with vacuum

ture. When this arrangement is used sublimation can be carried out either under an inert gas or in vacuo (Fig. 5). The simplest way to get a pure sublimate consists in cutting off that part of the tube containing the sublimate. A very simple arrangement for sublimation in a horizontal position, consisting of an electrical heating block and a glass tube sealed at one end, is represented in Fig. 6.

Numerous devices for microscale sublimation are commercially available. They have a water cooling device and are heated in a liquid bath (Fig. 7).

4. Extraction

During extraction the substance present originally in a solid or liquid phase concentrates gradually in the liquid phase. Particular methods of extraction differ slightly from each other and have their own names; numerous outfits have been developed and their descriptions may be found in catalogues of firms producing laboratory equipment.

If a solid substance is extracted repeatedly with several portions of cold solvent, the operation is called maceration, if the same operation is carried out with hot solvent, it is called digestion; and if the procedure has a countercurrent character, it is called percolation. So-called Soxhlet extractors, in which a solid substance is extracted with a solvent and the extract becomes more and more concentrated, function automatically. If a dissolved substance is extracted with one portion or more of an immiscible solvent, the operations is called simple extraction or multiple extraction, respectively. If the extraction is carried out by countercurrently-moving solvents, we deal with a countercurrent distribution technique. A continuous extraction of a liquid with another liquid immiscible with the first is called perforation; continuous countercurrent extraction using two liquid phases is carried out in special countercurrent columns. If one of the phases is fixed on a carrier, partition chromatography is the name given to the operation. The choice of the given type of extraction method will depend on the character of the analyzed substances and on the required separation efficiency.

Although rather complicated apparatus are available for the performance of the above-mentioned operations, we first describe a most simple, but much used, device for extraction—a simple separatory funnel. The analyst must always be supplied with a sufficient amount of separatory funnels of various sizes and must be acquainted with how to deal with them and take care of them, to keep them at all times in perfect condition;

otherwise, the stopcocks might get "stuck," or the valuable sample might get lost if the stopcocks or stoppers are not tight enough and even slip out. Normalized separatory funnels have various shapes; spherical, conical, or cylindrical. For small-scale work various separatory funnels of 10−20 ml volume are used. They have to be provided with tight, well-fitting ground-glass stoppers and stopcocks; the control of the tightness is carried out with ether. The stoppers and stopcocks should be bound to the funnel to prevent their falling out and breaking during washing and cleaning, or their being exchanged. They must also be protected from slipping out during shaking or separation. The stems of the separatory funnels should be cut off to make them as short as possible, and the ends must be ground. Smaller faults in fitting can be eliminated by grinding, but it is not correct to try to eliminate leakage by the application of grease. Lubricants should never be used for analytical work. Recently, new types of stopcocks for separatory funnels have been produced, the mobile parts of which are produced from Teflon or polyethylene. They have a number of advantages, but the authors have not yet had the opportunity of testing them personally. The extraction itself is carried out in the usual way.

Ether is often used for extraction in organic analysis. This makes the drying of the extracts before distillation necessary. Sometimes it is convenient to substitute benzene for ether (if emulsions are not formed) because on the distillation of benzene, water is distilled off first in the form of an azeotropic mixture. When the free organic acid is extracted from an ethereal or benzene solution of the derivative with aqueous alkali (for example, in the case of 3,5-dinitrobenzoate preparation) it often happens that a weakly soluble salt of the organic acid separates. In such a case the extract should either be diluted with water, or the extraction carried out first with a dilute alkali solution and only afterward with a more concentrated one. Very often, when derivatives are prepared, the ether or benzene solutions must first be freed from inorganic acids or organic bases. This is done with water, sodium hydrogen carbonate solution, and with dilute acids. As for ether, it should be kept in mind that it is appreciably soluble in concentrated aqueous solutions of inorganic acids (sulfuric, hydrochloric) and vice versa.

When working with small volumes t he two phases can be separated in a small test tube by means of a syringe.

To extract solid substances with hot organic solvents, a commercial Soxhlet extractor is fully sufficient. For microscale extraction the apparatus according to Blount (Fig. 8) can be used. Microscale extraction with high-boiling solvents is carried out in a Haanen and Badum extractor (Fig. 9). Among the apparatus for perforation, two types need to be considered:

those for the extraction of lighter and those for heavier liquids. The greatest obstacle to smooth extraction is the formation of emulsions and a bad separation of layers. A number of instruments have been proposed for such cases (1) (p. 26).

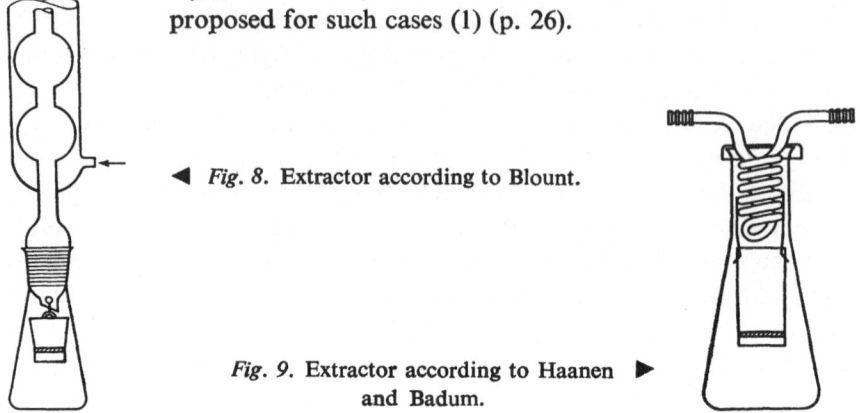

◄ *Fig. 8.* Extractor according to Blount.

Fig. 9. Extractor according to Haanen ► and Badum.

When substances having close partition coefficients are to be separated countercurrent distribution (Craig's method) is indicated. It serves for the determination of the identity and the purity of substances, as well as for the separation of complicated mixtures. This method, mathematically well founded, can be carried out in various ways, either in a series of separatory funnels or in partly or fully automated apparatus and machines.

5. Determination of Physical Constants

The following basic physical constants are used: For the identification of organic compounds—for solids, melting point, eutectic temperature (mixed melting point), and refractive index of the melt; for liquids, density, refractive index, boiling point, optical rotation, and critical demixion temperature. In this chapter attention will be given only to the melting and boiling points. For the determination of other constants see (1) and (2).

Melting Point

The melting point is the temperature at which the solid and the liquid phases of a substance are in equilibrium. For crystalline substances it represents the most important identification constant and the criterion of purity. For analytical purposes, it is determined predominantly on a microscale by four different methods: in a capillary, on the so-called Kofler bank (or Maquenne block), under the microscope, and by various automatic procedures. A correct interpretation of the melting point is dependent on the values found in the literature. However, these data may differ quite appreciably, they are not always quite correct, and may lead to incorrect conclusions. The discrepancies and inaccuracies in the literature data may be caused by the fact that certain measurements had been made on impure samples, by different methods (it is not correct to compare melting points determined in a capillary with those determined under the microscope, for example), by the use of an improper procedure (uncorrected melting points, a bad thermometer, too fast heating, etc.), by the existence of several crystalline modifications, and, finally, to a certain extent, by an uncritical acceptance of melting-point values by certain authors (the highest value for a melting point in the literature need not necessarily be the most correct one). It is generally advisable to collect as much data as possible on the identified substance, and if the found value differs by more than $2-3$ °C from the tabulated data, it is best to prepare an authentic sample and to check the identity by comparing the melting points and measuring the mixture melting point. It is rather surprising that the majority of melting-point values in the literature are uncorrect, in spite of the fact that they originated in well-known laboratories, and that it is quite simple to create the conditions necessary for obtaining corrected values.

Melting-Point Determination in a Capillary

The interval between two temperatures is determined as the interval between the point at which the liquid phase first appears and the temperature when the solid phase disappears completely. The determination itself is best performed in the Roth apparatus (Fig. 10), in which corrected values can be obtained directly without the necessity of making corrections for the extruding part of the thermometer. The apparatus is filled with high-boiling liquids such as glycerol, sulfuric acid, silicone oil, or chlorinated hydrocarbons. The capillaries are prepared by drawing out thoroughly clean thin-wall glass tubes to capillaries and cutting them to pieces $10-12$ cm long and 1.5 mm inner diameter. The capillaries for melting-point determinations are prepared from these by melting them through in the middle over a small flame to obtain capillaires $5-6$ cm long having their bottom ends as thin as the sides. They should be stored in a test tube over anhydrous calcium chloride.

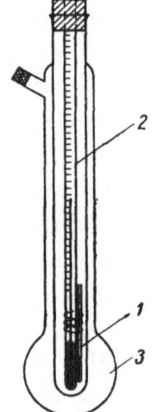

Fig. 10. Roth's device for melting-point determination: (1) capillary with the sample, (2) thermometer, (3) a suitable liquid.

The substance whose melting point is to be determined is ground on a rough plate with a spatula to a fine powder, and this is taken up directly with the capillary, which is then gently patted to help the powder to settle down to the bottom. The small column of the substance should be approximately 2 mm high. The patting can be substituted by allowing the capillary to fall vertically onto a glass plate through a glass tube $1 - 1.5$ m long. The capillary is fastened to the thermometer with a rubber band or with a steel spiral in such a way as to ensure that the 2 mm column of the substance is at the level of the center of the thermometer bulb. The heating is regulated so as to make the rate of temperature increase in the neighborhood of the melting point 1 °C/min. In the case of unknown substances it is recommended to carry out an orienting determination first. For substances which decompose easily on prolonged heating, the apparatus has to be prewarmed first to a temperature 20 °C below the expected melting point, and only then may the thermometer with the capillary be inserted into it.

When melting points of two substances are compared, or when mixture melting point is to be determined, it is possible to fix all the capillaries with the respective substances to the thermometer at once, so that the determination is carried out under identical conditions.

At a temperature close to the melting point a small column of the substance begins to separate from the walls of the capillary and to contract; on further heating it crumples and forms small droplets on the sides; and, eventually, the melting and the disappearance of the solid phase takes place.

The accuracy of the melting-point determination by this method depends on the graduation of the thermometer, i.e., on the spaces between the marks for 1 °C, and on the rate of heating. The use of a correct thermometer is a fundamental requirement. It is best to calibrate it personally, using standard substances. If this is done, the melting points measured on a Roth apparatus are already corrected. The thermometers (for a range up to 360 °C) are first heated at 300 °C for 8 hr and then calibrated using at least six standards melting in the range from 0° to 300 °C. The determined melting points, when plotted in a graph against the known melting points, should form a smooth curve, without appreciable deviations. If a thermometer is used every day, it is convenient to control it monthly by checking at least three temperatures on it. When the melting point of an especially important substance has to be determined, the melting point of the analyzed substance and of the standard should be determined simultaneously. The following substances are suitable as standards for calibration: water-ice 0 °C, menthol 42.5 °C, benzophenone 48.1 °C, p-nitrotoluene 51.7 °C, naphthalene 80.3 °C, acetanilide 114.2 °C, benzoic acid 122.4 °C, urea 132.8 °C, salicylic acid 158.3 °C, succinic acid 182.8 °C, anthracene 216.2 °C, phthalimide 233.5 °C, p-nitrobenzoic acid 241 °C, phenolphthalein 265 °C, anthraquinone 286 °C, and N, N'-diacetylbenzidine 317 °C.

Melting-Point Determination on a Kofler Bank

This is a rapid method enabling melting points to be determined with an accuracy of $1-2$ °C. It is used for rapid orientation and for serial work. The apparatus consists of a metal plate of rectangular shape heated electrically at one end. This causes a temperature gradient along the bank in the interval from 50° to 250 °C. The substance is spread over the bank and the formation of a sharp boundary between the solid and the liquid phase, which takes place rapidly, is observed. The determination requires a larger amount of substance and is not suitable for substances corroding a metallic surface. The apparatus described is commercially available and delivered with detailed instructions.

Melting-Point Determination under the Microscope (according to Kofler)

This method of determination has many advantages and can be recommended especially for analytical laboratories. The substance can be constantly controlled and its behavior determined before, during, and after the melting point is reached as well as during cooling. In addition, the consumption of the material is negligible, the manipulation simple and not time-consuming, and the measured melting points are already corrected. It is unnecessary to go into details at this point; it is in every analyst's interest to get acquainted with this method by studying the excellent monographs describing the whole apparatus, methods, and interpretations in detail (3, 4). The whole apparatus, including the standard substances for testing and the preparations for practical exercises, is commercially available and accompanied by appropriate instructions. The introduction of this technique is of the utmost value for every analyst. The method permits, in addition to the melting points, the determination of a series of other important constants – for example, determination of eutectic temperature at mixture melting point and refractive indices of melts. Several weeks of training in these methods are worthwhile in each laboratory.

Automatic Melting Point Determination

In an effort to mechanize and automate analytical methods, several proposals have been published recently for automatic melting-point determination. This can be based, for example, on the observation of light transmittance as dependent on the temperature (5), on the recording of the movement of a small rod supported by a column of the substance in a capillary, depending on temperature [at the melting temperature the small rod penetrates the melt (6)], or by observing the inflection of the time/temperature plot (7). Equipment for automatic melting-point determination is available commercially.

Mixture Melting Point and Eutectic Temperature

The determination of a mixture melting point is based on the assumption that a mixture of two different substances (with the exception of isomorphic substances) melts at a substantially lower temperature than each substance separately. If two substances having identical melting points are also identical, their mixture will not show a depressed melting point. The method can be utilized even when the melting point of the analyzed substance is several degrees lower than that of the pure standard; in such a case the mixture melting point lies between the two temperatures. The determination is carried out best under a microscope. If the analyzed substance is mixed with some standard substance and the mixture melting point determined, an additional identification constant is obtained — eutectic temperature, which is determined again by a method according to Kofler. Standard substances are part of the equipment of the manufactured apparatus, and they can be used especially for the differentiation of substances of various types displaying identical melting points. Only in the case of homologs having close melting points have they been found unsuitable; in such a case it is more rational to use as standards some of the compounds which are to be differentiated.

Boiling Point

The boiling point is defined as that temperature at which, at a given pressure, the liquid phase is in equilibrium with the vapor phase. For analytical purposes the approximate boiling point should first be determined by heating the liquid with a boiling stone in a test tube in which a thermometer is inserted in such a way as to be ~ 2 cm above the liquid surface. The boiling point determined in this manner is $1-3$ °C lower than the correct one.

Boiling-Point Determination According to Siwoloboff (8, 9)

A capillary of $3-4$ mm inner diameter and $50-100$ mm long is sealed at one end and $2-3$ drops of the substance are introduced into it with a capillary pipette to make the height of the liquid column $6-8$ mm. Another smaller capillary sealed at the other end, ~ 1 mm in diameter and 80 to 100 mm long, is immersed in the liquid (Fig. 11). The capillaries are then fixed to the thermometer and immersed into a bath as in the case of melting-

Fig. 11. Capillary for boiling-point determination according to Siwoloboff: (1) inner capillary, (2) external capillary, (3) the tested liquid.

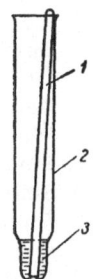

point determination. On heating, bubbles escape from the inner capillary. As soon as their rate increases to such an extent that they escape uninterruptedly (in a chain) the heating is interrupted and the temperature allowed to drop by $5-10$ °C. The stream of bubbles then stops and the liquid enters into the capillary. The bath is heated again and the rate of temperature increase is regulated to 1 °C/min. As soon as a rapid stream of bubbles starts to come out of the capillary the temperature is read and the heating stopped. The second reading is made at the moment when the liquid enters the capillary again. This temperature interval represents the boiling point, and in the case of pure substances it is small. The determination is then repeated and carried out simultaneously with a standard having a very close boiling point; the latter determination serves to make the correction of the first. The following substances may serve as standards (bp at 760 mm): ethyl bromide 38.4 °C, acetone 56.1 °C, chloroform 61.3 °C, carbon tetrachloride 76.8 °C, benzene 80.1 °C, water 100.0 °C, toluene 110.6 °C, chlorobenzene 131.8 °C, bromobenzene 156.2 °C, cyclohexanol 161.1 °C, aniline 184.4 °C, methyl benzoate 199.5 °C, nitrobenzene 210.9 °C, methyl salicylate 223.0 °C, *p*-nitrotoluene 238.3 °C, diphenylmethane 264.4 °C, 1-bromonaphthalene 281.2 °C, and benzophenone 306.1 °C.

For small amounts of substances (less then one drop) Emich's (10) method can be applied for boiling-point determination. Devices are also described allowing the determination of boiling points under reduced pressure (11, 12). The disadvantage of all these methods consists in the fact that the presence of even a small amount of impurity greatly influences the result. It is therefore advisable to combine this determination with a check of purity by gas-liquid chromatography.

If a sufficient amount of a sample is at the analyst's disposal, the boiling-point determination can be combined with the distillation of the sample, or with its fractionation (distillation curve).

References

1. Jureček, M.: Organická analysa I, Nakladatelství ČSAV, Prague 1955.
2. Müller, E., Editor: Houben Weyl: Methoden der organischen Chemie, Vol. 2, Analytische Methoden, Thieme, Stuttgart 1954.
3. Kofler, L., and Kofler, A.: Mikromethoden zur Kennzeichnung organischer Stoffe und Stoffgemische. Wagner, Innsbruck 1948.

4. Kofler, L., and Kofler, A.: Mikroskopische Methoden, in Handbuch der mikrochemischen Methoden. Vol. I, part 1. Springer, Vienna 1954.
5. Überreiter, K., Orthmann, H. J.: Kunststoffe **48**, 525 (1958).
6. Berhenke, L. F.: Anal. Chem. **33**, 65 (1961).
7. Walisch, W., and Eberle, H. G.: Mikrochim. Acta **1967**, 1 031.
8. Siwoloboff, A.: Ber. **19**, 795 (1886).
9. Smith, A., and Menzies, A. W. C.: J. Am. Chem. Soc. **32**, 897 (1910).
10. Emich, F.: Monatsh. **38**, 219 (1917).
11. García, C. R.: Ind. Eng. Chem., Anal. Ed. **15**, 648 (1943).
12. Rosenblum C.: Ing. Eng. Chem., Anal. Ed. **10**, 449 (1938).

6. Color Reactions

Reactions during which a strong color is formed, changed, or disappears are called color reactions. They represent a common means for the detection of organic compounds and an indispensable supplement to chromatographic methods, and form the basis of colorimetric methods. Their quantitative performance is usually very simple, and unfortunately this often means that they are not carried out with sufficient care. This results in the failure of the experiment, and some investigators therefore unjustly consider them as insufficiently reliable. In actual fact, color reactions have their own rules which must be respected. In addition to the chemical problems of the reaction, for example, reaction mechanism, which can be influenced by a number of structural factors and by the medium, color reactions are complicated by another factor, the problem of the color itself, i.e., by the relationship between the structure and the coloration (p. 79).

From the point of view of the process taking place and the cause of the color, color reactions of organic compounds can be classified into the following groups:

a) Group 1. The carrier of the color is the substance itself.
b) Group 2. The carrier of the color is the reagent.
c) Group 3. Formation of complex compounds from the reaction of an organic compound with inorganic compounds.
d) Group 4. The formation of π-complexes.
e) Group 5. The formation of new colored compounds.
f) Group 6. Catalytic reactions.

a) Group 1

A number of processes during which a reversible or irreversible change in the colored organic substance being tested itself takes place belong to this group. Reversible changes can occur, for example, under the influence of heat, pressure, or adsorption forces of the solvent; irreversible changes can be caused by chemical changes of the original substance, for example, as a consequence of reduction, oxidation, or deeper degradation. Specific cases will be discussed with examples.

Thermochromism (1, 2). This is a phenomenon characterized by the fact that certain almost colorless substances become intensely colored, or that colored substances become decolorized, when heated either in solution or in the solid state. The process is reversible and can be repeated at will if the substance is not simultaneously thermally decomposed and permanently changed. For example, dehydrodianthrone when heated in decaline changes its color from yellow to blue-green; in the case of thymol blue it was observed (3) that its blue-violet solution in benzene is decolorized on boiling, and that the color is restored on cooling. This is explained by a reversible change of the lactone form to the betaine form.

Piezochromism (1). This phenomenon has no bearing on the analysis, but it may be of general interest to point out that color can be induced in certain substances, for example, in phenolphthalein (4), if they are submitted to high pressure.

Photochromism (1, 5). A reversible change in color can be caused in certain types of substances by light energy. For example, triphenylformazan gives a red solution in benzene, which changes to light yellow when exposed to light, and turns back to red in darkness. This process was explained by the formation of cis-trans isomeric forms and by closing and opening of a hydrogen bridge (6).

Adsorptiochromism (7, 8, 9). In some instances color changes take place when substances are adsorbed on active surfaces. For example, it is known that p-dimethylaminoazobenzene with a red color is adsorbed from a yellow benzene solution on activated barium sulfate (7), colorless triarylcarbinols are adsorbed on active surfaces with the formation of color corresponding to the respective colored cations, and polynitrated aromatic hydrocarbons are adsorbed with a violet green color (10). This phenomenon can be explained by a double type of adsorption of these compounds: physical, caused by van der Waals forces, and chemical, so-called chemisorption, during which the adsorption bond is much more stable and more similar to a chemical bond. Depending on the nature of the adsorbent, i.e., depending on its ability to accept or to donate electrons, the interaction with the adsorbed substance may differ. This process is reversible, and the adsorbed substances can be eluted in their original form, i.e., the eluates display the coloration of the original solution. However, adsorptiochromism should not be confused with the coloring of substances on adsorbents caused by irreversible changes. So, for example, anthracene is partly oxidized on a column of alumina to give yellow anthraquinone. Derivatives of 1,4-

naphthoquinone (11) are colored irreversibly on alumina in the same manner.

Solvatochromism (12, 13). In solvatochromism the dissolution of a substance is accompanied by color formation, but the cause is not the formation of salts. On the other hand, if strongly colored salts are formed under the influence of mineral or Lewis acids, the process is called halochromism. The difference between the two phenomena is also apparent in the changes of the type of absorption spectra. In the case of halochromism absorption curves are radically changed (old bands disappear and new ones are formed), but in the case of solvatochromism the character of absorption curves is usually unchanged, and they are only shifted (as a whole) to shorter or longer wavelengths. Halochromism will be treated below.

In the case of solvatochromism a shift in the equilibrium between two types of molecules can occur (for example, keto $\rightleftharpoons$ enol equilibrium, associated $\rightleftharpoons$ nonassociated form, or the stabilization of polarized or the unpolarized form of the dye can take place under the influence of the polarity of the solvent. For example, the behavior of 4-phenylazo-1-naphthol (I) (12, 14) represents a clear example of the shift in the tautomeric equilibrium under the influence of the solvent. When dissolved in pyridine this compound gives a yellow solution and contains exclusively azonaphthol form (Ia),

(Ia)   (Ib)

while the red solution in acetic acid contains quinonehydrazone (Ib). Dyes containing an easily shifting electron system can be polarized under the influence of solvent, as, for example, in the case of 4-nitro-4'-dimethylamino-azobenzene containing both the electron-attracting nitro group and the electron-donating dimethylamino group. The true state of the molecule cannot be represented by either of the two extreme structures, but the basic state of the molecule is very close to formula (IIa), while the excited state of the molecule is closer to formula (IIb). According to Förster (15), the excited states of predominantly nonpolar dyes are predominantly polar, and vice versa.

(IIa)

$$\longleftrightarrow \quad \overset{|\overline{O}|}{\underset{\underset{\ominus}{|\overline{O}}}{>}} N = \langle \rangle = \overset{-}{N} - \overset{-}{N} = \langle \rangle = \overset{\oplus}{N} \overset{CH_3}{\underset{CH_3}{\diagdown}}$$

(IIb)

4-Nitro-4'-dimethylaminoazobenzene gives a yellow solution in the nonpolar hexane. With increasing polarity of the solvent the color changes through yellow-orange and orange to red in ethanol. The increasing polarity of the solvent increasingly stabilizes that form of the molecule which is closer to the type (IIb), which absorbs at longer wavelengths. On the other hand, substances are known which give a blue solution in benzene and a yelloworange one in water.

The Influence of pH on the Formation or Change in Color. The influence of pH on organic compounds is often observed in analytical chemistry during work with indicators. Acido-basic indicators are weak organic acids or bases which change their color within certain pH intervals, i.e., under the influence of the change of the hydrogen ion concentration. The change in color of acido-basic indicators is connected with the dissociation equilibria of their acidic or basic group. Depending on whether they are bases or acids, we can write

$$\text{Ind OH} \ \rightleftharpoons \ \text{Ind}^+ + \text{OH}^-$$
$$\text{H Ind} \ \rightleftharpoons \ \text{H}^+ + \text{Ind}^-$$

and use the Henderson-Hasselbalch equation:

$$\text{pH} = \text{p}K'_{\text{Ind}} + \log \frac{\text{Ind}^-}{\text{H Ind}}$$

in which $\text{p}K'_{\text{Ind}}$ represents the acid dissociation constant of the indicator. If the anion Ind^- has a different color than the undissociated molecule H Ind, the observed change in color takes place within the range of two pH units. With certain indicators, however, the situation may be more complicated.

For example, phenolphthalein exists in acid medium and below pH 8.3 in lactone form (IIIa), and at pH above 10.0 it is already completely changed to the red salt (IIIb) (16).

(IIIa) ⇌ (IIIb) + 2 H⊕

Eriochrom black T (IV) exists under the influence of pH in three forms: At about pH 6 it is transformed from wine-red to blue, and at pH 12

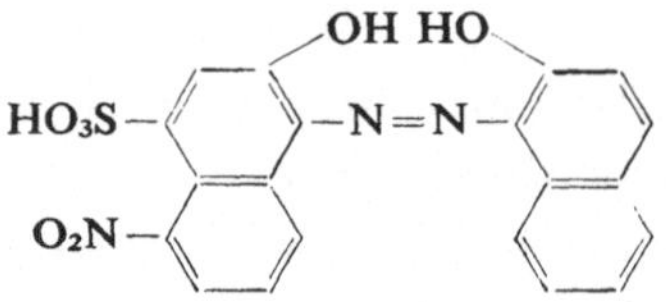

(IV)

it turns orange. The change in color is attributed to the formation of various dye anions. If the indicator is represented as anion H_2Ind^-, its color changes are caused by further dissociation:

$$H_2Ind^- \rightarrow HInd^{2-} \rightarrow Ind^{3-}$$
wine-red   blue   orange

Halochromism. Certain colorless or only very weakly colored organic substances are soluble in mineral acids with the formation of strongly colored solutions; this phenomenon is called halochromism. The pale-yellow dibenzalacetone (V), which dissolves in concentrated sulfuric acid with a strongly red-orange color, can serve as an example of this phenomenon. Gillespie and Leisten (17) have proved that during this process protonation of the carbonyl group takes place (Va). The intense color is then due to the extreme form (Vb) of the protonated form:

(V)

(Va)

(Vb)

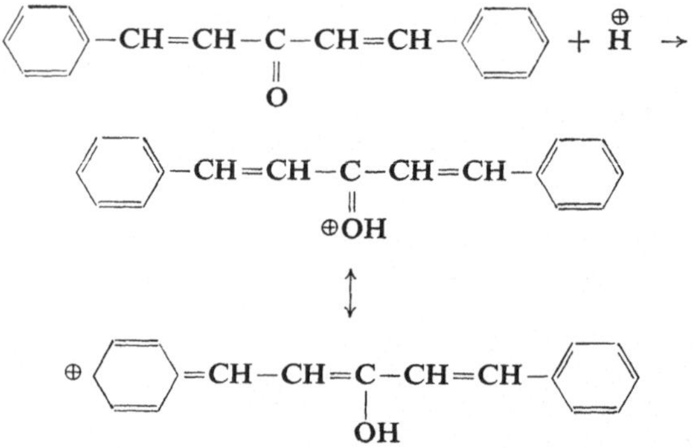

According to Brönsted-Lowry's ideas, dibenzalacetone behaves as an uncharged base:

$$B + H^\oplus \rightleftharpoons BH^\oplus$$

Depending on the weakness of the base B, acids of various strengths will be necessary to bring about a protonated form. As many of these compounds are actually indicators for the measurement of the function H_0, i.e., for the measurements of extreme acidities, we can take some data from the literature on these compounds.

Hammett and Deyrup (18) defined the acidity function as

$$H_0 = pK_{BH\oplus} - \log \frac{C_{BH\oplus}}{C_B}$$

where $C_{BH\oplus}/C_B$ is the directly measured ratio of indicator concentrations in its different colored forms, and $K_{BH\oplus}$ is the thermodynamic ionization constant of the conjugated acid. For example, the formation of the conjugated acid of benzalacetone takes place within the range of $H_0 = -4$ to -5.6 (i.e., $55-70\%$ H_2SO_4). Hence, with a concentration above 70% only the $BH^\oplus$ form occurs.

The analysis of this problem is important for color reactions because many color reactions are carried out in concetrated sulfuric acid and the resulting substances are colored just at a certain acidity.

The second reason for the formation of the color upon dissolution in concentrated acids can be demonstrated by the example of triarylcarbinols (19):

$$(C_6H_5)_3COH + 2\,H_2SO_4 \rightarrow (C_6H_5)_3\,C\oplus + H_3O\oplus + 2\,HSO_4\ominus$$

Strongly colored triarylcarbonium ions are formed. The corresponding trityl chlorides behave in the same manner.

The so-called Lewis acids ($AlCl_3$, $SbCl_5$, $HgCl_2$, $SnCl_4$, etc.) have an effect similar to that of mineral acids. In the case of trityl chloride the effect of aluminum chloride can be represented by the following equation:

$$(C_6H_5)_3C-Cl + AlCl_3 \rightarrow (C_6H_5)_3C\oplus + (AlCl_4)\ominus$$

From the analytical point of view, the halochromism of Schiff's bases, formed in the reaction of primary aromatic amines with aromatic aldehydes, play an important role:

$$-CH=N- + HCl \rightarrow CH=\overset{\oplus}{N}H- + Cl\ominus$$

In the case of azo compounds the addition of the proton in acid media was also studied, predominantly on 4-dimethylaminoazobenzene; it was shown that a mixture of two possible forms occurs (VIa, b and VII) (20, 21).

(VIa)

$$CH_3\diagdown \overset{\oplus}{N}= \langle\ \rangle =N-\overset{H}{\underset{|}{N}}-\langle\ \rangle$$
$$CH_3\diagup$$

(VIb)

$$CH_3\diagdown \overset{\oplus}{N}-\langle\ \rangle -N=N-\langle\ \rangle$$
$$CH_3\diagup \underset{H}{\underset{|}{}}$$

(VII)

The formation of colored salts can also take place in alkaline media. For example, the yellow azo dye (VIII) formed in the colorimetric determination of phenol dissolves in alkali with a red color with the formation of ions (IX). Addition of alcohol produces additional solvatochromic color changes to violet (22).

$$O_2N-\langle\ \rangle -N=N-\langle\ \rangle -OH \xrightarrow{\text{NaOH}}$$

(VIII)

$$\left[O_2N-\langle\ \rangle -N=N-\langle\ \rangle -\overset{\ominus}{O} \longleftrightarrow \overset{\ominus}{O_2N}=\langle\ \rangle =N-N=\langle\ \rangle =O \right] Na^{\oplus}$$

(IX)

Other Influences. The changes of colored substances possessing the properties of indicators can be caused by surface-active substances (23 – 26). Anion-active substances shift the coloration of indicator acids to the red region and of bases to the basic region. This effect is explained by the formation of micellary structures with positive or negative charges and by the absorption of indicator ions, or by the formation of insoluble salts or of various aggregates of the dye and the surface-active substance. In view of the fact that surface-active substances are widely used in technological practice, it is important that the analyst be well acquainted with these effects.

Irreversible changes can be caused by a chemical change of the substance, because of which the conditions for the coloring of the substances may cease to exist, or vice versa: for example, the reduction of azo compounds with stannous chloride to colorless products (p. 364). and the reactions of anthraquinone with sodium dithionite solution leading to an intensely red coloration (p. 300).

b) Group 2

All color reactions in which the carrier of the color is the reagent itself belong to this group. Here the reagent is changed during the reaction either reversibly or irreversibly from the colored to the colorless form or vice versa; a colored salt may result or a derivative of the tested substance without a change in color. This group includes, for example, the reaction with acido-basic indicators and redox indicators; reactions which lead to the decoloration of bromine water (p. 108) or potassium permangante (p. 108); the change in color of the Fehling reagent (p. 112); or the formation of color in reductive reactions with phosphomolybdic acid. We must also include in this group a series of color reactions used predominantly for colorimetry in extraction methods. Their principle consists in the formation of a colored salt on reaction of an organic cation or anion with a colored anion or cation, which can be extracted with a solvent other than that in which the reaction was carried out and in which the reagent is insoluble (27). The determination of certain cation-active substances with anion-active dyes (28) may serve as an example, as may the determination of certain alkaloids by the reaction with dye anions (29). The colored reagent or the reagent capable of a colored reaction can form a derivative with the tested substance, the skeleton of the original reagent (30, 31) remaining the carrier of the color reaction.

c) Group 3

In these reactions the tested organic substance and an inorganic salt give a more or less stable strong color due to a complex salt. In Part 2 of this monograph a series of such reactions is described, such as, for example, the reaction of hydroxamic acids with ferric salts (p. 276), the reaction of phenols with ferric chloride (p. 188), the reaction of molybdenum with o-dihydroxybenzenes (p. 191), of diacetyl dioxime with nickel salts (p. 227), of phenols with Millon's reagent (p. 196), alcohols with ceric ammonium nitrate (p. 170), alcohols with vanadium-hydroxyquinoline complex (p. 171), and the reaction of cis-enols of β-dicarbonylic compounds and α-dicarbonyl compounds with ferric chloride. (p. 294.)

d) Group 4

π-Complexes are formed on interaction of an organic compound capable of donating π-electrons (electron donor) with a reagent functioning as an electron acceptor. The interaction is based on the transfer of π-electrons to the acceptor, which brings about the appearance of a new absorption band in the electron spectrum of the π-complex, occurring usually at higher wavelengths than the first absorption band of the donor and acceptor themselves. The energy of the absorption-band maximum of the π-complex depends on the strength of the corresponding donor and acceptor. Often it happens that this absorption band is in the visible part of the spectrum, which results in the coloration of these complexes (32). The reaction of aromatic hydrocarbons with tetracyanoethylene may serve as an example, as may the detection of the aromatic nucleus with nitrosyl-sulfuric acid.

e) Group 5

The majority of color reactions of organic substances belong to this group. Reactions are classified here in which quite new, intensely colored organic compounds are formed from the reagent and the organic compound, either by condensation, or by some other chemical reaction.

In these cases it is usually possible to explain why the color is formed. The reactions can be further classified, on the basis of the type of colored compound formed, into reactions leading to the formation of azo dyes, di- or triphenylmethane dyes, xanthene dyes, polymethine dyes, indophenols, etc. Azo dyes are formed, for example, in the reaction of diazonium salts and phenols (p. 192) or amines (p. 324), azomethines in the reaction of primary aromatic amines with aromatic aldehydes (p. 215), di- and triphenyl-methane dyes in the reaction of aromatic aldehydes with aromatic hydro-carbons in concentrated sulfuric acid (p. 213), triphenylmethane dyes in the reaction of phenols with aromatic aldehydes or oxalic acid (p. 196), xanthene dyes in the reaction of anhydrides of dicarboxylic acids with resorcinol (p. 196), polymethine dyes are formed after the cleavage of the pyridine ring in the reaction of the glutaconaldehyde formed and barbituric acid (p. 378), indophenols on reaction of phenols with Gibbs reagent (p. 195), or 4-aminoantipyrine according to Emerson (p. 194), or on the Liebermann reaction (p. 195).

f) Group 6

In reactions of this group two organic reagents react with the formation of a colored compound in the presence of a catalyst, the catalyst being the substance to be tested or detected. The reaction of p-phenylenediamine with hydrogen peroxide, catalyzed by aldehydes (p. 239), may serve as an example.

Certain known color reactions cannot be included unequivocally into these groups, because they represent a combination of several of the processes mentioned. For example, in the reaction of primary aromatic amines with aromatic aldehydes (p. 215), Schiff's bases are formed which give strongly colored salts (halochromism) in the presence of acids, the color of which deepens on heating (thermochromism). We should also keep in mind that the color reaction with the organic substance can take place either directly, or it may be neccessary to transform the tested substance to a reactive species. For example, the diazonium salts react with phenols with the formation of azo dyes. The reaction can be used for the detection of primary aromatic amines after their diazotization (p. 341), but also for the detection of aromatic nitro compounds after their reduction to amines (p. 357), and even for the detection of aromatic hydrocarbons after their nitration and reduction, etc. (p. 123). In certain cases a reaction can take place in which a compound capable of giving a color reaction is only formed in the reaction medium during the course of the reaction. For example, formaldchyde reacts with chromotropic acid with the formation of a violet color (p. 214). This reaction is also obtained with compounds which liberate formaldehyde as part of the reaction (hexamethylenetetramine, dimethylolurea, etc.).

Practical Notes Concerning the Procedure of Carrying Out Color Reactions

The preceding discussion on color reactions aimed to demonstrate which factors can bring about or influence the formation and the change of color. We also wanted to show that during color reactions not only must the laws of chemical reaction be respected, but also the rules correlating color with structure. For example, when a reaction is carried out in concentrated sulfuric acid a particular concentration of sulfuric acid will be necessary to make the reaction (condensation) possible, and, further, a particular concentration of the acid to get optimum coloring of the product. It is therefore clear that by adding reagents, solvents, or an excess of substances accompanying the tested one in an unsuitable way, either the course of the required reaction or the development of the coloration, may be strongly negatively influenced. It should be borne in mind that the color reactions are usually elaborated for pure substances and that a positive or a negative result is reliable only if the suitability of such use is also checked. For example, if the analyst has to detect a small amount of phenols in glycol, he should not only carry out the reaction of the sample with Gibbs' reagent (p. 195), but he should also carry out the following control experiments with this reagent (especially if he is carrying out this test for the first time): (1) with pure glycol, (2) with an aqueous phenol solution, and (3) with pure glycol to which a small amount of phenol has been added, (4) with the sample and (5) with the sample to which a small amount of phenol has been added. Should he find with the first and the third controls that the test cannot be carried out, another reagent should be tried in the same manner.

It should be further kept in mind that each color reaction has its own optimum conditions, both with respect to the concentration of the tested substance and to the amount of the reagent. The Gibbs' reagent mentioned above is, for example, most suitable for the detection of small amounts (several micrograms) of phenols in aqueous solutions, but it is impossible to carry out the reaction by putting 1 g of phenol into a test tube and adding the reagent.

This monograph gives both the color reactions which the authors have found suitable during their long practice, and a series of reactions taken from the literature. The procedures are formulated so that the reaction could be carried out with pure compound. It is not excluded that many analysts will be compelled to adjust the procedures to suit their own conditions. We have compiled a very large number of color reactions, as it is often necessary to detect even a common substance in a medium which does not allow the use of any of the common reagents. Very often references to colorimetric applications are also given. This is because in very delicate or controversial cases it is best to carry out a quantitative procedure, i.e., a colorimetric procedure (but without the colorimetric evaluation).

Color reactions are carried out either in test tubes of various sizes, volumetric flasks, and graduated cylinders, or in the form of spot tests on an analytical filter paper, or on white porcelain spot-test plates (provided with small hollows). The procedure depends on the nature of the reaction, on the amount and character of the sample, and on the skill of the analyst. The majority of reactions can be adjusted in various ways. As regards reagents and auxiliary solvents, etc., the same remark is valid as that on p. 55 for reagents and derivatives. The scrupulous cleanness of the vessels and of all auxiliary substances and solvents (which can be checked in control experiments), as well as very accurate and neat work should be a matter of course for every analyst.

For a more detailed discussion of color reactions along the lines indicated in this chapter see (33).

References

1. Kortüm, G.: Angew. Chem. **70**, 14 (1958).
2. Day, J. H.: Chem. Revs. **63**, 65 (1963).
3. Wetz, E., Schmidt, F., and Singer, J.: Z. Elektrochem. **46**, 222 (1940).
4. Larsen, H. A., and Drickamer, H. G.: J. Phys. Chem. **62**, 119 (1958).
5. Luck, W., and Sand, H.: Angew. Chem. **76**, 463 (1964).
6. Hausser, I., Jerchel, D., and Kuhn, R.: Chem. Ber. **82**, 515 (1949).
7. Kortüm, G., Vogel, J., and Braun W.: Angew. Chem. **70**, 651 (1958).
8. Weitz, E., Schmidt, F., and Singer, J.: Z. Elektrochem. **47**, 65 (1941).
9. Weitz, E., and Schmidt, F.: Ber. **72**, 2099 (1939).
10. Ling, Ch.-S.: J. Chinese Chem. Soc. **18**, 135 (1951); C. A. **46**, 2965 (1952).
11. Green, J. P., and Dam, H.: Acta Chem. Scand. **8**, 1093 (1954).

12. Dimroth, K.: Marburger Sitzungsberichte **76**, No. 3, 3 (1953).
13. Dimroth, K., and Reichardt, Ch.: Palette **1962**, 28.
14. Fischer, E., and Frei, Y. F.: J. Chem. Soc. **1959**, 3159.
15. Förster, T.: Z. Elektrochem. **45**, 548 (1939).
16. Davies, M., and Jones, R. L.: J. Chem. Soc. **1954**, 120.
17. Gillespie, R. J., and Leisten, J. A.: J. Chem. Soc. **1954**, 1, 7.
18. Hammett, L. P., and Deyrup, A. J.: J. Am. Chem. Soc. **54**, 2721 (1932).
19. Lavrushin, V. F., and Verkhovod N. N.: Dokl. Akad. Nauk SSSR, **115**, 312 (1957).
20. Cilento, G., Miller, E. G., and Miller, A. J.: J. Am. Chem. Soc. **78**, 1718 (1956).
21. Lewis, G. E.: Tetrahedron **10**, 129 (1960).
22. Gasparič, J., and Beranová O.: Collection Czech. Chem. Commun. **26**, 3173 (1961).
23. Klotz, I. M.: Chem. Revs. **41**, 373 (1947).
24. Colichman, E. L.: Anal. Chem. **19**, 430 (1947).
25. Hiskey, C. F., and Downey, T. A.: J. Phys. Chem. 58, 835 (1954).
26. Bennewitz, R.: Fette, Seifen, Anstrichm. **58**, 832 (1956).
27. Pellerin, F.: Bull. Soc. Chim. France **1961**, 1071.
28. Ruf, E.: Z. Anal. Chem. **204**, 344 (1965).
29. Konyushko, V. S.: Zhur. Anal. Khim. **19**, 1012 (1964).
30. McIntyre, F. C., Clements, L. M., and Sroul, M.: Anal. Chem. **25**, 1757 (1953).
31. Kämmerer, H.: Chimia **19**, 61 (1965).
32. Briegleb, G.: Elektronen — Donator — Acceptor — Komplexe; p. 37. Springer Verlag, Berlin—Göttingen—Heidelberg, 1961.
33. Gasparič, J.: Chem. Listy — in the press.

7. Identification by Formation of Derivatives

The preparation of derivatives of the substance to be identified represents the most important and final step in identification. We consider a substance as identified by means of a derivative if the melting point of the prepared derivative does not differ more than $1-2$ °C from the values published in the literature, and if the mixed melting point (with the derivative of an authentic sample) is undepressed. If the amount of the analyzed substance is limited, it is advisable to first learn how to cary out the identification procedure by using an authentic sample, or at least with an analogous compound. The possibility of losing the valuable sample is thus limited in case of failure. For laboratories devoted to the identification of organic compounds it is advisable to gradually build up a collection of small amounts of authentic derivatives of basic groups of substances (in accordance with the problems and aims of the laboratory) with controlled melting points. The derivatives are thus immediately available for the determination of mixed melting points, X-ray analysis, chromatographic or other comparisons (the preparation of these derivatives is a very good and useful training for the laboratory worker). For a given case a derivative should be chosen, on the basis of literature data, which fulfills a maximum of the following requirements:

1. *Melting point of the derivative.* The chosen derivative should be a crystalline compound with a melting point between 80 and 250 °C, and it should melt abruptly and without decomposition, even when overheated. Melting points of the derivatives of isomers and a whole homologous series should be within a large range of temperatures and the melting point of the derivative of the tested substance should differ from that of the parent substance and of other derivatives of substances coming into cosideration by at least 5 °C. We should not forget that substances with low melting points are purified with difficulty, when crystallized they separate as oils, and they cannot be induced to crystallize easily. Substances with high melting point, on the contrary, are weakly soluble, and the difference in solubility in hot and cold solvents is often small, which causes great losses on crystallization. High-melting substances often have a diffuse melting point at which they decompose.

In a homologous series the first member often melts at the highest temperature and the melting point drops with increasing molecular weight until a certain final temperature is reached, so that the derivatives of a given type cannot be used for higher homologs. With an increasing number of reacting functional groups, the melting point increases, and it is therefore possible to use certain derivatives of polyfunctional derivatives which would be inconvenient for monofunctional compounds. For a general discussion on melting points also see p. 36.

2. *Preparation of the derivative.* The procedure of preparing a derivative should be as simple as possible and the reaction should be straightforward and unambiguous, it should give a high yield of the derivative and no by-products, its mechanism and kinetics should be known, and the product should be easily isolable from the reaction mixture. The requirements should rule out complicated apparatus, a multistep synthesis (low yields and a number of impurities), the transfer of the reaction mixture from one reaction vessel to another (losses), and the tiresome purification of the derivative. Theoretical knowledge of the reaction will then permit the correct choice and arrangement of optimum conditions for the reaction.

3. *Purification of the derivative.* The purification of the crude derivative is an important concluding operation of identification and it can be the source of numerous failures. For a correct choice of purification procedure it is important to understand the course of the reaction and to know which substances may be present in the sample, in order to be able to concentrate

on their elimination. Well-elaborated identification methods take the problem of the purification of the derivative into account and prescribe a correct procedure. Crystallization is used most often for purification (p. 26); other methods are applied less frequently, such as, for example, sublimation (p. 32) or preparative chromatography (p. 57).

4. *Reagent.* The reagent recommended for the preparation of the derivative should be easily accessible, best commercially, or its preparation should be simple and easy. Special attention should be paid to its purity. It should be as pure as possible in order to prevent the introduction of impurities present in the reagent. It always pays to crystallize or to distil the reagent before use, and in the case of a material of unknown origin the preparation of a derivative of a simple known substance should be tried first. It is also convenient to have a special collection of tested reagents, even if they are common chemicals (for example, acetic anhydride).

5. *Second series of derivatives.* It is convenient when a second series of derivatives can be prepared from the derivatives in a simple way, and when the identification constants of the second series prove definitely the identity of the tested substance. For example, alcohols are identified as esters of 3,5-dinitrobenzoic acid, and these give addition compounds with 1-naphthylamine, etc.

6. *Analysis of derivatives.* The presence of suitable functional groups in the molecule of the derivative makes a simple analysis possible, and hence the determination of the molecular weight. It is possible, for example, to determine the presence of a carboxyl group in p-carboxyphenylhydrazones of aldehydes and ketones by alkalimetry, the substances containing nitro groups can be analyzed titanometrically, in salts the acid or the basic component can be titrated in nonaqueous media, etc. It is more convenient to analyze the substances in this way, because the analysis is more specific and usually affords more information on the composition of the preparation than, for example, elemental analysis.

7. *Polymorphy.* The disadvantage of certain derivatives is that they occur in two or more modifications. This disadvantage is characteristic of some 2,4-dinitrophenylhydrazones which from other points of view are almost ideal for identification purposes. Therefore, the possibility of polymorphy should always be kept in mind.

8. *Regeneration of the original compound.* If the original substance can be easily regenerated from the derivative, this is a great advantage. Such a procedure can be used for a simultaneous purification of small amounts of substances, for example, liquids, which can be purified by crystallization after having been converted into a solid derivative. Derivatives having the character of salts are almost ideal in this respect, because the original com-

ponents can be regenerated from them by mere filtration of a solution through an ion-exchanger column.

9. *Literature data*. A chosen derivative can be considered as advantageous if the literature gives a maximum number of identification constants for a maximum number of substances of the considered type of compound. We should not forget that the final stage of identification is the interpretation of the data obtained. Such determined values as melting point, eutectic temperature, and refractive index of the melt are compared with the values given in the literature, and a substance is sought the constants of which correspond to known data. In case we are unable to find the values for the substance under consideration in the literature, it is necessary to prepare a derivative of an authentic substance and to compare its properties with those of the analyzed substance.

10. *Suitable chromatographic properties*. Suitable chromatographic properties should be an indispensable feature of the derivative. It should contain such functional groups in the molecule which allow easy detection on chromatography. This makes the control of purity during the purification of the derivative possible, as well as its identification by paper or thin-layer chromatography, even in such cases when the preparation of a solid derivative is not feasible. It often happens that the original sample contains two or more isomeric or homologous compounds which cannot be separated well, and in such a case the chromatographic methods mentioned are most suitable for the identification (p. 57).

11. *Work on semimicro and micro scale*. If the chosen derivative fulfills the majority of the preceding requirements, no obstacles should exist for semimicro or micropreparative work which is not only quicker but also indispensable if only a minute amount of the sample is at our disposal.

8. Organic Quantitative Analysis

On p. 24 we indicated that in a qualitative analysis the analyst also utilizes certain results of quantitative organic analysis. For example, the result of elemental analysis of the pure substance when combined with the determination of the molecular weight gives an accurate idea of the molecular formula of the unknown substance. Sometimes elemental analysis can be of great importance even for those substances which cannot be sufficiently purified. In such a case the calculation of the ratio of given elements is helpful. However, those elements should be chosen which are really decisive. On the other hand, it is sometimes very advantageous to already determine certain elements in the original sample even if this is a mixture. For example, when, in analyzing a technical product, the presence of sulfur or nitrogen (or some other element) is detected in the original sample, a quantitative analysis can establish whether the sulfur atom is constitutive, whether sulfur is present in the sample as an integral part of a component, or whether it is there only as an impurity. Sometimes a small amount of some element may indicate the presence of a small amount of a certain technologically important component. When samples containing an appreciable proportion of inorganic material are analyzed, it is advisable to complete the test for carbon. For mixture analysis it is possible to follow the course of separation and purification by the determination of some element. The same is true of the determination of a functional group. In combination with elemental analysis, it can thus be determined whether the all of the given element is present in the functional group. By determining the functional group, it is possible to follow the separation of the substance from a mixture and its purification. For the determination of functional groups in derivatives see p. 55.

When mixtures are analyzed it is sometimes indispensable to combine qualitative analysis to a certain extent with quantitative analysis, i.e., with the determination of single components. Often the analyst can only check whether he was able to account for all components of the sample by making a material balance. We strongly recommend the determination of water, dry residue, ignition residue, sulfate ashes, etc. whenever the sample requires it.

9. Chromatographic and Electrophoretic Methods

Nowadays chromatographic methods are inseparably connected with organic analysis and they should be commonly introduced in organic analytical laboratories. With chromatography a very efficient separation is attained on the basis of differential migration based on adsorption, ion exchange, partition between two phases, etc. Thereotical bases of chromatographic separations of all types have frequently been described. For more details specific monographs should be consulted (1 − 5).

All these types of chromatography are of special importance as methods for the separation of mixtures and as purification techniques on a macroscale and also a microscale and up to ultramicroscale operations.

If the chromatographic behavior can be expressed by a suitable quantity and the analyzed substance can be compared with an authentic sample, the method can be used for identification purposes. Such is the case with paper chromatography, thin-layer chromatography, and gas chromatography. Detailed descriptions of the mentioned techniques are beyond the scope of this chapter; the present book is merely intended to show how these methods are used by the analyst and what simple apparatus is required for it. It is in the interest of every analyst to get more closely acquainted with specific techniques as described in the corresponding works on paper chromatography (3, 6 – 11), thin-layer chromatography (12 – 19), column chromatography (20, 21), ion-exchange chromatography (21 – 24), and gas chromatography (25), as well as on electrophoretic methods (36 – 39). Developments in chromatographic methods can be followed in the Journal of Chromatography, which also publishes an extensive bibliography, and in the Journal of Chromatographic Science, as well as in articles from this field in Chromatographic Reviews (M. Lederer, editor, published by Elsevier) and in Advances in Chromatography (J. C. Giddings and R. A. Keller, editors, published by M. Dekker, New York). The literature on paper and thin-layer chromatography from 1957 to 1965 has been summarized in two bibliographic compilations (40, 41).

Generally, it can be said that all chromatographic methods are exceptionally efficient and enormously valuable for every organic analysis. It is impossible to say that any one of them is better than another. The following factors decide whether we choose for a given case a method based on adsorption, partition, or some other mechanism, or whether we carry out the separation on a column, on a thin-layer, on paper, or in the gas phase: the character of the problem itself, the properties of the sample, the scale of work, equipment in the lab, the number of analyses and the required rate of analyses, erudition and preferences of the worker, and often also the economic aspects of the problem. The success of an organic analyst in using chromatographic methods will depend on his ability to master these methods and to know how to use each one of them in the proper place, or to know how to combine them.

Paper Chromatography

In paper chromatography the separation of substances is based on partition between two liquid phases: one fixed on the cellulose fibers of the paper (stationary phase) and the second one moving (mobile phase). Under certain circumstances adsorption or ion exchange plays a certain role, occasionally a predominant one.

In doing chromatographic work suitable equipment and paper are needed. Detailed descriptions of different equipment are given in the literature cited as well as in the advertising literature of the companies which produce the equipment. Figure 12 represents a simple chromatographic chamber. If need be, it is possible to make the necessary apparatus in the laboratory;

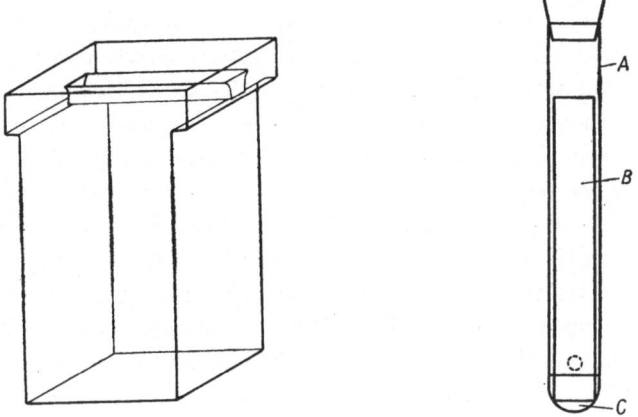

Fig. 12. A simple tank for descending chromatography according to Macek.

Fig. 13. Paper chromatography in a test tube: (A) test tube, (B) chromatogram, (C) solvent.

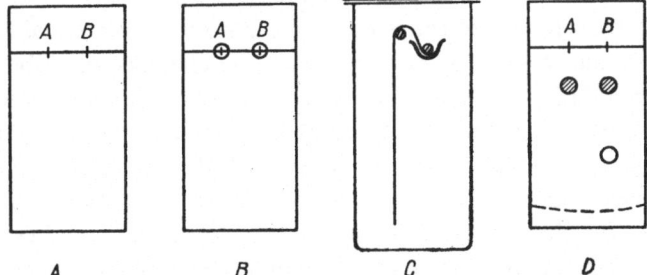

Fig. 14. Phases of paper chromatography: (A) strip of chromatographic paper with marked starting line and points of application, (B) chromatogram after the samples have been applied, (C) chromatography in the tank (side view), (D) detected chromatogram.

a closed chamber with a trough in it is needed, from which a strip of chromatographic paper with samples on it is hung, and which is filled with a solvent which moves downward by capillarity; an alternative consists in a glass cylinder with the solvent at the bottom and a chromatogram hanging in it with its lower end immersed in the solvent in such a way as to allow the solvent to move upward by capillary ascent. In the most simple case, such as, for example, in rapid qualitative tests, the ascending method in a test tube can be used (Fig. 13). Both the ascending and descending arrangements are fully satisfactory. They give equally good results, and the choice of one of them is usually influenced by local possibilities, individual preferences of the worker, and very often by the time factor, the descending arrangement usually being quicker. Ascending chromatography sometimes has to be used, because in a descending chromatogram the substance could, for example, run off the paper if left to run overnight. Micropipettes for putting the samples on the paper and atomizers for detection are also supplied with standard chromatographic equipment available commercially.

Usually, a strip of chromatographic paper 24—40 cm long and as broad as necessary is used, on which the start line is drawn 3—7 cm from the upper edge of the paper with an ordinary pencil; the points for putting the samples on are indicated on this line (Fig. 14). Then the solutions of the chromatographed substances are put on the start line with a micropipette. The amount of the sample is kept between 1 and 50 µg. This amount varies with the type of substance and is dependent on the sensitivity of the detection reaction and on the solubility of the chromatographed compounds. Samples for application are usually 0.5—2% solutions in volatile solvents if possible, which can be easily eliminated from the paper. If aqueous solutions have to be applied on the start line, the evaporation is enhanced by a stream of warm air. However, this method cannot be used if papers impregnated with relatively volatile stationary phases are used. After the samples have been applied the paper is put into the chamber and the solvent poured into the trough for development. The chamber must be saturated with vapors of all components of the solvent used, i.e., both stationary and mobile phases. When the solvent front has reached the required distance the chromatogram is taken out, the front marked immediately with a pencil, and the paper allowed to dry under a hood in a stream of air or in a drying oven. When dry the testing of the spots is carried out in a suitable way.

Chromatographic paper can be purchased from a number of companies. There is a wide choice. Best known are the following chromatography papers: Whatman (Balston and Balston); Schleicher and Schüll; Ederol; and Macherey, Nagel, and Co. For normal work with organic compounds all of them can be used equally well; if a special problem described in Part 2 requires a specific sort of paper, this will be pointed out. Different kinds of paper differ mainly in their flow rate and thickness. The latter factor plays a role especially when papers impregnated with organic stationary phases are used. Thicker papers retain larger amounts of the stationary phase and are also more resistant mechanically when wet with impregnating solvent.

Equipment for paper chromatography is, as shown, very simple, and there is no problem in introducing this method in every laboratory. However, there are certain specific requirements which become obvious only when such work is done on a large scale. It is advantageous, therefore, that larger institutes or factories have special laboratories for paper chromatography and also chromatography specialists. The description of a modern chromatography laboratory has been recently published (42).

If paper chromatography is to be used successfully in organic analysis, the following factors must be considered carefully: method and aim of the work, selection of solvent systems and experimental conditions, and choice of detection methods. In view of this, we shall try in the following paragraphs to give some practical advice.

The Use of Paper Chromatography in Qualitative Organic Analysis

In order to chromatograph a certain substance, it must fit the following requirements: it must be detectable, and also soluble, if we wish to find a suitable solvent system (p. 63). It should not be volatile and it should be reasonably stable under the conditions of chromatography. If the substance does not suit any of these requirements, it must be changed chemically to a suitable derivative (43). The chosen derivative should offer itself to sensitive detection (or it may be colored itself), it should have good chromatographic properties, i.e., it should be sufficiently soluble, and it must be preparable by a simple reaction in which all relevant components react without leading to side reactions, which would prevent reliable detection and orientation on the chromatogram. In certain cases the preparation of suitable derivatives can increase the structural differences of a pair of unresolvable substances so much that the derivatives are separable. In Part 2 a number of special cases are given where derivatives were successfully used, for example, for the separation of alcohols (p. 154), aliphatic and aromatic amines (p. 342), mercaptans (p. 387), and carbonyl compounds (p. 222).

Complicated molecules, on the other hand, can often be broken down by chemical (or some other) means to simpler but characteristic fragments easily identifiable by paper chromatography. This method was used for the analysis of polysaccharides (p. 314), azo dyes (p. 363), and so on.

Paper chromatography can be used for the following problems:

1. *Identification.* Every substance can be characterized by its position on the chromatogram, expressed as an R_F value. By definition, this value represents the ratio $R_F = a/b$, where a is the distance of the center of the spot from the point of application (start) and b is the distance of the solvent front from the start. The R_F value indicates the position of the spot on the chromatogram, but it cannot be considered as a true constant, because very often it changes slightly even though the experimental conditions are

kept rigorously constant. Its significance consists in the fact that it shows how the given substance is separated from other substances and, generally, how it behaves in a given system. The identity of two substances can be checked only if the two are compared on the same chromatogram in at least two completely different systems. We must endeavor to use such a solvent that the substances appear in the middle of the chromatogram, not near the front or the start. If we want to compare the analyzed compound A with the standard B, the sample should be applied on the same chromatogram as follows: (1) compound A, (2) compound B, (3) a mixture of A + B, their amount being only one half of that used for other runs, (4) compound A, and (5) compound B. In addition to the position on the chromatogram, detection also helps identification. In actual fact, the detection represents a color reaction with a pure substance. In order to further check the result obtained, the colored spots may be eluted and their spectra compared.

If no standard is available, one can sometimes utilize for an approximate identification of the analyzed substance the relationships between the chromatographic behavior and the structure (44). These relationships are based on Martin's assumption (45, 46) that the chromatographic mobility of substances expressed in terms of the R_M values is composed additively of contributions characteristic of their single parts, i.e., of functional groups (so-called group constants or ΔR_M values). According to this rule, the introduction of a new functional group into the molecule of an organic compound should produce a change in its chromatographic mobility which is characteristic of the given system and constant for this group irrespective of the other parts of the molecule. It was shown later that these relationships are not general, and that the influence of the rest of the molecule cannot be neglected (47, 48). This is why these relationships can be used with advantage only in closed, smaller groups of substances characterized by identical interactions inside and outside the molecule, and if all unfavorable external influences are eliminated. We recommend, therefore, that these relationships be used only by experienced workers and with great caution.

2. *Checking the purity.* Paper chromatography can also be used for detecting small amounts of impurities in analyzed samples. Success depends on a suitable choice of solvent system and color reaction. For example, all purification procedures can be followed chromatographically by checking the purified compound as well as the concentration of impurities in mother liquors, mixed fractions after column chromatography, etc. The preparation of derivatives can also be followed by this method.

3. *Analysis of mixtures.* Paper chromatography is most effective in analyses of mixtures. It is able not only to separate submicro quantities of a sample into its components, but of enabling them to be identified at

the same time. The success of such an analysis depends on the proper choice of a solvent system and a suitable detection scheme. The effect can be increased by combining paper chromatography with other types of chromatography or with paper electrophoresis. The use of paper chromatography for following the course of chemical reactions can be highly recommended. This can be done by subtracting minute aliquots from the reaction mixture at certain time intervals and chromatographing them. The chromatogram shows how the starting component disappears, while the reaction product or even by-products appear and their concentrations increase.

4. *Systematic analysis.* An analyst can utilize paper chromatography systematically in several ways. When analyzing complicated mixtures he can carry out a series of preliminary chromatographic experiments with them without consuming a weighable amount of the sample. The sample can be chromatographed in a series of solvents, and well-selected reagents are then used for detection. Very useful information can often be obtained in this way. For example, the behavior of a certain substance in different solvents can be indicative of its solubility, polarity, and the possibilities of its being separated from impurities. The relationships between the chromatographic behavior and the structure of substances can be studied and utilized. The analyst can also submit the substances to various systematic chemical changes, for example, to cleavage or hydrolysis, and transform them in this way to simpler known substances, which can be then identified chromatographically.

The Choice of Solvent Systems

A successful application of paper chromatography depends on the correct choice of the solvent system. A knowledge of the solubility of organic substances (see also p. 105), and of the fact that the chromatographed substance should be well soluble in the stationary phase and correspondingly less so in the mobile one, is of great help (49). If the substance is only slightly soluble in the mobile phase, it remains on the start line; if it is too soluble, it moves with the solvent front. On the basis of these considerations the following three cases of the stationary phase can be distinguished:

1. Untreated papers; i.e., the system cellulose plus water as the stationary phase. This is suitable for water-soluble substances. As the mobile phase, mixtures of organic solvents with water can be used (butyl acetate or ethyl acetate, amyl alcohol, n-butanol, n-propanol, ethanol, methanol). The enumerated organic solvents are put in order so that with increasing polarity the R_F values also increase. This means that if a substance remains on the start line in water-saturated n-butanol, a lower alcohol will have to be used instead, and vice versa, if the substance moves with the solvent front, it will be

necessary to add a less-polar solvent to the system. An addition of water also increases R_F values; therefore, the R_F values in water-saturated butanol will be higher than in water-saturated amyl alcohol, because amyl alcohol not only decreases R_F values, but it also can dissolve much less water in itself.

Sometimes it is necessary to make use of a basic or acid medium; for example, chromatography is carried out in the presence of ammonia, pyridine, ethylamine, acetic acid, formic acid, or hydrochloric acid. In some cases the substances mentioned function only as solvents; in other cases they have their own important role. For example, it might be necessary to transform volatile acids or bases to water-soluble salts, and this is done just by acting on them with bases or acids. In some cases the addition of strong acids or bases is used either for the suppression of the dissociation of acids or bases, or on the contrary, to enhance their complete dissociation. If the chromatographic behavior of substances is dependent on pH, the chromatography is carried out in the presence of buffers.

2. Papers impregnated with polar solvents other than water, for example formamide, dimethylformamide, acetamide, ethylene glycol, propylene glycol, N-methylformanilide, or methanol, serving as the stationary phase. If the paper is impregnated with formamide, then hexane, cyclohexane, benzene, toluene, carbon tetrachloride, chloroform, ethyl acetate, and their mixtures can be used as the moving solvent. The R_F values will increase with the polarity of the mobile solvent from hexane to ethyl acetate. Glycols behave as formamide, on dimethylformamide the R_F values are lower than on formamide. These systems are suitable for substances of medium polarity, and they will often be used in this monograph. The indication "25% dimethylformamide/hexane" means that the paper is impregnated with a 25% dimethylformamide solution (p. 65), and that hexane is the moving solvent.

3. Papers impregnated with nonpolar solvents, for example, with paraffin oil, kerosene, soya oil, olive oil, lauryl alcohol, or 1-bromonaphthalene, forming the stationary phase. The mobile phase is always a mixture of water and an organic solvent (e.g., alcohols, acetone, acetic acid, formamide, dimethylformamide). If paraffin oil is fixed on the paper, even glacial acetic acid can be employed, as well as concentrated methanol, for they are immiscible with the stationary phase. In other instances, especially in the case of lauryl alcohol and 1-bromonaphthalene, the moving phase must be thoroughly saturated with the stationary phase, in order to prevent the elution of the stationary phase during the chromatographic process. The addition of an organic solvent to the mobile phase increases the R_F values, while the addition of water decreases them. Exactly as in the case of untreated papers, here, too, a base or an acid is added to the mobile

phase if necessary. In this monograph the indication "1-bromonaphthalene/ /80% acetic acid" means that the paper is impregnated with 1-bromonaphthalene see below and that the mobile phase is 80% acetic acid saturated with 1-bromonaphthalene.

In addition to common chromatographic papers, specially treated papers are also available, for example, acetylated, ion-exchange papers, glass-fiber papers, and papers impregnated with adsorbents. If acetylated cellulose is used, mixtures of organic solvents with water are used as the mobile phase (see above). The use of acetylated papers is advantageous only if they are commercially available, because their laboratory preparation is very tedious. Papers impregnated with adsorbents make adsorption chromatography on paper possible. To a certain extent, they are analogous to thin layers of adsorbents (see below).

Work with Impregnated Papers

The preparation of impregnated papers requires a certain attention and skill. It is carried out by dipping and drawing a strip of paper (marked with indications in pencil) through a solution of the stationary phase in a volatile solvent. This is done by taking the paper at the opposite end to the start and drawing it through the solution in a dish. The wet strip is then hung in the hood by clipping the end opposite to the start to a glass rod. The chromatogram is allowed to stand under the hood until the auxiliary volatile solvent is evaporated, approximately 15 min. In the case of dimethylformamide and ethylene glycol the drying must be watched carefully because both stationary phases evaporate from the paper to an appreciable extent. Therefore, in summer the drying time should be shortened to 10 min or less, if necessary. The R_F values on impregnated papers depend very much on the amount of the stationary phase; the more of it the paper contains, the lower the R_F values are (see Fig. 27 p. 155). It is important, therefore, to keep the concentration of the impregnation solutions constant, as well as the drying time. The appropriate concentrations of the impregnation solutions which should be employed for the cases described in Part 2 are: 5% solution of paraffin oil in hexane or cyclohexane, 5% solution of lauryl alcohol in ethanol, 10% solution of 1-bromonaphthalene in chloroform, 20% solution of dimethylformamide in ethanol or acetone, 25−50% solution of dimethylformamide in benzene, and 20% solution of ethylene glycol in ethanol.

Preparation of Silica-Gel-Impregnated Paper (50)

A strip of chromatographic paper is drawn three times through a freshly prepared mixture of 100 parts of an ∼7% water-glass solution and 30 parts of a 5% ammonium chloride solution. The excess mixture is eliminated from both sides of the paper with a glass rod, and the paper is allowed to hang freely overnight. It is then washed three times with water and dried for 30 min at 80 − 90 °C. Papers so treated can be kept in the laboratory indefinitely.

Detection

If an organic compound is to be chromatographed successfully, a suitable method of its detection on the chromatogram must be available. Detection is carried out by applying a color reaction in the following manner: the dried chromatogram is sprayed with a fine spray of the detection reagent from an atomizer (usually making use of compressed air). When doing this the chromatogram is either placed on a sheet of filter paper or hung vertically in a suitable frame. The spraying is carried out under the hood or in a room from which the exhalations and the mist of the reagent can be drawn off. In some cases the spraying can be substituted by immersing the chromatogram in or drawing it through the reagent solution. A very common method of detection is the inspection of the chromatogram under UV light. Color reactions used for detection (51) can be quite general and nonspecific, but can also be specific or even selective for a given group of substances or an individual substance. The value of universal color reactions consists in the fact that they can detect a large amount of substances, which is very advantageous during orientation analyses of an unknown sample. Selective or specific reactions help in identification. When a varied mixture of organic compounds is analyzed, it is recommended that it be chromatographed several times in parallel sections on the same chromatogram, that the latter be cut into single strips which are then each detected with a different reagent, and that they then be compared.

Thin-Layer Chromatography

With this method of chromatography the actual process takes place on a thin layer of adsorbent fixed on a glass or aluminum plate, or on a plastic foil. The separation takes place on the basis of a partition mechanism, adsorption, or ion exchange, etc. Individual mechanisms are often combined or overlap, and it is not easy to determine the boundaries between individual processes. Basic forces playing a decisive role during the separation are ionic forces, further coordination forces, chelate formation, association of dipoles, hydrogen-bond formation, and dispersion (van der Waals forces).

Two modifications of the technique are known. In the first a suspension of the adsorbent, often containing some kind of binder to bind the layer more firmly to the plate, is poured on a glass plate, and in the second modification the adsorbent is poured onto the plate in a dry state. In the first instance we speak of chromatography on fixed or adhering layers (with binder or without it), and in the second we speak of loose or nonadhering layers. To carry out the chromatography, a suitable place to work as well as suitable equipment and a supply of powdered material for the preparation of layers are necessary. For a detailed description of various types of equipment see the corresponding specialized monographs (11 – 21). The

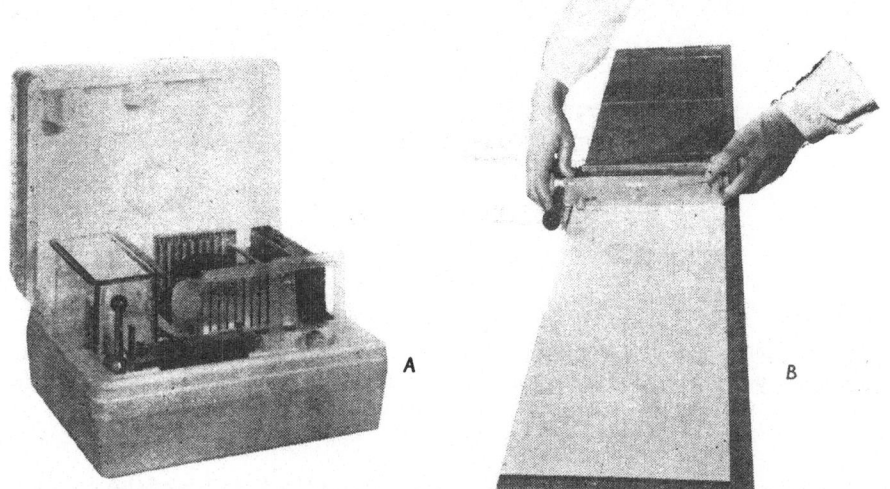

Fig. 15. Equipment for thin-layer chromatography offered by Desaga and Co.: (A) general view, (B) application of a suspension of adsorbent.

preparation of adhering layers is carried out either manually or by using simple devices – a simple glass rod with two plastic rings on its ends – or by means of more complicated commercial devices. Figure 15 shows the equipment for the preparation and work with adhering layers available from the firm Desaga, Heidelberg. Plates of 5, 10, or 20×20 cm size are most commonly used, the thickness of the layer is 250 μ to 1 mm, and the ascending development is preferred. To obtain very rapid information, the process can be carried out on microscope slides of 75×25 mm size coated with an adsorbent layer (52). The coating of the plates is carried out in exactly the same manner as in the case of normal-size plates. This type of chromatography gives extremely valuable results in those cases where the course of a reaction is followed, or when a suitable solvent system is looked for. The chromatography (development) itself is carried out in a 50 ml beaker covered with a watch glass, and the operation usually takes only several minutes. It is very convenient to use various foils with ready layers, which are supplied by various firms and which can be used directly for analysis.

The principle of the preparation of nonadhering or loose layers consists in an even spreading of the dry adsorbent over a suitable plate. The size of such plates is in no way limited; plates 20 – 40 cm long are often used, their breadth depending on various factors. They can be made from simple window glass. Sometimes photographic plates freed from emulsion are recommended.

Spreading of the adsorbent (alumina, silica gel, of 0.1 mm and smaller particle size) is carried out with the above-mentioned glass rod, conveniently

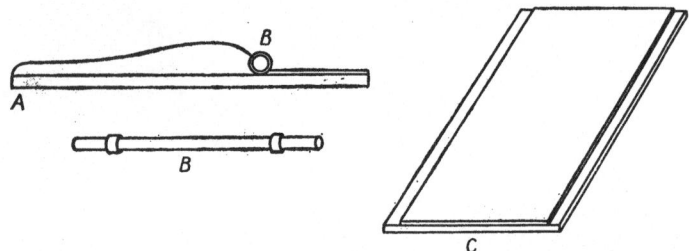

Fig. 16. Preparation of a dry nonadhering layer of adsorbent: (A) glass plate with the spread adsorbent, (B) device for making layer smooth, (C) layer ready for application of samples.

1 – 1.5 cm in diameter, and 2 – 3 cm longer than the breadth of the prepared chromatogram (plate). Both ends of the rod are provided with ~ 1.5-cm-long pieces of rubber or polyethylene tubing. The distance between the inner edges of the rings determines the breadth of the layer, and the size of the walls of the tubing determines the thickness of the layer. The thickness is usually 0.6 – 2.0 mm. The layer is formed by spreading an evenly distributed layer of adsorbent (2 – 3 mm thick) over a horizontally fixed plate. The glass rod is then taken with both hands and laid on one end of the plate, and the excess of the adsorbent is eliminated by sliding evenly with the device over the plate. An even, smooth layer is so formed (Fig. 16). The sliding movement and the pieces of tubing on the ends of the glass rod produce two strips of empty space on the longitudinal edges of the plate. The application of a loose, nonadhering layer can be carried out inside an adapted plastic dish used for photographic development, provided with suitable stops (Fig. 17). The plates with loosely spread adsorbent are immediately ready for use, and should not be stored. During the manipulation they should be protected from sudden motion, strong streams of air, and similar mechanical influences. The plates can be inclined up to 40°, so that the development can

be carried out with safety at an inclination of 20°. It is carried out in an ascending manner in a closed glass vessel, as can be seen in Fig. 18.

During the work itself care should be devoted to the application of samples on the layer. In contrast to paper chromatography, this operation requires a certain skill and feeling in order to preserve the integrity of the layer. As for adhering layers, they should be touched carefully with the tip

Fig. 17. Polyvinyl dish adapted for the application of thin layers.

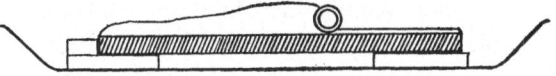

of a capillary pipette and the solution allowed to be sucked into the layer. As for nonadhering layers, they should not be touched with the tip of the pipette, but only with the outgoing drop of solution. The application is best carried out by using special templates made from transparent plastic.

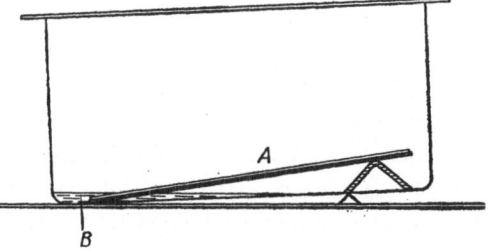

Fig. 18. Running the chromatogram with a nonadhering layer: (A) plate with the layer, (B) solvent.

It is also important to keep the applied volumes constant and to respect the fact that the solvent in which the sample was dissolved, if not eliminated from the adsorbent, can impair its activity and appreciably influence the R_F values. In chromatography on thin layers of nonadhering alumina we have observed that the use of various solvents for the application of the sample on the same chromatogram, such as, for example, benzene, alcohol, or pyridine, results in appreciable differences in R_F values of the same substance.

The amount of substance applied is often much smaller than in paper chromatography, and it is also determined by the thickness of the layer.

Application of Thin-Layer Chromatography in Organic Analysis

The use of thin-layer chromatography in organic analysis is much the same as that described for paper chromatography. Its use is broader and also includes those cases where the substances can be detected only by aggressive reagents (conc. sulfuric acid), or when the substance does not give color reaction and when it is necessary (on using a mineral adsorbent) to localize the spots by carbonization, which is carried out by sweeping the plate with the direct flame of a gas burner. Some volatile substances are also more firmly bound to adsorbents than on paper.

Choice of Adsorbents and Solvent Systems

When using thin-layer chromatography, either on adhering or nonadhering layers, the choice of a suitable adsorbent is of major importance. For adhering layers the following materials are commercially available: silica gel, alumina, kieselguhr, magnesium silicate, polyamide, cellulose, acetyl cellulose, and ion exchangers. They can be purchased from various companies, for example, E. Merck, Fluka AG., Serva, M. Woelm, Macherey, Nagel and Co., Schleicher and Schüll, and Excorna. It is important to stress here that due attention should be given to the preparation and the manipulation of the adsorbent. First, it must have the required activity, acidity, or alkalinity; impregnation with suitable reagents may be necessary in order to direct separations on it (for example, impregnation with silver nitrate for the separation of unsaturated compounds); impregnation with fluorescent indicators is often useful, as is impregnation with nonvolatile organic solvents functioning as a stationary phase if partition chromatography is more desirable than a mechanism based on adsorption. A standard procedure has been proposed (52a) for the testing of activity. The choice of a solvent system is equally important. This is based on the same principles as in classic adsorption, partition, and other types of chromatography depending on the principle involved in the separation process.

Paper or Thin-Layer Electrophoresis

Paper electrophoresis is a separation process based on the differential migration of electrically charged molecules in an electric field. It is carried out on a strip of paper impregnated with a suitable electrolyte. Solutions of substances to be analyzed are applied on it (as in paper chromatography) and the ends of the strip are connected with the positive and negative poles,

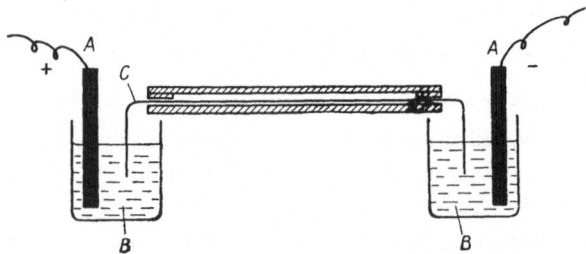

Fig. 19. Apparatus for paper electrophoresis: (A) electrodes, (B) electrolyte, (C) paper strip impregnated with electrolyte.

respectively, of a dc electric field. A potential drop is thus obtained, the magnitude of which depends on the applied voltage and on the electrolyte levels around both electrodes (Fig. 19). Ions charged positively move toward the cathode, negatively charged ions move toward the anode, and neutral particles do not move at all. After a suitable time the process is interrupted and the separated substances — unless colored — are detected as in paper chromatography.

Several factors influence the movement of particles in electrophoresis: primarily, the properties of the particles themselves (for example, charge, size of the molecule and its shape, and degree of dissociation), but also the properties of the electrolyte (for example, its concentration, pH, ionic

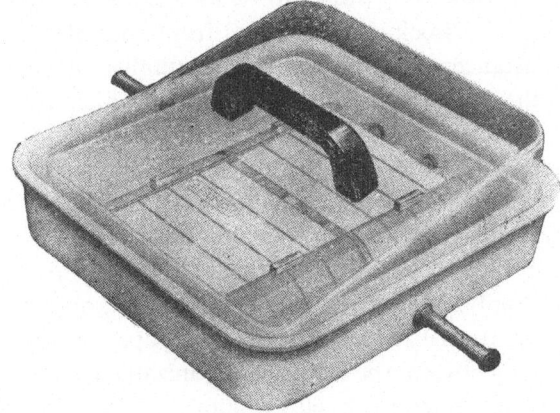

Fig. 20. Equipment for paper electrophoresis offered by Shandon and Co.

strength, conductivity, temperature, viscosity, dielectric constant of the solvent, and the possible presence of nonelectrolytes) and those of the electric field (for example, its intensity, homogeneity, the possible presence of an alternating component in the direct current used, and the possible creation of heat and changes in pH caused be the influence of the electric current). Generally, it can be stated that the rate of migration increases with the charge of the particles and with the potential difference, and that it decreases with the increasing size of the molecule and with increasing electrolyte concentration.

For the analysis suitable equipment is necessary: a source of direct current and a chamber for the electrophoresis itself, solutions of electrolytes, and chromatographic paper. The apparatus for electrophoresis is very simple and can be improvised in every laboratory. However, many kinds of both simple and more complicated apparatus are commercially available which are satisfactory for routine work with organic compounds. Figure 20 shows the newest equipment offered by Shandon and Co. For a more detailed description of the various apparatus and operating directions, as well as for the choice of optimum conditions, see the appropriate monographs (8, 11, 36 – 39). Here we shall only briefly summarize several principles for the use of paper electrophoresis in organic analysis.

Paper electrophoresis is of enormous importance for the separation of mixtures of organic substances and for their identification and the checking of their purity. As the separation of substances during electrophoresis is based on completely different principles than in chromatography, this method is complementary to chromatographic methods, and can often help in those cases where the chromatographic behavior of a pair or a group of substances is not sufficiently different. Its use will be of major importance for the separation of dissociable substances from neutral ones and for the separation of substances carrying a different number of charges, or substances carrying the same number of charges but varying in their molecular size; it will also separate mixtures of anions and cations which could interfere in paper chromatography even of neutral substances. In such a case the efficiency of separation can be increased by combining electrophoresis with chromatography. It is obvious, however, that the comparison of an unknown substance with a supposed authentic specimen for identification can be carried out only under identical conditions, i.e., on the same electropherogram, as was described for the case of paper chromatography on p. 61. The relation between the number of ionizable functional groups in the molecule and its mobility at electrophoresis can be utilized for the determination of the number of such groups in the molecule.

Analogously to paper electrophoresis, thin-layer electrophoresis can also be performed. Corresponding equipment is also commercially available.

Gas Chromatography

In gas chromatography the organic substance in the gaseous state is conducted by a stream of a carrier gas through a column filled with a solid carrier usually impregnated with a suitable high-boiling liquid, the so-called stationary phase. When moving through the column different components of the mixture move at different speeds, depending on their affinity to the stationary phase, and are so separated. At the end of the column a suitable detector is attached to measure the concentration changes of substances in the effluent gas. A scheme for such a device is shown in Fig. 21.

For the actual work suitable apparatus is essential. Gas chromatographs are relatively expensive, but they are easily available in the market. The choice of the apparatus should be made on the basis of the work planned, the applicability of the apparatus as described by the producer, of experience from other laboratories, and financial considerations. The economical use of gas chromatographs depends on their correct maintenance and repair.

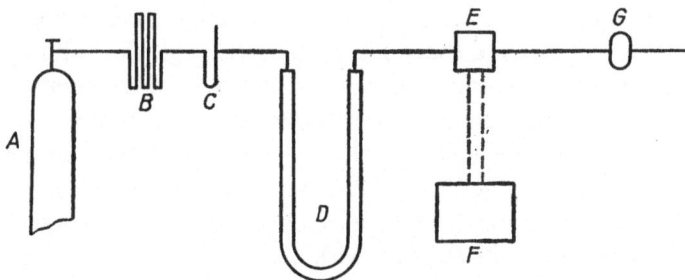

Fig. 21. Scheme of a gas chromatograph: (A) carrier gas tank, (B) manostat, (C) manometer, (D) chromatographic column, (E) detector, (F) flow meter, (G) recorder.

The Use of Gas Chromatography in Qualitative Organic Analysis

In organic qualitative analysis gas chromatography is used for identification, the analysis of mixtures, and the preparative separation and purification of organic substances. A prerequisite is a suitable volatility of the substance, its chemical stability under the given experimental conditions, and the possibility of detection by available detectors. If the substance does not possess these qualifications, it can be modified, chemically, by changing it to a suitable derivative or breaking it down to simpler substances. When a very complicated mixture is analyzed it is sometimes advisable to first carry out a preliminary separation of the mixture by distillation, preparative gas or column chromatography, or by chemical means, as, for example, by separating the mixture into acid, neutral, and basic components. Unknown substances can generally be identified in six ways (53), which are often combined:

1. *Identification by comparison with an authentic sample.* After obtaining the first record of the tested substance or mixture, substances believed to be identical or similar to the analyzed one are gradually added to it, and it is observed whether one of the components of the mixture gives the same peak as the added substance, i.e., whether after the addition of the standard to the mixture the peak of a component of the mixture increases. If the added substance is not identical with any of the components, a new peak appears in the record. Of course, two different substances having the same elution volume can also occur in the mixture. It is therefore necessary to compare the tested mixture with standards using several stationary phases of different polarity, or even at different temperatures.

2. *Identification by comparison of elution data.* The basic quantity characterizing the compounds qualitatively is the elution volume V_R, which represents the volume of gas necessary for the elution of the analyzed component. For the sake of comparison the specific elution volume V_g is also used; it is calculated from the elution volume by correcting it for the amount of carrier gas necessary for passage through the column and the pressure drop in the column, and it is calculated for a temperature of 0 °C and for 1 g of the stationary phase. In practice, this is done as follows: the value of the specific elution volume is calculated for one component of a series, and this value is taken as unity, the other values then being calculated with respect to it (the so-called elution rations r are obtained).

To eliminate the difficulties with poorly accessible standards, Kovats (54) and Evans and Smith (55) have proposed a system of retention indexes, the members of n-paraffin series being taken as standards.

3. *Identification based on the relationship between the structure and the chromatographic behavior of a substance.* From the rules governing the behavior of substances in gas chromatography, a series of relationships has been deduced which can be utilized for identification:

a) The logarithm of the elution volume for each member of a structurally identical homologous series is linearly proportional to the number (n) of carbon atoms in the molecule (log $V_R = kn$), and the logarithm of the ratio of the elution volume of neighboring (vicinal) members of a series (so-called separation factor q) are constant for the whole series.

b) The logarithm of the elution volume varies linearly with the reciprocal of the absolute temperature T [log $V_R = k(1/T)$]. Graphical representation of this relationship gives straight lines for various groups of substances (within a temperature range of 100 °C), the gradients of which are characteristic for groups or series of substances.

c) Elution volumes of a homologous series of substances on two different stationary phases are proportional ($V_R = KV_R'$). In a graphical representation this means that straight lines are obtained issuing from the same origin, but having different gradients. In actual practice, experimentally obtained data for standard compounds are usually plotted and it is observed whether the unknown substance satisfies the given relationship within the limits of error. In such instances it is not necessary to have a standard at one's disposal, but the method has only a limited application because it can be employed only for the members of a homologous series.

4. *Reaction gas chromatography.* Identification of a series of groups of substances can be performed by carrying out a chemical reaction with the sample either before it enters the column or after its passage through it. A reaction should be chosen which occurs selectively with only a certain

group of substances. Comparison of the analysis record of the original sample with that after the reaction has been carried out show immediately that any missing peaks belonged to the corresponding (reactive) group of substances. It is also possible to carry out a reaction with the mixture before it enters the column, by cleaving it to simpler compounds and so transforming to substances easily identifiable by gas chromatography. Thus, for example, on reaction with hydriodic acid, alkyl groups can be split off from alkoxy groups in the form of alkyl iodides, and the latter can be identified by gas chromatography. These procedures are usually carried out in a small reactor to which a gas chromatograph is attached. The procedure of Beroza (56), in which the tested substance is catalytically hydrogenated, is very interesting; during this reaction the functional groups are eliminated, and, depending on the type of substance being analyzed, either the parent paraffin or a paraffin shorter by one carbon atom is formed; the number of carbon atoms indicates the length of the carbon-atom chain in the original aliphatic compound. This method can be used for determination of the basic skeleton of aliphatic as well as aromatic compounds. In special cases the identification of organic compounds is carried out by combining pyrolytic decomposition of the organic compound with analysis of the pyrolysis products by gas chromatography. The method is employed for nonvolatile substances of a complicated nature (for example, polymers), which can be characterized by other means only with difficulty. The reader should refer to some recent reviews (57, 58).

5. *Identification by means of "Responsegrössen."* Similarly to paper chromatography, where, in addition to the R_F value, a color reaction also serves for identification, in gas chromatography various detectors can be used for a more-detailed identification, which can indicate preferentially or even specifically substances belonging to certain groups.

6. *Identification on the basis of other methods.* A mixture of substances can be separated by gas chromatography and the components isolated and identified by other means, for example, by infrared spectroscopy or mass spectrometry.

For the isolation of substances, larger columns are used, 10 – 25 mm in diameter. For the detection of vapors streaming out from the columns a device should be used in which the analyzed substance is not chemically changed (for example, the detection may be based on the decrease in thermal conductivity). On a column of 1 m length and 25 mm diameter, 2 g of mixture can be worked up (depending on the weight of the analyzed substance in it). The choice of the stationary phase and of experimental conditions is the same as given below; a flow rate of 200 to 300 ml/min is usually employed.

The Choice of Optimum Conditions for Gas Chromatography

The successful application of gas chromatography to a given problem depends on a suitable arrangement and choice of working conditions. The quality of separation depends on the following factors:

1. *Temperature of the column*. Optimum temperature is at about 30 − 50 °C below the boiling point of the components. The increase in temperature causes a logarithmic decrease in elution time and the separation deteriorates; at low temperatures the elution curves broaden and the separation is also impaired.

2. *Gas flow*. For columns 0.4 − 0.5 mm in diameter a flow of 20 − 40 ml is suitable. With an increase in flow, the elution volumes decrease.

3. *Length of the column*. The separation increases with the square root of the column length.

4. *Stationary phase*. This is the most important variable in gas chromatography. The separation of substances which are closely related chemically, for example, homologs having different boiling points, does not present difficulties. As stationary phase a substance is chosen whose polarity is not too different from the polarity of the separated substances. The separation of substances with close or even identical boiling points is more difficult the closer they are chemically. In such cases the separation can be enhanced by increasing the efficiency of the column by taking a capillary column (also by prolonging the column, diminishing the amount of the stationary phase and of the sample), or by employing selective stationary phases, which would make the separation possible even on short columns. When choosing selective stationary phases the considerations are based on the thermodynamics of miscible phases. From these considerations it can be deduced that gas chromatography can even separate substances whose vapor pressures are practically identical if it is possible to find a specifically interacting phase which will cause the separated substances to show deviations from ideal behavior: substances showing positive deviations from ideal behavior will be eluted earlier than substances behaving ideally, and vice versa. The deviations from ideal behavior may be caused by interactions between the chromatographed substance and the stationary phase through van der Waals forces, hydrogen bonds, complex formation, etc. For example, a mixture of benzene and cyclohexane will be eluted on a nonpolar phase in the given order, while on a polar phase the order will be reversed because in this case the effective tension of benzene is decreased in consequence of an interaction with its π-electrons; this interaction causes a retention of benzene in the column.

In the choice of particular substances as phases an empirical approach is the one usually taken. The choice is done by analogy. The greatest efficiency is attained by combining the use of selective stationary phases with a prolongation of the column.

The several hundreds of substances for stationary phases described in the literature can be classified into the following four groups:

1. *Nonpolar substances:* paraffin oil, squalene, apiezon oils, silicone oils, polyethylene, etc.

2. *Substances of medium polarity:* phthalic acid esters, and ethyl hexyl sebacate, etc., polyesters.

3. *Polar substances:* dimethylformamide, dimethylenesulfonal, β, β'-oxydipropionitrile, ethylene glycol-bis (propionitrile) ether, polyethylene glycol.

4. *Special substances:* these contain several polar groups (sugars, 3,5-dinitrobenzoates), or they act by the configuration of their molecules (benzoquinolines).

In Part 2 of this monograph optimum conditions for individual groups of substances will be given; as regards a more detailed study of the problem, the original literature and the appropriate monographs should be consulted.

References

1. Cassidy, H. G.: Adsorption and Chromatography. Interscience Publishers, New York, 1951. 477 pp.
2. Abbott, D., and Andrews, R. S.: An Introduction to Chromatography. Longmans, London, 1965.
3. Heftmann, E. (Editor): Chromatography. Reinhold, New York, 1961. 753 pp.
4. Lederer, E., and Lederer, M.: Chromatography. II. Ed. Elsevier Publishing Company, Amsterdam, 1957, 711 pp.
5. Strain, H. H.: Chromatographic Adsorption Analysis. Interscience Publishers, New York, 1942.
6. Hais, I. M., and Macek, K. (Editors): Paper Chromatography. A Comprehensive Treatise. Academic Press, New York and Publishing House of the Czechoslovak Academy of Sciences, Prague, 1963. 955 pp.
7. Hais, I. M., and Macek, K. (Editors): Handbuch der Papierchromatographie. Vol. 1. Grundlagen und Technik. 2 nd Ed. G. Fischer Verlag, Jena, 1963. 1069 pp.
8. Block, R. J., Durrum, E. L., and Zweig, G.: A Manual of Paper Chromatography and Paper Electrophoresis. Academic Press, New York-London, 1958. 710 pp.
9. Cramer, F.: Papierchromatographie. 5th Ed. Verlag Chemie, Weinheim, 1963. 216 pp.
10. Waldi, D.: Chromatographie. 2nd Ed. E. Merck, Darmstadt, 1959. 189 pp.
11. Smith, I., and Feinberg, J. G.: Paper and Thin Layer Chromatography and Electrophoresis. A Teaching Level Manual. 2nd. Ed. Shandon Sci. Co., London, 1965. 241 pp.
12. Stahl, E.: Dünnschichtchromatographie. Springer-Verlag, Berlin, 1962.
13. Stahl, E. (Editor): Thin-Layer Chromatography. A Laboratory Handbook. Springer-Verlag, Berlin — Heidelberg — New York, 1969. 1041 pp.

14. Truter, E. V.: Thin-Film Chromatography. Cleaver-Hume Press, London, 1963. 205 pp.

15. Bobbitt, J. S.: Thin Layer Chromatography. Reinhold, New York, 1963, 250 pp.

16. Marini-Bettolo, G. B. (Editor): Thin-Layer Chromatography. Elsevier, Amsterdam, 1965. 232 pp.

17. Randerath, K.: Thin-Layer Chromatography. Academic Press, New York, 1963. 250 pp.

18. Randerath, K.: Dünnschicht-Chromatographie. 2nd Ed. Verlag Chemie, Weinheim, 1965. 291 pp.

19. Lábler, L., and Schwarz, V. (Editors): Chromatografie na tenké vrstvě. (Thin-Layer Chromatography.) Publishing House of the Czechoslovak Academy of Sciences, Prague, 1965. 465 pp.

20. Hashimoto, Y.: Thin-Layer Chromatography. Hirokawa Publishing Co., Tokyo, 1963.

21. Ishikawa, N., Hara, S., Furuya, T., and Nakazawa, Y.: Thin-Layer Chromatography. Fundamentals and Applications. Nanzando Co., Tokyo, 1963.

22. Keil, B., Herout, V., and Protiva, M.: Laboratoriumstechnik der Organischen Chemie. Akademie-Verlag, Berlin, 1965.

23. Hrapia, H.: Elutionschromatographie. Akademie-Verlag, Berlin, 1965.

24. Helferich, F.: Ion Exchange. McGraw-Hill, New York, 1962.

25. Samuelson, O.: Ion Exchange Separations in Analytical Chemistry. Wiley, New York, 1963.

26. Griessbach, R.: Austauschadsorption in Theorie und Praxis. Akademie-Verlag, Berlin, 1957.

27. Dorfner, K.: Ionenaustausch-Chromatographie. Akademie-Verlag, Berlin, 1963.

28. Desty, D. H. (Editor): Vapor Phase Chromatography. Academic Press, New York, 1957.

29. Bayer, E.: Gas-Chromatographie. Springer-Verlag, Berlin, 1962.

30. Keulemans, A. I. M.: Gas Chromatography. 2nd Ed. Reinhold, New York, 1959.

31. Kaiser, R.: Gas Phase Chromatography. Vol. 1 − 3. Butterworths, London, 1963.

32. Kaiser, R.: Gas-Chromatographie. Akad. Verlagsges. Geest und Portig, Leipzig, 1962.

33. Lewis, J. S.: Compilation of Gas Chromatography Data. Special Technical Publication No. 343. ASTM, Philadelphia, 1963.

34. Littlewood, A. B.: Gas Chromatography. Academic Press, New York—London, 1962.

35. Knox, J. H.: Gas Chromatography. Methuen, London, Wiley, New York, 1962.

36. Lederer, M.: An Introduction to Paper Electrophoresis and Related Methods. Elsevier Publishing Company, Amsterdam—New York, 1957.

37. Smith, I.: Chromatographic and Electrophoretic Techniques. Vols. 1 and 2. W. Heinemann Medical Books, Inc., London, 1960.

38. Wunderly, C.: Principles and Applications of Paper Electrophoresis. Elsevier, New York, 1961.

39. Ribeiro, L. P., Mildieri, E., and Alfonso, O. R.: Paper Electrophoresis. Reviews of Methods and Results. Elsevier Publishing Company, Amsterdam, 1961.

40. Macek, K., Hais, I. M., Gasparič, J., Kopecký, J., and Rábek, V.: Bibliography of Paper Chromatography 1957 − 1960 and Survey of Applications. Publishing House of the Czechoslovak Academy of Sciences, Prague, 1962.

41. Macek, K., Hais, I. M., Kopecký, J., and Gasparič, J. (Editors): Bibliography of Paper and Thin-Layer Chromatography 1961—1965 and Survey of Applications. Elsevier Publishing Company, Amsterdam, London, New York, 1968.
42. Von Arx, E., and Neher, R.: J. Chromatog. 8, 145 (1962).
43. Gasparič, J.: Chem. Listy 55, 1439 (1961).
44. Procházka, Ž.: Chem. Listy 58, 911 (1964).
45. Martin, A. J. P.: Biochem. Soc. Symposia (Cambridge) 3, 4 (1949).
46. Martin, A. J. P., and Synge, R. L. M.: Biochem. J. 35, 1358 (1941).
47. Gasparič, J., Gemzová, I., and Šnobl, D.: Collection Czech. Chem. Commun. 31, 1712 (1966).
48. Marcinkiewicz, S., Green, J., and McHale, D.: J. Chromatog. 10, 42 (1963).
49. Gasparič, J., and Večeřa, M.: Mikrochim. Acta 1958, 68.
50. Borecký, J.: Collection Czech. Chem. Commun. 30, 2549 (1965).
51. Anfärbereagenzien für die Dünnschicht- und Papier-Chromatographie. E. Merck, Darmstadt.
52. Naff, M. B., and Naff, A. S.: J. Chem. Educ. 40, 534 (1963).
52a. Heřmánek, S., Schwarz, V., and Čekan, Z.: Collection Czech. Chem. Commun. 26, 3170 (1961).
53. Schomburg, G.: Zeit. anal. Chem. 200, 359 (1964).
54. Kovats, E.: Helv. Chim. Acta 41, 1915 (1958).
55. Evans, M. B., and Smith, J. F.: J. Chromatog. 6, 293 (1961).
56. Beroza, M.: Anal. Chem. 34, 1801 (1962); J. Assoc. Offic. Agr. Chemists 47, 1 (1964).
57. Perry, S. G.: J. Gas Chromatography 1964, 54.
58. McKinney, R. W.: J. Gas Chromatography 1964, 432.

10. Spectral Methods

Absorption spectroscopy in the visible, ultraviolet, and infrared regions are among the methods used very frequently for the identification and structural analysis of organic compounds. The absorption of light in the visible or ultraviolet parts of the spectrum corresponds to changes of electronic states of the molecule. If a quantum of light energy hv is absorbed, certain electrons are transferred (see below) from their ground state in the molecule of energy E_0 to the excited state of energy E_1. The increase in energy is directly proportional to the frequency and inversely proportional to the wavelength of the light:

$$\Delta E = E_1 - E_0 = hv = \frac{hc}{\lambda},$$

where h is Planck's constant, v the frequency of the radiation, c the light velocity, and λ the wavelength. The energy change ΔE comprises the whole vibrational and rotational energy of the electron.

Absorption in the ultraviolet and the visible regions is based on energetic changes of valence electrons (organometallic compounds are an exception, where the transfer of electrons takes place in the outer sphere of the metal atom). Spectral properties then depend on the type of valence electrons.

σ-Electrons, responsible for the formation of simple $C-C$ or $C-H$ bonds, can be brought to the excited state only under the influence of high-energy photons, and, consequently, substances with such bonds absorb only in the far-ultraviolet part of the spectrum (Fig. 22). Groups with multiple

Ultraviolet		Visible		Infrared		
vacuum	quartz		near	middle		far
1000 Å	4000 Å	8000 Å	10000 Å			
		0.8 μ	2 μ		16 μ	300 μ
	electronic spectra			vibrational spectra		

Fig. 22. Part of the electromagnetic spectrum with regions of vibrational and electronic spectra.

10 Å = 1 mμ (millimicron) = 1 nm (nanometer)
1 μ (micron) = 1 μm (micrometer)

bonds, for example, $C=C$, $C=O$, and $N=N$, contain, in addition to σ-electrons, π-electrons, the steric position of which is outside the axis of the bond, and also sometimes contain so-called *free-electron pairs*.

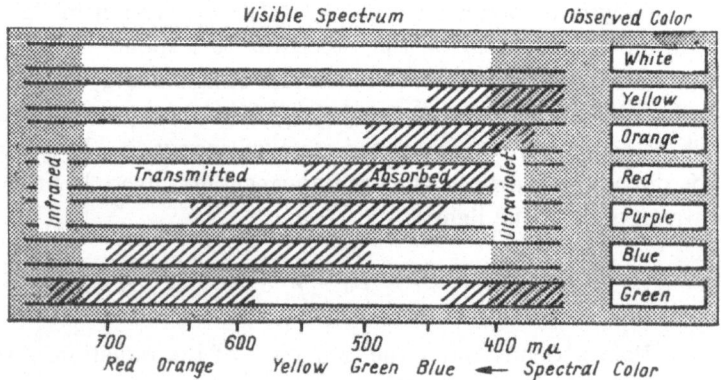

Fig. 23. Graphical representation of the relationship between the color of a substance and its absorption in the visible region.

To excite π-electrons to higher energy states, photons with lower energy suffice. When one or more such groups are introduced into the molecule, absorption takes place in the longer-wavelength regions, and eventually the absorption maxima can be shifted to the visible region and the substances appear colored with a deepening hue. When the absorption passes from the ultraviolet region to the visible region absorption in the blue region appears first. The deepening colors are then green-yellow, followed by yellow, orange, red, violet, blue, and green, as apparent from Fig. 23.

An organic compound can contain, of course, several unsaturated functional groups in one molecule (earlier, these groups were called chromophoric groups). In this case their relative position plays an important role. The following cases can occur:

$$\text{Chromophoric groups: } \textit{isolated} \quad -C=C-CH_2-C=C-$$
$$\textit{conjugated} \quad -C=C-C=C-C=O$$
$$\textit{cumulated} \quad >C=C=C<; \; >C=C=O$$
$$\textit{cross-conjugated} \quad -C=C-C-C=C-$$
$$\overset{\|}{O}$$

From the point of view of color the second case is of greatest importance, because the conjugation of π-bonds can connect several chromophoric centers and create a sufficient condition for absorption in the visible region. In addition to groups with double bonds, atoms and groups with free-electron pairs also play a very important role in the molecules of colored substances, because they can take part in mesomeric shifts in the molecule and thus have a substantial influence on the color (auxochromes). However, groups can also exist which produce an opposite effect (so-called anti-auxochromes).

In contrast to atomic spectra, which give narrow absorption lines, molecular spectra form broad bands in consequence of a large number of rotation and vibration levels under normal conditions. In the case of spectra of dissolved compounds, a broadening of the band is also caused by such intermolecular interactions as solvation and by associations of the molecules.

We assume that the analyst in the organic laboratory always collaborates with a specialized spectroscopist. It is important, therefore, that he know the possibilities and the extent of the application of spectral methods for his work.

Absorption spectroscopy in the visible and ultraviolet regions is useful to the analyst in the following cases:

1. *Comparison of the identity of two substances.* Electronic spectra in the ultraviolet region are less complex than infrared spectra, and, hence, less suitable for identity proof. If two substances are really identical, their

spectra must be identical too. However, two substances with identical spectra need not necessarily be identical.

2. *Detection of certain types of compounds* (for example, of an aromatic skeleton). If the compound trasmits radiation ($\varepsilon < 1$) between 220 and 800 nm, it may well be an aliphatic or alicyclic hydrocarbon, amine, nitrile, alcohol, ether, alkyl chloride, or alkyl fluoride, but the presence of conjugated double bonds, the carbonyl group, nitro group, bromine, or iodine is excluded. The detection of the aromatic nucleus in an unknown sample is carried out by measuring the absorption in the ultraviolet region of ethanolic, cyclohexane, or *n*-hexane solutions of various concentrations ($10^{-2} - 10^{-4}$ mole/liter).

3. *In certain cases the absorption in the ultraviolet region can serve as a valuable indicator.* For example, during column chromatography it can indicate the presence of the substance sought or of impurities with multiple bonds in the eluate.

Particular attention should be devoted to the purification and the preparation of samples for spectral measurements. In view of the order-of-magnitude differences in extinction coefficients of various substances, even a small amount of an impurity with a high extinction coefficient can cause quite erroneous conclusions.

Vibrational and rotational changes in the molecule lead to the absorption of electromagnetic radiation of wavelength $2.5 - 50\,\mu$ (100 to $4\,000\ \text{cm}^{-1}$), i.e., in the infrared region (Fig. 22). Organic compounds display very characteristic spectra in this region, so that the latter can be considered as one of the most characteristic physical attributes (especially in the interval between 1000 and 1400 cm^{-1}, the so-called fingerprint region) of organic molecules; this attribute (spectrum) can therefore be utilized to a certain extent even for mixtures (unless these react and form addition compounds). The infrared spectrum of a mixture is the sum (superposition) of the spectra of single components. The smaller the molecule of a substance, the more characteristic is the vibrational spectrum. With increasing molecular weight, infrared spectra become less characteristic. In such a case a large number of indistinct peaks occur in the spectrum and the spectral curve does not represent single absorptions of definite vibrational modes, but the general shape of a large number of superimposed bands (for example, the spectra of polysaccharides, polymers, or rotational isomers of higher n-parrafins).

Normal vibrations, represented by peaks in the infrared region, are specific for every substance. The frequencies of vibration are determined by the geometrical arrangements of the atoms in the molecule, and by the character of the bonds between atoms, and the atomic masses. Every normal vibration in the molecule is dependent in a different way on the vibrations of other atoms. Those vibrations which are dependent on the other parts of the molecule and which occur repeatedly in a series of compounds within a certain short frequency interval are called characteristic vibrations; this is, however, a relative conception and is significant only for a certain limited group of substances with identical inter- and intramolecular interactions.

The infrared spectrum of a low-molecular-weight compound is usually composed of a few strong absorption bands belonging to some of the basic vibrations, and a larger number of weaker peaks which are formed by combination (sum or difference) of basic vibrations, or which are the multiple of their values. Absorption in the infrared represents changes in rotational and vibrational states of atoms, and, therefore, substances with similar structures have differing spectra in certain regions (o-, m-, and p-isomers; cis- and trans- isomers; optical isomers; isotopic molecules).

Groups of atoms or particular kinds of bonds which differ from groups or bonds in other parts of the molecule in their mass or type, respectively, or in their position in the molecule, display, in a series of similar compounds, absorption bands which occur in approximately the same part of the spectrum. The frequencies of certain absorptions can be designated as characteristic of the appropriate groups or bonds. This fact, observed empirically and also accounted for theoretically, is the basis of qualitative analysis by means of infrared spectra, which makes it possible to identify certain groups of atoms or even whole molecules of unknown structure. Characteristic frequencies of certain bonds are usually listed in monographs or in work in the literature dealing with infrared and Raman spectroscopy.

In qualitative organic analysis infrared spectra are used primarily in the identification of an unknown molecule or in detecting certain bonds or groups in the molecule. Certain absorptions or the whole spectra are compared with the data from the literature or with the spectra of known substances. In doing this, it is important to respect the method of measurement and the parameters of the apparatus used. In some cases it is also possible to use infrared spectra for the determination of the purity of substances (detection and identification of impurities, control of the purifications, etc.), for the identification of the components in a mixture, and in structural analysis. In the last case, which aims at the determination of the total molecular structure, infrared spectroscopy is only one of the methods necessary for the solution of this often difficult task. The use of infrared spectroscopy is very suitable for so-called structural diagnosis because it can sometimes give a maximum of preliminary information regarding the type of substance being analyzed and the presence of certain bonds or groups in the molecule while consuming a minimum amount of the substance itself (3 − 50 mg); for example: the detection of the aromatic skeleton (by comparing the infrared spectrum with the ultraviolet one) of a paraffinic or olefinic hydrocarbon, the detection of the OH or NH_2 group on the basis of certain absorption bands, and the detection of $C=O$, $C=C$, and $C\equiv C$ groups.

Infrared spectroscopy is also much used for the quantitative analysis of multicomponent mixtures in the case of very similar organic compounds (position isomers, and cis- and trans-isomers).

When a substance is given to the infrared spectroscopist for identification, it is important that it be known if the substance is a single chemical compound; it is also advisable to give the spectroscopist as much information as possible. If published spectra of the substance are not available, it is possible to interpret the spectra of at least the simpler compounds on the basis of their comparison with the spectra of synthetic models. It is important to indicate the origin of the sample, the method of its preparation, and the hypothetical structures if possible. At this point we want to stress the importance of close collaboration between the spectroscopist and the organic analyst and the synthesist. The sample must be absolutely dry, it should contain neither polar solvents, because they could damage the windows of the cells, nor mechanical impurities. The elemental composition and the molecular weight of the sample must also be known. In some cases it is necessary that the substance be sufficiently soluble in nonpolar solvents (carbon tetrachloride, for the detection of hydrogen bonds).

To measure a spectrum, 3 − 50 mg of substance are usually necessary, depending on the intensity of absorption. If the substance is a liquid, the spectrum can be measured either directly (without dilution), or after dissolving it in a suitable solvent (CS_2, CCl_4, $CHCl_3$); the sample may also be prepared in a very fine suspension in Nujol, or in perfluorinated paraffin (fluorolube). In certain regions of the infrared spectrum the solvents used have a strong absorption of their own, and if we wish to have a spectrum of a substance over its whole range, we must measure the spectrum in at least two solvents. Solid or poorly soluble substances can also be measured after compressing their homogeneous mixture with powdered KBr to form a thin disk (approximately 1 mm); the usual concentration is 3 mg of substance per 1 g of KBr.

Raman spectra have an importance similar to infrared spectra, and they can also be used with advantage for identification purposes, as well as for qualitative and structural analysis. They are especially useful for the determination of the symmetry of the molecule, and Raman and infrared spectroscopy are complementary.

In addition to the spectral methods mentioned, the method of nuclear magnetic resonance (or proton magnetic resonance, NMR or PMR) is used more and more in analytical practice. This method makes it possible to study and trace all isotopes the mass numbers of which are odd and which possess a magnetic moment. For the organic chemist, ^1H comes into primary consideration, followed by ^{13}C, ^{14}N, ^{19}F, and ^{31}P.

In NMR spectroscopy the absorption of high-frequency energy (up to 10^2 MHz) corresponds to energetic transitions of the magnetic isotopic nuclei between states which were formed on the splitting (separation) of the ground energy state under the influence of a strong magnetic field (up to 10^4 G). The following relationship holds between the frequency of the absorbed quanta and the intensity H_0 of the magnetic field acting on the nucleus:

$$v = \frac{\gamma}{2\pi} H_0$$

where γ is the gyromagnetic ratio characteristic of each type of nucleus. As the γ values of the nuclei of various isotopes differ considerably, it is possible at a given adjustment of the apparatus to measure the NMR spectra of a single type of isotope only.

The differences between the nuclei of the same isotope bound in different ways are created as a consequence of the shielding of the external magnetic field by the bonding electrons. The change of the intensity of the magnetic field in the vicinity of the nucleus is therefore proportional, in the first approximation, to the density of bonding electrons, and hence to the polarity of the bond. Therefore, the shifts of the frequencies of absorption maxima caused by the shielding on the nucleus are called "chemical shifts." In order to make the values of these shifts, measured at varying working frequencies of the apparatus, comparable, a dimensionless unit is introduced,

$$\delta = \frac{\Delta v}{v_0} \omega^6$$

where Δv is the difference between the frequencies of the measured and standard absorption maxima, and v_0 is the working frequency of the apparatus. Chemical shifts of protons are mostly within the 10-ppm range, less frequently they attain values of up to 20 ppm, but in the case of heavy isotopes the range is broadened to $100-1000$ ppm.

Hence, isotopes of the same kind, but bound in different ways, can be distinguished on a recording of the NMR spectrum according to the positions of the absorption maxima which correspond to them. However, it is also possible to determine the character of the position at which the studied group is bound. This can be done by resolving the maximum into multiplets caused by the spin coupling of mutually chemically bound atoms. In the case of two neighboring groups of protons, the mutual chemical shift of which appreciably exceeds group interactions, a very simple rule applies: the multiplicity of the maximum of a group is equal to the number of protons in the neighboring group plus one.

The regularities described can be clearly seen on a proton spectrum of ethylbenzene (Fig. 24).

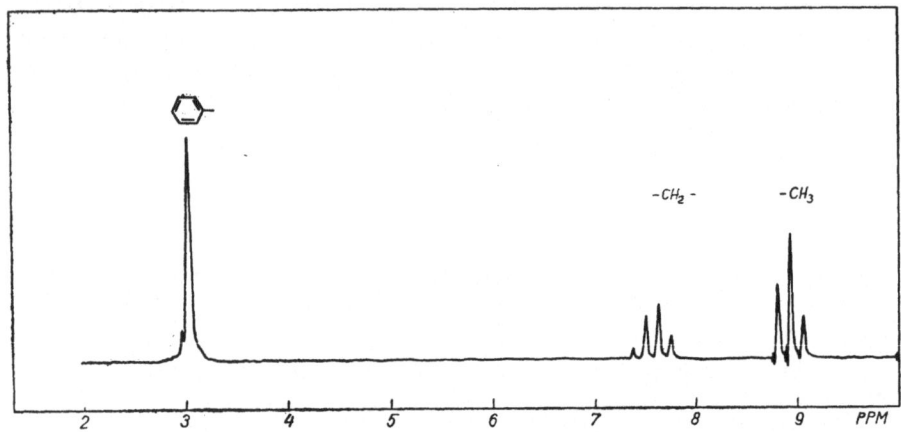

Fig. 24. NMR spectrum of ethylbenzene.

For qualitative organic analysis only those NMR spectra are meaningful which are recorded with high resolution. This is attained only in the case of liquid samples (pure liquids or solutions). For the recording of a spectrum usually 0.5 ml of a $2-20\%$ solution is necessary.

Those liquids can serve as solvents which give in the spectrum a single maximum which does not coincide with the maxima of the sample [H_2O, CH_2Cl_2, C_6H_{12}, $(CH_3)_2SO$, H_2SO_4]. At low concentrations it is necessary to use aprotic solvents (CCl_4) or totally deuterated solvents ($CDCl_3$, D_2SO_4, $CD_3 . CO . CD_3$, D_2O, etc.).

NMR spectroscopy represents one of the most useful methods of identifying groups containing protons, especially if the elemental composition of the sample is known. The detection of alkyls and alkoxyls bound to aliphatic or aromatic residues can be performed easily and unambiguously.

As the area under the recorded peaks is practically directly proportional to the number of protons of the group to which it corresponds, and because the number of protons in a group is always a whole number, it is possible to determine, after identification, the number of individual groups in the molecule without a special measurement. In suitable cases the molecular weight is thus also determinable. The determination of the number and type of double bonds can also be very convincing. NMR spectra are also useful for the identification of the aliphatic or the aromatic skeleton, and for deciding whether the aliphatic chain is branched or how many secondary or tertiary groups it contains. NMR spectra are also suitable for identification purposes, because it is hardly possible that two compounds would have identical spectra. The utilization of NMR spectra for quantitative analysis is especially convenient when the content of various isotopes is determined, for the measurements of tautomeric equilibria, and in kinetic measurements. NMR spectra are also used in detailed studies of the structure and conformation of molecules, of hydrogen bonds, and of very rapid equilibrium exchange reactions.

These three spectral methods, very significant for modern organic analysis, do not exhaust the number of instrumental methods of analysis in use. We could add, for example, the measurement of X-ray spectra or the use of mass spectrometry, the description of which is beyond the scope of this monograph. Specialized monographs should be consulted for the individual spectral methods (1 − 30).

References

1. Emsley, J. W., Feeney, J., and Sutcliffe, L. H.: High Resolution Nuclear Magnetic Resonance Spectroscopy, Pergamon Press, Oxford, Vol. I. 1965; Vol. 2. 1966.
2. Pople, J. A., Schneider, W. G., and Bernstein, H. J.: High Resolution Nuclear Magnetic Resonance, McGraw-Hill, New York, 1959.
3. Jackman, L. M.: Applications of Nuclear Magnetic Resonance Spectroscopy in Organic Chemistry, Pergamon Press, London, 1959.
4. Roberts, J. D.: An Introduction to the Analysis of Spin-Spin Splitting in High Resolution Magnetic Resonance Spectra, W. A. Benjamin, New York, 1961.
5. Bhacca, N. S., and Williams, H. D.: Applications of NMR Spectroscopy to Organic Chemistry: Illustrations from the Steroid Field, Holden-Day Inc., New York, 1964.
6. Bible, R. H.: Interpretation of NMR Spectra, an Empirical Approach, Consultants Bureau, Plenum Press, New York, 1965.
7. Bellamy, L. J.: The Infrared Spectra of Complex Molecules. 2nd Ed. Methuen and Wiley, London − New York, 1958.
8. Brügel, W.: Einführung in die Ultrarotspektroskopie. 2nd Ed. D. Steinkopf, Darmstadt, 1962.
9. Braude, E. A., and Nachod, F. C.: Determination of Organic Structures by Physical Methods. Academic Press, New York, 1965.
10. Weisberger, A.: Technique of Organic Chemistry. Vol. I. Physical Methods of Organic Chemistry. Part II; Vol. II. Chemical Applications of Spectroscopy. Interscience Publishers, New York, 1946, 1956.

11. Houben-Weyl: Methoden der Organischen Chemie. Bd. III. Physikalische Methoden. Teil 2. Verlag G. Thieme, Stuttgart, 1955.
12. API-Sammlung (American Petroleum Institute — Research Project 44). Carnegie Institute of Technology, Pittsburgh.
13. Sadtler-Catalogue S. P. Sadtler, Philadelphia. This catalogue comprises about 3000 IR spectra.
14. DMS Sammlung (Dokumentation der Molekülspektroskopie). Verlag Chemie, Weinheim and Butterworths Scientific Publishers, London, from 1956 up to now, 12, 000 spectra.
15. Hershenson, H. M. (Ed.): Infrared Absorption Spectra. Academic Press, New York—London, 1959.
16. Lang, L.: Absorption Spectra in the Ultraviolet and Visible Regions. I—VII. Vol. Akadémiai Kiadó, Budapest, 1959—1966.
17. Kössler, I.: Methoden der Infrarotspektroskopie in der chemischen Analyse. Akad. Verlagsges. Geest und Portig, Leipzig, 1961.
18. Brandmüller, J., and Moser, H.: Einführung in die Ramanspektroskopie. D. Steinkopf, Darmstadt, 1962.
19. Rao, C. N. R.: Ultraviolet and Visible Spectroscopy. Chemical Application. Butterworth's Scientific Publishers, London, 1961.
20. Rao, C. N. R.: Chemical Application of Infrared Spectroscopy. Academic Press, New York, 1963.
21. Gross, A. D.: An Introduction to Practical Infrared Spectroscopy. Butterworths and Co., London, 1964.
22. Colthup, N. B., Daly, L. H., and Wiberly, S. E.: Introduction to Infrared Spectroscopy. Academic Press, New York—London, 1964.
23. Jaffé, H. H., and Orchin, M.: Theory and Application of Ultraviolet Spectroscopy. J. Wiley and Sons, New York, 1962.
24. Baumann, R. P.: Absorption Spectroscopy. J. Wiley and Sons, New York, 1961.
25. Otting, W.: Spektrale Zuordnungstafel der Infrarot-Absorptionsbanden. Springer Verlag, Berlin, 1963.
26. Nakanishi, K.: Infrared Absorption Spectroscopy — Practical. Holden Day Inc., San Francisco—Tokyo, 1962.
27. Davies, M.: Infrared Spectroscopy and Molecular Structure. An Outline of the Principles. Elsevier Publishing Comp., Amsterdam, 1963.
28. Potts, W. J., Jr.: Chemical Infrared Spectroscopy. 2 Volumes. J. Wiley and Sons, New York, 1963.
29. Borsdorf, R., and Scholz, M.: Spektroskopische Methoden in der organischen Chemie. Akademie Verlag, Berlin, 1964.
30. Geppert, G.: Experimentelle Methoden der Molekülspektroskopie. Akademie Verlag, Berlin, 1964.

CHAPTER IV

A REVIEW
OF ANALYTICAL LITERATURE

The analyst must regularly follow the relevant literature, which can be divided into the following groups:

1. Textbooks of organic analysis, presenting basic knowledge to university students (1 – 10).

2. For practical use in the laboratory, books by the following authors can be used: Bauer and Moll (11), Houben-Weyl (a thorough compendium of organic analysis in the 2nd volume) (12), Cheronis et al. (13), Feigl (for color reactions) (14), and Peséz and Poirier (15). A special place is reserved for Mulliken's work (16, 17), in which the analytical properties of a large number of pure compounds containing C, H, and O, and compounds containing C, H, O, and Cl are listed.

3. In addition to the monographs mentioned on general organic analysis, analysts have at their disposal a number of publications on analytical problems of certain specialized industrial fields. Monographs on the analysis of fats and oils (18), monomers (19) and polymers, and mineral-oil products (20) may serve as examples.

4. Publications containing data on a great number of organic substances are of great practical value, as, for example, physical properties (21), melting points (22, 23), and other properties which are not immediately connected with organic analysis (24 – 30). As the analyst, when identifying an unknown organic substance, must know its molecular formula at a certain point of investigation, all works containing indexes or tables of these formulas [for example, (24)], and Chemical Abstracts, in which it is possible to look for a compound using a formula index and find more information in the original paper, are of great value.

5. In solving certain problems, the analyst must get well acquainted with the chemistry of given classes of substances, and he must therefore devote his attention to the organic chemical literature.

6. To acquire knowledge of special fields or organic analysis, the appropriate literature quoted in the preceding chapters is of great importance.

7. Periodicals. To workers in the field of organic analysis we suggest

and recommend the following 11 analytical journals, in which they will be able to find original papers, review articles, and also abstracts (A): Analytica Chimica Acta (Holland), Analytical Chemistry (USA), The Analyst (Great Britain), Chimia Analityczna (Poland), Chimie Analytique (A) (France), Microchemical Journal (USA), Mikrochimica Acta (Austria), Talanta (Great Britain), Zavodskaya Laboratoriya (USSR), Zeitschrift für Analytische Chemie (A) (German Federal Republic), and Zhurnal Analiticheskoi Khimii (USSR). Among journals which contain abstracts only, the following are of primary importance: Analytical Abstracts (Great Britain), Chemical Abstracts (USA), Chemisches Zentralblatt (German Democratic Republic), and Referativnyi Zhurnal, Khimiya (USSR), as well as the Index Chemicus (USA).

References

1. Jureček, M.: Organická analysa I. 2nd Ed. Nakladatelství ČSAV, Prague 1955.
2. Kamm, O.: Qualitative Organic Analysis. 2nd Ed. Wiley, New York 1947.
3. McElvain, S. M.: Characterization of Organic Compounds. Macmillan, New York 1953.
4. Middleton, A. R.: Systematic Qualitative Organic Analysis. 2nd Ed. Arnold and Co., London 1943.
5. Shriner, R. L., Fuson, R. C., and Curtin, D. Y.: Systematic Identification of Organic Compounds. 5th Ed. Wiley, New York and London 1964.
6. Smith, F. J., and Jones, E.: A Scheme of Qualitative Organic Analysis. Blackie and Son, London 1948.
7. Veibel, S.: The Identification of Organic Compounds. GAD Publisher, Copenhagen 1961.
8. Wild, F.: Characterization of Organic Compounds, 2nd Ed. Cambridge University Press, New York 1961.
9. Schneider, F. L.: Qualitative Organic Microanalysis. Springer Verlag, Wien; Academic Press, New York—London 1964.
10. Meyer, H.: Analyse und Konstitutionsermittlung organischer Verbindungen. 6th Ed., Springer Verlag, Vienna 1938.
11. Bauer, H., and Moll, H.: Die organische Analyse. Akademische Verlagsges. Leipzig 1960.
12. Müller, E. (Editor): Houben-Weyl: Methoden der organischen Chemie. Vol. 2. Analytische Methoden. Thieme, Stuttgart 1954.
13. Cheronis, N. D., Entrikin, J. B., and Hodnett, E. M.: Semimicro Qualitative Organic Analysis. 3rd Ed. Wiley, New York and London 1965.
14. Feigl, F.: Spot Tests. Organic Analysis. 6th Ed. Elsevier, Amsterdam 1960.
15. Pesez, M., and Poirier, P.: Méthodes et réactions de l'analyse organique. Vol. III. Réactions colorées et fluorescences. Masson et Cie, Paris 1954.
16. Huntress, E. H., and Mulliken, S. P.: Identification of Pure Organic Compounds. Wiley, New York 1946.
17. Mulliken, S. P.: Organic Chlorine Compounds. Wiley, New York 1946.
18. Kaufmann, H. P.: Analyse der Fette und Fettprodukte. 2 Volumes. Springer Verlag, Berlin—Göttingen—Heidelberg 1958.

19. Kline, G. M. (Ed.): Analytical Chemistry of Polymers, 3rd Vol. Interscience Publishers, J. Wiley and Sons, New York—London 1959—1962.
20. Finke, M., and Leipnitz, W.: Moderne Methoden der Erdölanalyse. Akademie Verlag, Berlin 1964.
21. Timmermans, J.: Physico-Chemical Constants of Pure Organic Compounds. Elsevier, Houston 1950.
22. Kempd, R. K., and Kutter, F.: Schmelzpunkttabellen zur organischen Molecular-Analyse. Edward Bros., Ann Arbor, Mich. 1944.
23. Utermark, W., and Schicke, W.: Schmelzpunkttabellen organischer Verbindungen, 2nd Ed. Academie Verlag, Berlin 1963.
24. Beilsteins Handbuch der organischen Chemie. Springer Verlag, Berlin 1918.
25. Radt, F. (Editor): Elsevier Encyclopedia of Organic Chemistry. Elsevier, Houston 1946.
26. Heilbron, I., and Bunbury, H. M.: Dictionary of Organic Compounds. 2nd Ed. New York 1953.
27. Rodd, E. H. (Editor): Chemistry of Carbon Compounds. Elsevier, Houston 1952.
28. Hodgman, Ch. D., Weast, R. C., and Selby, S. M. (Ed.): Tables for Identification of Organic Compounds. 2nd Ed. Chemical Publishing Co., Cleveland 1964.
29. Hodgman, Ch. D., Weast, R. C., and Selby, S. M.: Handbook of Chemistry and Physics, 45th Ed. Chemical Publishing Co., Cleveland 1964.
30. Lax, E., and Synowietz, C. (Ed.): Taschenbuch für Chemiker und Physiker (D'Ans-Lax). Vol. II. Organische Verbindungen, 3rd Ed. Springer Verlag. Berlin—Göttingen—Heidelberg 1964.

PART 2

CHAPTER V

TESTS FOR ELEMENTS

Tests for elements in organic substances are carried out after oxidative or reductive mineralization by common methods of inorganic qualitative analysis.

The most efficient method of mineralization of organic substances used for the detection of elements is the sodium fusion method (1, 2). The procedure is very simple and does not require any special equipment. It is suitable primarily for the detection of nitrogen, halogens, and sulfur. The test for nitrogen is especially suitable when combined with the sensitive detection of the cyanide formed by its conversion to cyanogen chloride (3), which reacts with pyridine with the formation of glutaconic dialdehyde. The latter reacts with dimedone and gives rise to strongly colored polymethine dyes (4). The high sensitivity of the reaction (0.02 µg CN^-/ml) makes detection also possible in such cases where the conversion of cyanide to Prussian blue (2) or benzidine blue (5) fails. In addition, it is possible to detect nitrogen without taking special measures in the presence of sulfur in the analyzed sample.

A suitable method of mineralization of organic compounds consists, according to Schöniger [for reviews, see 6, 7)], in their combustion in an oxygen filled flask; this process is quick, and in it the elements to be detected are not present in a large excess, of other products of mineralization, as in the case of the sodium fusion method.

Oxidation with the decomposition product of silver permanganate is also very rapid. The procedure makes possible the detection of C, H, N, S, Cl, Br, I, P, and Hg. If combined with the method of Vašák and Šedivec (8), As can also be detected.

The procedure for the detection of carbon and hydrogen should be considered as preliminary. In special cases the procedure according to Jureček (9) is suitable for the detection of carbon. To carry out tests for Cl, Br, and I in parallel in quantities below 1 µg, spot-test reactions according to Weisz (10) are suitable. A very sensitive test for N, halogens, S, As, and P was described by Widmark (11).

Silicon can be detected according to the method given by Filman and Smart (12). The test for oxygen is not usually carried out. The so-called "ferrox test" has a limited use (13), and it fails with substances containing nitrogen or sulfur, and also in the case of higher ethers and nitro compounds.

The tests for metals are usually carried out with the residues after annealing using common reactions of inorganic qualitative analysis or also emission spectroscopy; the latter method is often carried out using the original sample.

Generally, the test for elements is carried out only if a sufficient amount of the sample is available. If only a small amount is in the analyst's possession, we recommend carrying out quantitative elemental microanalysis directly, in which approximately the same amount of sample is consumed, but which leads, in addition to identification, to quantitative determination of the composition. This is true mainly of the detection and the determination of C, H, O, N, S, and halogens.

1. Mineralization by Fusion with Sodium

a) Decomposition of Solid Substances

The sample (10 — 15 mg) is placed on the bottom of a micro test tube (approximately 50×8 mm) and a piece of cleaned sodium the size of a pea is allowed to slide along the sides of the tube to its bottom. The sodium is pricked on a tiny glass rod approximately 1 cm long and 1 mm in diameter. It is then melted over a small flame until it fills the tube and the sodium drop slides to the bottom (protective goggles!). Rotation of the tube helps the mixing of the sodium with the sample. The temperature is slowly increased until the glass of the tube softens (dark red glow). The still-hot tube is immersed into 5 ml of water in a small porcelain dish (hood!). The carbon formed and the glass splinters are filtered off and washed with 5 ml of water. The colorless filtrate is submitted to further tests.

b) Decomposition of Liquid Substances

A piece of sodium is introduced onto the bottom of a micro test tube (approximately 50×8 mm) blown out to bubble (a small spherical flask) which is fastened vertically on a stand, and the sodium is melted slowly over a small flame. Heating is continued until its blue-green vapors have filled the expanded part of the tube. A capillary containing the liquid sample is introduced into the flask, but the hairlike part of the capillary should be broken beforehand. The liquid from the capillary distills and its vapors react with the vapors of sodium and with the melted sodium. When the main reactions have subsided the small flask is taken off the stand, is then heated to a red glow, and is thrown into a dish containing 5 ml of water. Further procedure is as described under (a).

2. Oxidation with the Decomposition Product of Silver Permanganate (14)

Preparation of the reagent: $KMnO_4$ (87 g) is dissolved in two liters of hot water, and 102 g of $AgNO_3$ are added with stirring until dissolved. The mixture is allowed to cool and the precipitated product is filtered off on a Büchner funnel, washed with 750 ml of water, and recrystallized from two liters of water. The crystals of $AgMnO_4$ are isolated by filtration on a sintered glass filter, washed with 500 ml of water, and dried at $60-70$ °C. For single experiments 2 g of $AgMnO_4$ are decomposed in a test tube by heating at ~ 150 °C.

Equipment: Glass tubes 13 cm long, 4 mm inner diameter, which are constricted to 0.3 mm at ~ 9 cm from the upper end and drawn out to a capillary at the lower end (Fig. 25).

Procedure: A piece of annealed asbestos is inserted and gently pressed into the tube (it is held by the constriction of the tube) and ~ 1 mg of the sample well mixed with 50 mg of the silver permanganate decomposition product is added on the filtration layer thus formed. The layer of the mixture is tightened with a lump of asbestos. Liquids are introduced with a capillary pipette on the

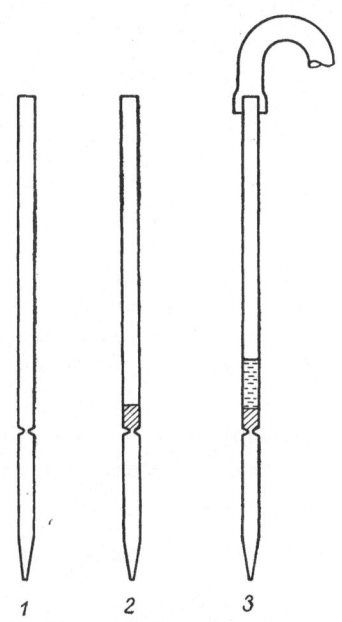

Fig. 25. Filtration tube: (1) empty, (2) full, (3) during washing.

silver permanganate decomposition-product layer, which is then covered with an additional 20 mg of the reagent. The decomposition of the organic substance takes place on heating of the layer with the flame of a micro-burner and it is complete within several seconds.

3. Oxygen Flask Method

A $250-300$ ml Erlenmeyer flask provided with a ground-glass stopper is used for the oxidation. A platinum wire of $0.5-0.7$ mm diameter with a loop at its end is sealed into the stopper (Fig. 26).

A folded piece of filter paper (ashless, halogen-free; a control experiment is necessary) containing 2−4 mg of the sample is inserted into the loop. Liquids are allowed first to soak into a piece of filter paper, which is then folded in the same manner as above.

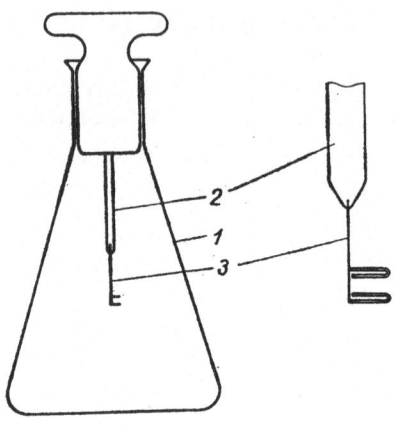

Fig. 26.

Flask for the combustion of organic compounds in oxygen atmosphere; 1-flask, 2-glass rod, 3-platinum wire.

The folded paper is inserted into the platinum wire loop in such a manner as to allow a strip of it to project upward freely. The appropriate absorption liquid is then introduced into the combustion flask and the flask is filled with oxygen by allowing a stream of the latter to pass through the flask for 5−10 sec. The free end of the filter paper in the wire loop is then ignited and the stopper is quickly put into the neck of the flask. The combustion of the substance in the flask lasts only a few seconds. It is recommended that the stopper be held lightly to prevent the expanding gases from pushing it out. After 10 min of absorption, which is enhanced by occasional shaking, the tests can be carried out.

4. Test for Carbon

Immediately after mineralization of the organic substance with the decomposition product of silver permanganate a rubber tube ending with a rubber ball (used, for example, for filling automatic burettes) is connected to the upper end of the tube and the gases are carefully expelled over a pure surface of a freshly filtered saturated barium hydroxide solution (∼1 ml in a micro test tube). If the sample contains carbon, an easily observable turbidity of $BaCO_3$ occurs.

5. Test for Hydrogen

The presence of hydrogen manifests itself immediately after the oxidation of the organic substance with silver permanganate decomposition product by the formation of small droplets of water condensed on the cool sides of the decomposition tube.

6. Test for Sulfur

a) After Mineralization with Sodium

After mineralization with sodium 1−2 drops of a freshly prepared 1% sodium nitroprusside solution $(Na_2[Fe(CN)_5NO] \cdot 3H_2O)$ are added to the filtrate (0.5 ml). If a red color is formed, proving the presence of sulfur in the tested sample, 2−3 drops of sodium plumbate are added to another 0.5 ml portion of the solution (to a 5% solution of lead acetate, 20% sodium hydroxide is added dropwise until the initially-formed precipitate is again dissolved). The formation of a black precipitate proves the presence of sulfur in the molecule of the substance tested, a brown precipitate indicates the presence of sulfur-containing impurities in the sample.

b) After Oxidative Mineralization

Several drops of water are introduced into a cooled decomposition tube by means of a micropipette, and the liquid is pushed with the rubber ball through the reagent layer into the constricted part of the tube. One to two drops are then pressed out into a micro test tube and mixed with a few drops of a 5% solution of barium nitrate. If sulfur is present in the substance tested a white precipitate of $BaSO_4$ is formed. A modification of this procedure was described by Ackermann (15).

c) After Mineralization in the Oxygen Flask

A 5% H_2O_2 solution (5 ml) is used as the absorption solution. After absorption 1 ml of the solution is boiled briefly and the presence of sulfur is detected in the form of $BaSO_4$.

7. Test for Nitrogen

a) After Mineralization with Sodium (15)

One milliliter of the filtrate is mixed in a 10-ml calibrated flask with 1 ml of a 1% chloramine T solution, and the mixture is acidified with 1−2 ml of N HCl and stirred. After 1 min, 3 ml of 3% dimedone solution are added (3,5-dimethyldihydro resorcinol is dissolved in 30 ml of pyridine and the solution is made up to 100 ml with water; it should be kept in a brown bottle). If nitrogen was present in the substance investigated, an orange color appears after shaking the mixture, which changes to red-violet and then a deep violet, which is stable for approximately 10 min. The color then changes to blue and its intensity diminishes gradually. When only trace amounts of nitrogen are present in the sample the traces of Cu from the water can disturb the reaction.

b) After Oxidative Mineralization

The part of the aqueous solution remaining in the tip of the decomposition tube after the test for sulfur is expelled onto a watch glass containing 1 ml of a 1% diphenylamine solution in conc. sulfuric acid. In the presence of nitrogen a blue color appears on the sites of contact of the two liquids, which is especially well visible on a white background.

8. Test for Chlorine, Bromine, and Iodine

a) Beilstein's Test

A copper wire gauze, 5×5 cm, is placed on a quartz triangle on a stand and a nonluminous flame of a Teclu burner, ~ 15 cm high, is put under the stand in such a way as to allow the wire gauze to cut the flame into two parts. The access of air to the burner is regulated so as to prevent the formation of the inner green conus. When the center of the wire gauze is glowing red and the flame passes through it, $1-2$ mg of the substance are taken on a spatula and immersed into the outer edge of the flame ~ 5 cm below the gauze. The flame immediately assumes a yellow color (caused by the burning carbon), and if the substance contains halogens, the flame over the wire gauze immediately turns green to blue-green. The volatile substances are tested for the presence of halogen by allowing a drop of the substance to vaporize in a test tube and by blowing out the air together with the vapor into the flame below the glowing wire gauze by means of a bent tube.

The arrangement described gives reliable results. However, very often a copper wire broadened at its end to the form of a small spatula (by beating the end with a small hammer) is also quite satisfactory for the performance of the test for halogens. A few crystals of the sample are placed on the broadened part of the wire and introduced into the flame. However, the test may fail in the case of certain very volatile substances.

b) Test for Halogens after Mineralization by Sodium Fusion

(i) In the absence of nitrogen and sulfur 1 ml of the filtrate (see above, p. 96) is acidifed with a few drops of nitric acid, and a few drops of a 2% silver nitrate solution are added to it. A white or yellow floculent precipipate proves the presence of halogen in the substance.

(ii) If the preceding tests have proved the presence of nitrogen or sulfur in the sample, the third part of the filtrate is acidified with 30% acetic acid and boiled gently, and, after cooling and acidification with HNO_3, $AgNO_3$ is added to detect the halogens.

(iii) Test for chlorine, bromine, and iodine simultaneously. After mineralization by melting the sample with sodium the filtrate is divided into two parts. To detect chlorine ion in the presence of bromine and iodine ions, one-half of the filtrate is acidified with HNO_3, and a 2% $AgNO_3$ solution is added dropwise until a precipitate is formed. This is then filtered off, washed with water, mixed in a test tube with 5 ml of an ammoniacal solution of ammonium carbonate [10% solution of $(NH_4)_2CO_3$ with the addition of ammonia in the ratio 4 : 1], and thoroughly shaken. The undissolved precipitate proves the presence of bromine or iodine. The undissolved precipitate is filtered off an the filtrate is acidified with HNO_3. A white, floculent precipitate represents a proof of the presence of Cl in the substance.

To detect bromine and iodine in the presence of chlorine, the second part of the filtrate is acidified with 30% acetic acid and the solution is boiled briefly to expel H_2S and HCN. After cooling, 1 ml of CCl_4 is added, followed by dropwise addition of freshly prepared chlorine water. After the addition of each drop the mixture is thoroughly shaken. In the presence of iodine the organic layer (carbon tetrachloride) becomes violet. After further addition of drops of chlorinated water the color vanishes, and in the presence of bromine a yellow-to-brown color of the CCl_4 layer appears.

To detect halogens in mixture (in filtrate), paper chromatography in acetone-water (4 : 1) can be employed. Detection on paper can be carried out by spraying with ammoniacal silver nitrate and irradiation with UV light until dark spots appear.

c) Test for Halogens after Mineralization with Silver Permanganate Decomposition Product

A few drops of ammonia (1 : 1) are introduced into the decomposition tube after the extraction with water and the solution is expelled into a micro test tube. On acidification with HNO_3, a turbidity or a precipitate due to silver halogenides is formed.

d) Test for Halogens after Mineralization in the Oxygen Flask (6)

Five milliliters of N NaOH are used for the absorption. After acidification the test for halogens is carried out by means of silver nitrate solution.

9. Test for Fluorine

The substance (3−5 mg) is ignited in the flask in the manner described above; water (10 ml) is used for absorption. To 3−5 ml of the solution after absorption 1 ml of 0.0005 M alizarine complexone solution is added followed by 0.2 ml of acetate buffer (pH = 4.3), water up to 7.5 ml, and

1 ml of 0.0005 M ceric nitrate solution [for the preparation of the basic compounds necessary for the reagent see (16)]. After making up to 10 ml with water the mixture is allowed to stand for 1 hr in the dark; in the presence of fluorine a change in color is observed, from red to violet and eventually deep blue.

10. Test for Boron

Five milliliters of conc. H_2SO_4 and 5 ml of a 1,1'-dianthrimide solution (0.1 g of 1,1'-dianthrimide is dissolved in 25 ml of conc. H_2SO_4 and the aliquot part is diluted with 20 parts of concentric sulfuric acid immediately before use) are added to the absorption solution (0.5 − 1 ml) from the preceding experiment. The solution is allowed to stand at least for 30 min at 90 °C. The presence of boron in the sample is manifested by an intensely blue color.

11. Test for Phosphorus

The substance (3 − 5 mg) is ignited in the flask in the manner mentioned above; a mixture of 5 ml of 0.5 N NaOH and 4 ml of bromine water is employed for the absorption. After absorption and shaking, the solution is acidified with 3 ml of N HCl, and bromine is expelled by gentle boiling, until the solution is colorless. A 2.5% ammonium molybdate solution (5 ml) is then added, followed by 10 ml of N H_2SO_4. After cooling, the solution is extracted with 10 ml of amyl acetate, the layers are separated, and the amyl acetate layer is filtered, after washing with water, through a cotton wool plug into a test tube. Added to this are 2.5 ml of a $SnCl_2$ solution (0.7 g of $SnCl_2$ in 15 ml of conc. HCl and 100 ml of water). If phosphorus was present in the sample, a strong blue color appears in the amyl acetate layer.

For the absorption 5 ml of a dilute HNO_3 solution (1 : 2) may also be used. After absorption the solution is boiled briefly. For the extraction of the yellow modification of the heteropoly molybdatophosphoric acid, butyl acetate or isobutyl acetate, n-butanol, isobutanol, n-amyl alcohol, isoamyl alcohol, or a mixture of isobutanol-benzene (1 : 1), isobutanol-chloroform (3 : 3), etc., can also be used. The reduction of the extracted heteropoly acid after acidification with 10 ml of N HCl or N H_2SO_4 can also be carried out by the addition of 1 ml of 0.2% hydrazine sulfate or 0.05 − 0.1 g of solid ascorbic acid. Both of the latter two reduction methods take place at 40 − 60 °C.

12. Test for Mercury

The decomposition product of silver permanganate is used for oxidation. The filling of the decomposition tube is the same as described earlier, but a gentle stream of air is blown through it by means of a rubber bulb during the heating. If the sample contains mercury, this condenses in the narrowed part of the capillary, where it can be observed with the naked eye. A solution of dithizone (1.3 mg in 100 ml of CCl_4) is sucked into the capillary tip, taking great care to keep the asbestos plug dry. In the presence of mercury the color of the dithizone solution changes from green to orange-yellow.

References

1. Lassaigne, J. L.: Ann. Chem. **48**, 367 (1943).
2. Kainz, G., and Schöller, F.: Mikrochim. Acta **1954**, 327.
3. Spěvák, A., Kratochvíl, V., and Večeřa, M.: Collection Czech. Chem. Commun, **26**, 887 (1961).
4. Kratochvíl, V.: Collection Czech. Chem. Commun. **25**, 299 (1960).
5. Jureček, M.: Organická analysa I, p. 100. Nakladatelství ČSAV, Prague, 1955.
6. Ayres, D. C., and Dawson, B. E.: J. Chem. Soc. **42**, 270 (1965).
7. MacDonald, A. M. G.: Analyst **86**, 3 (1961).
8. Vašák, V., and Šedivec, V.: Chem. Listy **46**, 341 (1952).
9. Jureček, M.: Microchim. Acta **1955**, 1088.
10. Weisz, H.: Microchim. Acta **1956**, 1225.
11. Wildmark, G.: Acta Chem. Scand. **8**, 246 (1954).
12. Gilman, H., and Smart, G. N. R.: J. Org. Chem. **15**, 720 (1950).
13. Feigl, F.: Spot Tests in Organic Analysis. Elsevier, Amsterdam, 1956.
14. Körbl, J.: Microchim. Acta **1956**, 1705.
15. Ackermann, G.: Microchim. Acta **1962**, 106.
16. Belcher R., Leonard, M. A., and West, T. S.: J. Chem. Soc. **1959**, 3577, 1959, 2390.

CHAPTER VI

GROUP AND CLASSIFICATION REACTIONS

Organic compounds can be suitably and systematically classified by group and classification reactions. Color and precipitation reactions lead to the determination of the presence of functional groups, and the analyst can thus choose a suitable means for further analysis. For practical procedure see the general remarks on p. 51 and Table 1.

A Review of Classes of Organic Compounds Grouped According   *Table 1*
to Their Solubility Characteristics, and a Review of Classification Reactions*

---

1. Substances soluble in water and ether (monofunctional compounds, usually up to C_5)
   (a) C, H, O: alcohols (5, 11, 27), aldehydes and ketones (9, 13, 16, 17, 31), carboxylic acids (2, 21), acid anhydrides (2, 21), acetals (12), esters (12, 21), certain glycols (11, 18), lactones (2, 21), polyhydric phenols (7, 17, 24)
   (b) C, H, O, N: amides (5, 12, 20, 29), amines (8, 10, 11, 20, 23, 29), nitrogen heterocycles (5, 10), nitriles (12, 21, 29), nitroparaffins (32, 34, 35), oximes (33)
   (c) C, H, O, halogens: halogenated compounds enumerated under (a)
   (d) C, H, O, S: heterocyclic hydroxylated sulfur compounds (6, 11), mercapto acids and thioacids (2, 6)
   (e) C, H, O, N, halogens: halogenated amines, amides, nitriles
2. Substances soluble in water, insoluble in ether
   (a) C, H, O: di- and polycarboxylic acids (2), hydroxy acids (2, 18, 27), polyhydric alcohols (11, 18), polyphenols (13), simple sugars (13, 16)
   (b) C, H, O, metals: salts of acids and phenols, various metalloorganic compounds
   (c) C, H, O, N: salts of amines and of organic acids (8, 10, 11, 20, 23, 29), amino alcohols (2, 11), ureas (8, 10)
   (d) C, H, O, halogens: halo acids (2, 21), acyl halogenides (21), chlorohydrins (11)
   (e) C, H, O, S: sulfonic acids (2, 6), alkylsulfuric acids (2, 6), sulfinic acids (2, 6)
   (f) C, H, O, N, halogens: salts of amines with halo acids
3. Basic, water-insoluble substances, soluble in 1.2 N HCl. Amines (except negatively substituted ones and diaryl- and triarylamines; these are in group 7) (8, 10, 11, 20, 23, 29), amino acids (7, 8, 20, 23), amphoteric compounds (aminophenols, aminothiophenols, aminosulfonamides) (20), arylsubstituted hydrazines (5, 16, 17)
4. Weakly acid substances insoluble in water and 1.5 N NaHCO₃, soluble in 2.5 N NaOH
   (a) C, H, O: acids (2, 21), acid anhydrides (21), phenols (7, 11, 19, 22, 24, 27), enols (7)

(b) C, H, O, N: amino acids (7, 8, 20, 23), nitrophenols (32, 34), amides (includ ng monoalkylamides; dialkylamides are in group 3) (5, 12, 20, 29), aminophenols (20), imides (5), aromatic N-monoalkylamines (19, 20), oximes (33), primary and secondary nitroparaffins (32, 34, 35)

(c) C, H, O, halogens: halophenols (2)

(d) C, H, O, S: mercaptans (2, 7, 11, 13), thiophenols (2, 11)

(e) C, H, O, S, N: sulfonamides (6) and their amino derivatives (2, 6, 20), aminosulfonic acids (2, 6, 20)

5. Strongly acid substances insoluble in water, soluble in 1.5 N NaHCO₃

(a) C, H, O: acids and acid anhydrides (2, 21)

(b) C, H, O, N: amino acids (2, 7, 8, 20, 23), nitro acids (2, 32, 34), carboxylic acids of nitrogen heterocycles (2, 21), polynitrophenols (2, 32)

(c) C, H, O, halogens: halo acids (2, 21), polyhalophenols (2)

(d) C, H, O, S: sulfonic and sulfinic acids (2, 6)

(e) C, H, O, N, S: aminosulfonic acids (2, 6, 20), sulfates of weak bases (2, 6, 10)

(f) C, H, O, S, halogens: halogenides of sulfonic acids (2, 6)

6. Neutral substances not belonging to the preceding groups, soluble in conc. H_2SO_4 and containing only C, H, and sometimes O or halogens. Alcohols (5, 11, 27), aldehydes and ketones (9, 13, 16, 17, 31), esters (12, 21), ethers (30), unsaturated acyclic (3, 4), and aromatic hydrocarbons easily sulfonated—for example, di- and polyalkylsubstituted benzenes (22, 26), acetals (12), anhydrides (2, 21), lactones (2, 21), polysaccharides (13, 16)—turn dark in H_2SO_4

7. Neutral substances also containing N (or also S and other elements) which do not belong to the preceding classes

(a) C, H, O, N: amides (5, 12, 20, 29), nitroarylamines (32, 34), nitrated hydrocarbons (32, 34, 35), aminophenols (7, 11, 19, 20, 22, 24, 27), azo compounds (32), azoxy compounds (32), hydrazo compounds (5, 16, 17), di- and triarylamines

(b) C, H, O, N, S: mercaptans (7, 11, 13), sulfonamides (6), N-dialkylsulfonamides (6). sulfides (6), disulfides (6), sulfones (6)

(c) C, H, O, N, halogens: halogenated amines, amides, nitriles

8. Inert substances insoluble in classification solvents. Hydrocarbons (the majority of cyclic and all saturated acyclic ones) (5, 22, 26) and their halogenated derivatives, diaryl ethers and their halogenated derivatives

---

* The numbers in parentheses refer to the group reactions described on pp. 108—116.

1. Solubility

On the basis of a suitably chosen system of solvents, and using an optimum ratio of solvent to investigated substance, organic compounds can be classified into several groups. Together with the determination of the elemental composition, this facilitates the identification procedure and enables a rational choice of subsequent tests.

Procedure: 30 mg of substance (solid substances are ground to a fine powder) are shaken in a test tube (volume ~8 ml) with 1 ml of solvent for 1 min. If the substance dissolves completely, it is considered soluble, if a part remains undissolved, we consider it insoluble. If the test is carried out with inert solvents, it is advisable to heat the solution mildly and then to cool it by vigorous shaking. In some cases, it is possible to inoculate with a trace of the solid substance.

Solvents: water, ether, 1.2 N HCl, 1.5 N NaHCO$_3$, 2.5 N NaOH, conc. H$_2$SO$_4$.

The test is carried out with solvents in the given sequence. Once the solubility group is determined (see below), no more experiments with subsequent solvents are made. If the substance is not soluble in water, its solubility in ether is not tested, but solubility in subsequent solvents is tested immediately.

The Solubility Groups

1. Substances soluble in water and in ether (monofunctional compounds, usually up to C$_5$).

2. Substances soluble in water and insoluble in ether (difunctional and polyfunctional compounds).

3. Basic substances, insoluble in water, soluble in 1.2 N HCl.

4. Slightly acid substances, insoluble in water and in 1.5 N NaHCO$_3$, soluble in 2.5 N NaOH.

5. Strongly acid substances, insoluble in water, soluble in 1.5 N NaHCO$_3$.

6. Neutral substances, containing C and H — sometimes also oxygen and halogenides — and which do not belong to the above-mentioned groups, and which are soluble in conc. H$_2$SO$_4$.

7. Neutral substances, also containing nitrogen (or sulfur and other elements as well), which do not belong to the preceding groups.

8. Inert substances, containing only C and H (or also oxygen and halogens), which do not belong to the preceding groups and which are insoluble in concentrated H$_2$SO$_4$.

Remarks on the Determination of Solubility

The definition of solubility chosen for the separation of organic compounds does not distinguish between solubility caused by physical properties of the compound and that caused by chemical reactions. For practical purposes it is recommended to weigh (or measure) the investigated compound with an accuray of ±10%; the weighing or measuring of the substances is recommended especially in cases when the solubility is on the boundaries of individual groups. When the determination of the solubility group is unambiguous, it is sufficient just to estimate the amount of the substance. Certain substances give, with reactive solvents (HCl, NaOH, H₂SO₄), reaction products which are insoluble in the solvent used, such as, for example, 1-naphthylamine, which gives an insoluble hydrochloride. Nonetheless, it is classified in group 3 as a base. In such cases the isolation of the product and a test of identity with the original substance is recommended.

Certain secondary and tertiary alcohols are dehydrated to olefins, under the influence of H_2SO_4. These can then polymerize to give polymers insoluble in concentrated sulfuric acid. In spite of this, such compounds are classified into groups 6 or 7. In homologous series of monofunctional compounds (ethers, esters, ketones, alcohols, nitriles, amides, acids, amines) substances with 5 C in the molecule are usually soluble in water; those with branched carbon chains are more soluble.

The tests of solubility are of course carried out with chemically pure compounds. Examples of the solubility of typical representatives of the given classes of compounds are shown in Table 2, where compounds of intermediary character, i.e., the positions of which are on the boundaries of individual solubility classes, are also listed (1).

Examples of the Classification of Some Compounds   *Table 2*
into Solubility Groups and a Review of Compounds
which are on the Boundaries of Solubility Classes*

Acids: chloroacetic 1, *n*-butyric 1, l-chloropropionic 1, crotonic 1, isovaleric 1—5, valeric 5

Alcohols: butanol 1, tert-amyl alcohol 6, isopropylmethylcarbinol 1—6, isoamyl alcohol 6, benzyl alcohol 6, cyclopentanol 6

Amides: formamide 1—2, acetamide 2, propionamide 1—2, isobutyramide 1, *n*-butyramide 1—7, formanilide 1—7, acetanilide 7

Amines: diethyl 1, isoamyl 1, *n*-amyl 1, benzyl 1, piperidine 1, cyclohexyl 1, di-*n*-propyl 1—3, di-*n*-butyl 3, aniline 3, tri-*n*-propyl 3

Esters: ethyl acetate 1, methyl propionate 1, *n*-propyl formate 1, isopropyl acetate 1—6, methyl isobutyrate 1—6, *n*-butyl formate 1—6, methyl isovalerate 6, sec-butyl acetate, *n*-butyl acetate 6, benzyl acetate 6, diethyl oxalate 1—6, dimethyl malonate 1—6, diethyl phthalate 6, di-*n*-butyl oxalate 6

Ethers: ethyl methyl 1, diethyl 1—6, ethyl isopropyl 1—6, di-isopropyl 6, di-*n*-butyl 6, di-*n*-amyl 6

Aromatic hydrocarbons: mesitylene 6—8, isodurene 6—8, cymene 8, *p*-xylene 8, *m/*xylene 8, naphthalene 8

Ketones: methyl ethyl 1, methyl isopropyl 1, methyl *n*-propyl 1—6, diethyl 1—6, cyclopentanone 6, cyclohexanone 6, acetophenone 6, benzil 6

Nitriles: propionitrile 1, isobutyronitrile 1—7, *n*-butyronitrile 7

Nitro compounds: nitromethane 1—4, nitroethane 4, nitrobenzene 7

Phenols: hydroquinone 1, chlorohydroquinone 1—4, phloroglucinol 2—4, phenol 4

---

* For the classification into groups 1—8 see p. 106. For example, the indication 1—5 means that the substance is on the limits of solubility in water and 1.5 N $NaCO_3$; compounds with a lower number of C will be in group 1, those with a higher number of C in group 6.

2. Acidity and Basicity

a) *A test with universal indicator.* A sample (5 – 10 mg) is mixed with 0.5 ml of water in a test tube and 3 drops of a universal indicator are added. The color obtained indicates the pH of the solution. If the substance is insoluble in water, a suitable solvent (methanol, ethanol, dioxane, approx. 0.5 – 1 ml) is added to its suspension in 0.5 ml of water. A control experiment with the same solvent mixture is carried out in parallel, and the changes of color are compared.

b) *Test with Congo red.* A small crystal or a drop of the substance is deposited on a small strip of paper impregnated with the indicator and moistened with water. The formation of a blue color indicates that the substance is acidic (a strong acid).

c) *Test with iodate and iodide* (for strong and also weak acids). To the substance in a test tube (~ 5 mg or two drops of a saturated ethanolic solution) two drops of a 2% solution of potassium iodide and two drops of 4% potassium iodate are added. The test tube is stoppered and dipped into boiling water bath for 1 min. After cooling, 1 – 4 drops of a 0.1% starch solution are added. If the substance is an acid, the solution turns blue.

d) *Test with Ni-dimethylglyoxime.* Two drops of an aqueous solution of the investigated substance or a few milligrams of the solid sample are put on a spot test plate (with indentations), and two drops of the reagent are added. If the substance is a base, a red precipitate of Ni-dimethylglyoxime is formed (weak bases, such as, for example, nitroanilines or diphenylamine, do not react).

The preparation of the reagent: To a solution of nickel sulfate (7.7 mg in 1 ml) the same amount of dimethylglyoxime in alcohol (8.1 mg in 1 ml) is added and the mixture shaken thoroughly. A clear filtrate of the solution is used for the test.

3. Potassium Permanganate

For procedure see p. 120.

Reacting substances: the majority of substances with a double or triple bond or easily oxidizable compounds.

4. Bromine in Carbon Tetrachloride

For procedure: see p. 120.

Reacting substances: the majority of compounds containing a double or triple bond. The decolorization of the bromine solution with the formation of hydrogen bromide (fumes above the liquid, especially upon a gentle blowing into the test tube) proves that substitution has taken place.

5. Sodium Test

For procedure see p. 174.

Reacting substances: alcohols, arylhydroxylamines, monosubstituted acetylenes (alkynes), methyl ketones, 1,3-diketones, β-ketoesters, and all other enolizable substances, for example, $CNCH_2CN$ and $RCOCH_2CN$ react in the cold. When heated, certain hydrocarbons (indene, fluorene, cyclopentadiene), amides, hydrazines, imides, and heteroxyclic compounds with an NH group also react.

6. Melting with Sodium Hydroxide

The substance $(20 - 30 \text{ mg})$ is heated in a silver or nickel crucible with $1 - 2$ pellets of solid sodium hydroxide and kept in the melted state for 1 min. During the melting the evolution of basic vapors is tested with wetted litmus paper. The cooled and solidified melt is extracted with 3 ml of water and filtered. The alkaline filtrate $(2 - 3 \text{ drops})$ is mixed with a few drops of 1% sodium nitroprusside solution to test the presence of sulfidic sulfur. The formation of a violet color proves the presence of bivalent sulfur in the tested substance.

Reacting substances: sulfides, thiophenols, thioesters, sulfilimines, thioanilides, thioureas, isothiocyanates, and xanthogenates. If the test for the sulfidic sulfur is negative, the main filtrate is mixed with three drops of 30% hydrogen peroxide, boiled, acidified with dilute hydrochloric acid (1 : 1), and additioned with three drops of a 10% barium chloride solution.

Reacting substances: sulfonic acids and their salts, sulfinic acids, sulfoxides, sulfones, and sulfonamides.

7. Ferric Chloride

For procedure see p. 189, and for more detailed discussion p. 188.

Reacting substances: phenols, enols, aliphatic amino acids, mercaptans, thiophenols, antipyrine, and pyramidon.

8. p-Dimethylaminobenzaldehyde

For procedure see p. 322.

Reacting substances: aromatic primary amines, amino acids, and proteins.

9. 2,4-Dinitrophenylhydrazine

For procedure see p. 221.

Reacting substances: aldehydes, ketones, carbonyl-group-containing compounds.

10. Sodium Tetraphenylboranate

For procedure and details see p. 331.

Reacting substances: basic compounds.

11. 3,5-Dinitrobenzoyl Chloride and Paper Chromatography

The substance (5 – 10 mg or a small drop) is mixed in a micro test tube with approx. 50 mg of 3,5-dinitrobenzoyl chloride and heated at 100 – 110 °C for 10 – 15 min in a sand, water, or some other bath. After cooling, 1 ml of benzene and five drops of 5% Na_2CO_3 are added, the mixture is stirred thoroughly with a glass rod, and the benzene layer (5 – 30 µl) is deposited on the start of a strip of chromatographic paper. Chromatography is carried out (it is sufficient to use test-tube chromatography) in the following solvent systems: paper A is impregnated with dimethylformamide in benzene (1 : 1) and developed in hexane, paper B is impregnated with a 20% solution of formamide in ethanol and developed in chloroform-ethyl acetate 1 : 4. Detection of substances on the chromatogram is carried out by reduction with stannous chloride and spraying with Ehrlich reagent (see p. 154). In this way aliphatic, cycloaliphatic, unsaturated and heterocyclic alcohols, arylalkyl alcohols, glycols, polyethylene glycols, monoethers of glycols and of polyethylene glycols, chlorohydrins, phenols, mercaptans, thiophenols, primary and secondary aliphatic and aromatic amines, and other compounds with functional groups capable of benzoylation can be detected. The chromatography is carried out first in system A. If, after detection, a spot appears between the start and the solvent front, it is considered as evidence of the presence of a compound of the above-mentioned classes. If a spot appears directly on the start, chromatography in the system B is carried out, in which 3,5-dinitrobenzoic acid has an R_F value of 0.03 and other derivatives always have a higher R_F value. The procedure may also serve for further identification on the basis of found R_F values, or after chromatography in other systems and comparison with standards. For R_F values of 150 derivatives of 3,5-dinitrobenzoic acid in 20 solvent systems see (2).

12. Hydrolysis

The substance (50 − 75 mg) is mixed in a micro test tube with 2 ml of 20% NaOH and shaken, and it is observed if the separation of phases takes place (or if the smell of amines can be noticed). The liberated amines are isolated by extracting them with ether or by steam-distillation. Additional tests can then be carried out. For the behavior of esters at hydrolysis see p. 264, for amides see p. 272, and for nitriles p. 369. When carrying out this test one should remember that concentrated alkali solutions, especially if hot, cause chemical changes of certain classes of organic compounds (aldehydes, diketones, halogenated compounds). Arylamines substituted in *o*- or *p*-position with a nitro or nitroso group are hydrolyzed in an alkaline medium, setting free ammonia or an amine and forming the sodium salt of the corresponding phenol. For the hydrolysis of acetals and ketals see p. 296, for nitriles p. 369, amides p. 272, and esters p. 265.

13. Tetrazolium Salt

To 0.1 ml of tetrazolium salt solution [1% aqueous solution of 2,2-diphenyl-5,5-(3,4-dimethoxyphenyl)-3,3-dimethoxy-4,4-diphenyleneditetrazolium chloride (3)] 0.3 ml of 0.3 N NaOH solution and 0.5 ml of water are added. The solution is boiled briefly, and if it turns colored, 1 − 2 drops of water are added and the solution heated again until decolorized. Then 0.1 ml of an aqueous solution of the tested substance is added and the mixture is heated to boiling and coloration is observed. In the presence of 1 − 5 µg of a reducing compound the solution turns blue, and at a higher concentration a dark precipitate of formazan is formed.

Reacting substances: compounds with reducing groups, aldehydes, sugars, and polyphenols, and compounds with SH groups.

14. Aluminum Chloride and Chloroform

For procedure see p. 130.
Reacting substances: aromatic hydrocarbons, for example, benzene and its homologs, arylhalogenides, naphthalene, anthracene, phenanthrene, etc.

15. Molisch Reagent

For procedure see p. 308.
Reacting substances: monoses, polyoses, and glycosides.

16. Fehling Reagent

For procedure see p. 210.

Reacting substances: aliphatic aldehydes, α-hydroxyaldehydes, α-hydroxyketones and reducing sugars, hydrazines, etc.

17. Tollens Reagent

For procedure see p. 210.

Reacting substances: aldehydes polyphenols, α-diketones, tartaric acid, reducing sugars, some aromatic amines, aminophenols, hydrazines, etc. Formic acid reacts at elevated temperatures.

18. Periodic acid

For procedure and discussion see p. 177.

Reacting substances: 1,2-glycols, α-hydroxyaldehydes, α-hydroxyketones, and 1,2-diketones.

19. Benzenediazonium Fluoroborate

For procedure see p. 192.

Reacting susbstances: monohydric phenols with an unsubstituted para position or, at least, with a free ortho position, polyhydric meta-phenols. For more information see p. 341.

20. Nitrous Acid

A solution of 0.1 g of substance in 4 ml of 2 N HCl is cooled to 5 °C and mixed with a 10% $NaNO_2$ solution, until a drop of the mixture gives a blue color with KI-starch paper. Nitrogen is set free by amides, aliphatic primary amines, and amino acids (amides are converted to acids, and amines to alcohols: yields are often small, but the products can be isolated). A yellow oil or a solid substance (nitrosamine) is formed in the case of secondary aliphatic and aromatic amines. Nitrosamines can be extracted with ether.

Primary aromatic amines give a clear solution on diazotization, and the formation of the diazonium salt is proved by coupling it with phenols; 3 ml of 1% 2-naphthol solution in 2 N NaOH are added to 1 ml of the diazonium salt solution. Only some negatively substituted primary aromatic amines do not react with nitrous acid (see p. 341).

Tertiary aliphatic amines and tertiary aromatic amines with substituted para position do not react with nitrous acid under these conditions. On alkalization the unchanged base precipitates, if insoluble in water. Tertiary amines with a free para position give nitroso derivatives which sometimes precipitate as insoluble hydrochlorides; in contrast to nitrosoamines, these hydrochlorides are insoluble in ether. Upon alkalization of the solution the green nitroso derivative is set free, which can be extracted with ether (a blue extract). m-Phenylenediamine reacts with HNO_2 under the formation of a brown dye (vesuvin), mercaptans with HNO_2 give strongly colored thionitrous acids (esters are usually red).

21. Hydroxamate Test

For procedure see p. 262.

Reacting substances: carboxylic acids, their esters, acyl halogenides, anhydrides, and lactones. Esters and anhydrides need not be converted to chlorides.

22. Picric Acid

For procedure see p. 128.

Reacting substances: polynuclear aromatic hydrocarbons, phenols, aryl ethers, and bases, including alkaloids and purines.

23. Ninhydrin

For procedure see p. 279.

Reacting substances: α- and β-amino acids, aliphatic primary and secondary amines.

24. Liebermann Reaction I

The substance (3 − 5 mg) is dissolved in 0.5 ml conc. H_2SO_4 and this solution is mixed with 1 ml of a 5% KNO_2 solution in conc. H_2SO_4. After a while a strong coloring occurs. If the reaction does not take place at room temperature, the solution is boiled briefly and then cooled. A careful slow dilution with water sometimes leads to a deepening of the color; alkalination of the solution (with cooling) very often changes the color.

Reacting substances: monohydric phenols with free para position, polyhydric phenols of the meta series; with warming, m-aminophenol and its derivatives, and thiophene and its derivatives also react.

25. Liebermann Reaction II

The substance (1 – 2 mg) is melted with a crystal of phenol. To the cooled melt a few drops of concentrated sulfuric acid are added. A dark, cherry-red color is formed. The reaction mixture is diluted with 2 ml of water and alkalized with a few drops of sodium hydroxide solution; the solution turns green, blue, and sometimes even violet.

Reacting substances: aromatic nitroso compounds.

26. Formaldehyde and Sulfuric Acid

For procedure see p. 130.

Reacting substances: aromatic hydrocarbons, ethers and halogenated derivatives, and unsaturated hydrocarbons.

27. Ceric Ammonium Nitrate

For procedure see p. 170.

Reacting substances: alcohols and phenols (approximately up to C_{10}), hydroxy acids, hydroxy aldehydes, and other hydroxy-group-containing compounds (see also p. 170).

28. Sodium Iodide

For procedure see p. 137.

Reacting substances: alkyl chlorides and bromides, acid chlorides and bromides, allyl chlorides and bromides, ketones, esters, amides and nitriles substituted with chlorine or bromine in α-position. For discussion see p. 136.

29. Fluorescein Chloride

For procedure see p. 324.

Reacting substances: primary and secondary aliphatic amines, amides, nitriles, and pyrrole derivatives; for specific hues and fluorescence see p. . . . .

30. 3,5-Dinitrobenzoic Acid
Anhydride + SnCl₄

For procedure see p. 203.

Reacting substances: ethers.

31. Iodoform Reaction

For procedure see p. 235.

Reacting substances: substances containing the CH_3CO group bound to a hydrogen or carbon atom, or compounds which can be oxidized by alkaline hypoiodite to a substance containing such a group.

32. Stannous Chloride and Hydrochloric Acid

The substance (30–50 mg) is dissolved or suspended in 2 ml 3 N HCl in a test tube, and to this solution approximately 0.1 g of granulated tin is added in small portions. If the reaction is slow, it should be enhanced by mild heating. After the dissolution of tin the test tube is heated for 10 min over a boiling water bath. Then 6 N NaOH is added until the tin hydroxides are dissolved. After cooling, the alkaline solution is extracted twice with 2 ml of ether. The ethereal extracts are then reextracted with 1 ml 6 N HCl and the aqueous layer is used for the test for amines (see p. 318).

Reacting substances: nitro compounds, nitroso compounds, azoxy compounds, and azo compounds.

33. Iodine and Sulfanilic Acid

The substance (3–5 mg) is wetted with three drops of conc. HCl and the mixture is then evaporated almost to dryness. A small amount of solid sodium acetate (on the tip of a knife) is added first, followed by two drops of a sulfanilic acid solution (1 g of sulfanilic acid dissolved in 75 ml of water and 25 ml of glacial acetic acid) and one drop of iodine-acetic acid solution (1.3 g of iodine dissolved in 100 ml of glacial acetic acid). After 3 min standing the excess iodine is eliminated by addition of thiosulfate (approx. 0.1 N $Na_2S_2O_3$), and one drop of a 1-naphthylamine solution is added (0.3 g of 1-naphthylamine dissolved in 70 ml of water and 30 ml of glacial acetic acid). A pink to purple color is formed.

Reacting substances: oximes of aldehydes and ketones.

34. Ferrous Hydroxide

In a narrow micro test tube approximately 20 mg (one drop) of the sample are mixed with 1.5 ml of a freshly prepared acidified 5% solution of ferrous ammonium sulfate (0.1 ml of conc. H_2SO_4 per 25 ml of the solution).

One drop of $3 \text{ N } H_2SO_4$ is then added, followed by 1 ml of $2 \text{ N } KOH$ in methanol. The test tube is immediately closed and thoroughly shaken. If the test is positive, a precipitate is formed, which changes its color to red-brown within 1 min.

Reacting substances: nitro compounds, nitroso compounds, quinones, nitrates, and nitrites.

35. Nitrole Reaction

Several drops of the substance are shaken for 1 min with 2 ml of a sodium nitrite solution (5% solution of $NaNO_2$ in $2 \text{ N } NaOH$). Then $2 \text{ N } H_2SO_4$ is added dropwise. A dark red color is formed (alkali nitrolate), disappearing on addition of acids and reappearing on repeated alkalization; sometimes a blue-green color is formed after acidification, which does not disappear after alkalization; extraction with chloroform gives a blue extract (pseudo-nitrole).

Reacting substances: primary nitroparaffins (red color); secondary nitroparaffins (blue-green color); tertiary aliphatic and aromatic nitro-hydrocarbons.

References

1. Shriner, R. L., and Fuson, R. C.: The Systematic Identification of Organic Compounds, 3rd Edition. Wiley, New York 1948.
2. Gasparič, J., and Borecký, J.: J. Chromatog. 5, 466 (1961).
3. Cheronis, N. D., and Stein, H.: J. Chem. Educ. 33, 120 (1956).

CHAPTER VII

HYDROCARBONS

1. Paraffins and Cycloparaffins

Paraffins can be separated from mixtures with other solvents almost quantitatively by using a mixture of 170 parts of 85% phosphoric acid ($\varrho = 1.7$) and 100 parts of concentrated sulfuric acid ($\varrho = 1.84$). Oxygen-containing compounds dissolve, and hydrocarbons remain undissolved in this mixture. Only some reactive olefins can partly form monoesters of sulfuric acid. The process is carried out with cooling, and the reduction in the volume of the upper layer, as well as the change of its refractive index, are followed. The procedure can be considered as completed when both the volume and the refractive index become constant. If a mixture of sulfuric acid and phosphorus pentoxide is used, olefins and aromatic eompounds also react, so that enly paraffins remain undissolved (1).

Detection of Higher Paraffins

In view of the fact that paraffins do not give any characteristic color reactions, their elementary analysis (C + H = 100%), molecular weight, and the fact that they lack absorption in UV light are fully sufficient for their detection.

If a mixture of aliphatic hydrocarbons and other substances (for example, paraffin oil in technical products) is being investigated, we can detect them best by chromatogaphing the mixture (0.5−1.0 g) on an alumina column (activity I−II) with hexane. Paraffinic hydrocarbons are eluted in the first fractions, the hexane solution does not absorb in the UV region, and, after evaporation of hexane, an oily residue is obtained the elementary analysis of which will prove that it is a hydrocarbon. To carry out the experiment, an alumina column of 2 cm diameter and 20 cm in height is sufficient.

Identification of Paraffins and Cycloparaffins

Paraffins and cyloparaffins do not form derivatives suitable for identification. Although paraffins can undergo substitution reactions (for example, halogenation) under suitable conditions, giving rise to a number of isomers the separation of which is difficult, these reactions are not suitable for identification. Therefore, the compounds of this class are identified by determining their physical constants (boiling point, refraction, density) and measuring their absorption spectra (2−4). Gas chromatography represents an improved and suitable method for the identification of paraffinic hydrocarbons (5−7). (The preparation of clathrates has only a limited use − see below.)

It is important to point out that pure compounds are necessary for the measurement of physical constants and spectra, and that the isolation of such substances is usually very difficult. Gas chromatography is very suitable for the determination of purity, because it can also be used for the isolation of a pure individual compound (see p. 133).

Addition Compounds − Clathrates

Of those compounds (clathrates) which during crystallization occlude other compounds in the cavities of their crystals, those with urea and thiourea (8, 9) have a certain importance for the identification of hydrocarbons.

The formation of clathrates is not specific for aliphatic hydrocarbons; a necessary condition is that the added substance have a particular given length of the unbranched hydrocarbon chain (usually C_6). The added compounds are usually not formed in molar stoichiometric ratios, and the amount of urea increases with the chain length of the added compound. For example, the clathrate of urea with n-heptane contains 6.1 mole of urea for 1 mole of n-heptane, while in the case of n-dotriacontane (n-$C_{32}H_{66}$) the ratio is 23.3 : 1. Addition compounds with urea usually crystallize in hexagonal systems, in contrast to urea itself, which crystallizes in a cubic system. In the hexagonal network of urea, free channel-like spaces are present in which the added substances are occluded. The diameter of the channels is approx. 5 Å, and straight-chain hydrocarbons enter into them easily. Branched chains cannot enter this space, and therefore only clathrates of straight-chain hydrocarbons can be formed.

In a similar manner, clathrates of hydrocarbons with thiourea are formed. The diameter of the channels is, however, 5.4 Å, which makes it possible for even molecules with branched chains to enter into them.

We must stress the fact that the formation of clathrates is not specific for hydrocarbons. A large number of monofunctional and difunctional compounds are known which form clathrates with urea or thiourea.

Urea forms addition compounds with paraffinic and olefinic hydrocarbons, with alcohols, ethers, carbonyl compounds, mono − and dicarboxylic acids and their esters, with alkyl halides, thioethers, amines, diamines, nitriles, and dinitriles. It is obvious that the chemical character (i.e., the presence of certain functional groups) is not decisive for the ability to form clathrates. The decisive factor is the length and the size of the chain.

The range of compounds which combine with thiourea is still more varied. As regards their formation, the rule is that clathrates are most easily formed from aliphatic compounds with two methyl groups or with substituents of approximately the same size (halogens) in the chain. A larger side chain (ethyl, isopropyl) prevents the formation of clathrates with thiourea. Thiourea also forms adducts with cycloparaffins; simple aromatic compounds (benzene, nahphthalene, anthracene) do not form clathrates with thiourea.

From the above it follows that the preparation of clathrates is not specific for the preparation of derivatives of hydrocarbons; in addition, clathrates do not possess properties suitable for identification purposes (unclear melting points, sometimes quite indistinguishable, and independent of the character of the occluded molecule). Their main importance consists in their use for isolating and purifying substances, especially hydrocarbons (10).

Their preparation is very simple. The hydrocarbon is mixed with excess urea (thiourea) in methanol.

The stability of clathrates in air is very varied. Clathrates with volatile components usually decompose on standing in air; clathrates of higher-molecular compounds are more stable. For example, the clathrate of thiourea and ω, ω'-dicyclohexyloctane is permanently stable at room temperature.

Clathrates dissociate in solution; the degree of dissociation is dependent on the polarity of the solvent. In water, decomposition takes place, so that the components can be easily separated and isolated.

Addition Compound of n-Decane and Urea

Reagents: urea, methanol.

Procedure: n-Decane (1 ml) is dissolved in 20 ml absolute methanol containing 3 g of urea. After 5 hr standing in a refrigerator the addition compound is filtered off in a filtration crucible. Yield of clathrate is 1.49 g. It decomposes slowly above 66 °C. Crystallization from 10 ml of methanol affords 184 mg of the compound; a slow decomposition begins above 55 °C.

Addition Compound of Cyclohexane with Thiourea

Reagents: thiourea, methanol.

Procedure: Cyclohexane (1 ml) is dissolved in 20 ml absolute methanol containing 3 g of thiourea. After 5 hr standing in a refrigerator the addition compound is filtered off with suction in a filtration crucible. Yield, 114 mg; the decomposition point cannot be determined.

2. Unsaturated Hydrocarbons

In order to detect unsaturated hydrocarbons, either addition or oxidative reactions are used. However, we must point out that no single reaction used for the detection of a double bond can give sufficiently reliable information in itself, because for each, a number of exceptions are known. Therefore, several reactions must be applied in order to detect the presence of a double bond.

Terpenic dienes give specific color reactions with 2,3,5,6-tetrachloro-p-benzoquinone (11). Hydration of acetylenic hydrocarbons gives rise to carbonyl compounds which can be converted to 2,4-dinitrophenylhydrazones (see p. 218) (12).

Reaction with Bromine

Compounds with double or triple bonds add bromine:

$$>C=C< \; + Br_2 \;\; \rightarrow \;\; >\underset{\underset{Br}{|}}{C}-\underset{\underset{Br}{|}}{C}<$$

The reaction is carried out in an indifferent solvent, for example, carbon tetrachloride, acetic acid, chloroform, or carbon disulfide. The decolorization of the solution takes place during the reaction. Hydrobromic acid is not set free. Its formation would prove that substitution had also taken place. The rate of this reaction is not equal with various compounds. Generally, the presence of negative substituents in the molecule slows it down.

Reagent: 5% solution of bromine in carbon tetrachloride.

Procedure: The tested substance (approx. 0.1 g) is dissolved in 2 ml of carbon tetrachloride and the reagent is added dropwise, with shaking. Decoloration is observed. If fumes of hydrobromic acid are formed, the test may be considered as dubious.

Reaction with Potassium Permanganate (13)

Unsaturated compounds decolorize potassium permanganate solution with the formation of a brown precipitate of manganese (IV) hydroxide. This reaction is more general and less specific than the addition of bromine. With water-soluble substances it is carried out in water solution. Among organic solvents, acetone or benzene were found suitable. Both solvents must be purified with permanganate and should not be decolorized with permanganate in a control experiment. A positive reaction is given by all easily oxidizable compounds.

Reagent: 2% aqueous potassium permanganate solution.

Procedure: To a solution of approximately 0.1 g of the tested substance potassium permanganate solution is added dropwise until a pink color persists.

Reaction with Ilosvay Reagent (14)

Compounds of the $-C \equiv CH$ type, carrying a hydrogen atom on the carbon atom bound with a triple bond, give a positive reaction with Ilosvay reagent (14).

Reagent: Crystalline copper sulfate or nitrate (1 g) is dissolved in a 50-ml graduated flask in the necessary amount of water, 4 ml of 20% ammonia are added, and the mixture is stirred. To the blue solution of the amo-complex which is formed, 3 g of hydroxylamine hydrochloride are added, and after dissolution the flask is filled with water up to the mark; the solution is colorless at this stage.

Procedure: To an aqueous or alcoholic solution of the tested substance an equal volume of the reagent solution is added and the mixture shaken thoroughly. In the presence of compounds of the mentioned type a black-yellow, red-violet, or brown precipitate of the corresponding copper acetylide is formed.

For the preparation of a more stable reagent see (15).

In spite of the fact that a double or triple bond is very reactive, there is no general method known for the preparation of solid derivatives. A series of unsaturated hydrocarbons (especially terpenes) can be identified by conversion them to nitroso chlorides (16). Crystalline derivatives are often obtained by addition of HCl or HBr:

$$R-CH=CH-R' + NOCl \rightarrow RCH(NO)CHClR'$$

Unsaturated hydrocarbons can also be identified after their oxidation to carboxylic acids. In the case of an asymmetrical unsaturated hydrocarbon two acids are formed, which can be identified by paper, thin-layer, or gas chromatography.

$$R-CH=CH-R \xrightarrow{O} 2 RCOOH$$

$$R-C \equiv C-R \xrightarrow{O} 2 RCOOH$$

The oxidation is carried out by shaking the hydrocarbon (heating is sometimes useful) with potassium permanganate in an alkaline aqueous solution. The MnO_2 is filtered off and the acids are isolated from the filtrate after acidification by extraction with ether (see p. 129). In this case the procedure also serves for the determination of the position of the double bond in the chain. In more complicated cases (for example, in the case of steroids) a combination of chemical and spectral methods is used [for example, oxidation with OsO_4 and $Pb(CH_3COO)_4$, and measurement of infrared spectra (17)].

1,2-Dibromostyrene

Identification of styrene: To a solution of styrene (0.2 ml) dissolved in 1 ml of CCl_4, 10 drops of Br_2 are added. After cooling, the dibromide separates, and is then redissolved by addition of 5 ml of methanol and heating. The hot solution is filtered through cotton, to give, after cooling, 259 mg of 1,2-dibromostyrene, mp 70−71 °C.

Monosubstituted derivatives of acetylene react with K_2HgI_4 according to the following equation:

$$2\,RC{\equiv}CH + K_2HgI_4 + 2\,KOH \rightarrow (RC{\equiv}C)_2Hg + 4\,KI + 2\,H_2O$$

The reaction is carried out by mixing an alcoholic alkyne solution with an aqueous alkaline mercuric iodide solution; the precipitated substance is crystallized from alcohol or benzene (18).

Among other reagents which add to the double bond (mercaptans, thiophenols, thiolic acids) and which are proposed for identification purposes, the most suitable seems 2,4-dinitrobenzenesulfenyl chloride (19), which easily forms crystalline sulfides with symmetrical olefins:

$$R-CH{=}CH-R + ClSC_6H_3(NO_2)_2 \rightarrow \underset{\underset{SC_6H_3(NO_2)_2}{|}}{RCHClCHR}$$

Asymmetrical olefins can give a mixture of isomers (for example, 2-pentene). Olefins substituted with negative groups (styrene) react unambiguously. A general method of preparation of sulfides consists in heating 200 mg of the reagent with 0.2−0.4 ml of the olefin in 5 cm of glacial acetic acid on a water bath. The end of the reaction (as a rule, after 10−20 min) is indicated by the cessation of liberation of iodine from an aqueous KI solution (spot test). The sulfide separates after cooling or upon pouring the reaction mixture onto ice. It is purified by crystallization from ethanol.

3. Aromatic Hydrocarbons

Reactions for detection and identification of aromatic hydrocarbons can be divided into substitution reactions (nitration, chlorosulfonation, acetylation), addition reactions (π-complexes), and oxidative reactions (oxidation of aliphatic chains, oxidation of aromatic hydrocarbons to quinones). The use of spectral methods is very common, usually in combination with separation methods, as, for example, column chromatography or counter-current distribution, or else column chromatography on silica gel combined with the use of fluorescing dyes (20−28) or gas chromatography.

For the application of spectroscopic methods for the identification and structure determination of polynuclear hydrocarbons see (29), and for the application of correlation analysis (chemical shifts of protons) for the identification of polyacetylnaphthalenes in mineral oil see (30).

Nitration of Aromatic Hydrocarbons

General remarks. Of a series of nitrating agents (HNO_3 + H_2SO_4; fuming HNO_3; HNO_3 in acetic acid, in acetic anhydride; alkali nitrates + H_2SO_4; nitrogen oxides), the nitration mixture HNO_3 + H_2SO_4 is most convenient for identification purposes; for micronitrations 100% nitric acid was proposed; it can be prepared by distilling HNO_3 with a double excess of sulfuric acid, preferably in vacuo (31). Nitration in sulfuric acid is, from the point of view of the reaction mechanism, an electrophilic aromatic substitution. It was found that when mononitrobenzene was nitrated in sulfuric acid, the reaction rate can be expressed by the equation

$$v = K\,[C_6H_5NO_2]\,.\,[HNO_3]$$

in which the terms in brackets correspond to the formal concentration, without regard to the form in which they are present in the solution. Later on it was found that in sulfuric acid nitric acid dissociates according to the equation

$$HNO_3 + 2\,H_2SO_4 \;\rightarrow\; NO_2^{\oplus} + H_3O^{\oplus} + 2\,HSO_4^{\ominus}$$

and the nitronium cation is the main and the most active nitration agent. The equilibrium of the above equation is shifted to the right by the increase in sulfuric acid concentration. Of importance for identification purposes is the orientation of the functional groups already present in the molecule (on the aromatic nucleus) at the moment of the introduction of the nitro group. Usually, ortho- and para-directing groups cause activation of these positions, while groups attracting electrons render nitration more difficult and cause meta orientation of the substituent because the reduction of the electron density of the carbon atoms of the aromatic nucleus is least in this position. In accordance with this, the groups OR, alkyl, NR_2, and halogens direct the entering nitro groups into positions ortho and para. The first three groups facilitate nitration, while halogens inhibit it. In contrast to this, polar groups, as for example, $-NO_2$, $-CN$, $-CHO$, $-COOH$, and $RCO-$, decrease the electron density at the aromatic nucleus, leaving the meta position most negative, because it is least influenced, and the nitronium cation seeks this position. If the aromatic nucleus carries two substituents, it is often very difficult to determine the direction of substitution beforehand because the influences mentioned can interfere. The situation can become still less clear if a greater number of substituents are present on the aromatic nucleus. Further, it is important to keep in mind that nitration always leads to a mixture of isomers. Typical examples are listed in Table 3, according to (34). From the table it is evident that for the identification of toluene, for example, mononitration cannot be utilized, but after the introduction of two nitro groups a product can be obtained which is suitable for identification (similar influence of the methyl and

Nitration of Monosubstituted Benzenes   *Table 3*

Group already present in the nucleus	Content of the isomer in the product in %		
	ortho	*meta*	*para*
CH_3	56.5	3.5	40
CH_2Cl	32.0	15.5	52.5
$CHCl_2$	23.3	33.8	42.9
CHl_3	6.8	64.5	28.7
F	12.4	—	87.6
Cl	29.6	0.9	69.5
Br	36.5	1.2	62.4
I	38.3	1.8	59.7
NO_2	6.4	93.2	0.4
COOH	18.5	80.2	1.3
$COOC_2H_5$	28.3	68.4	3.3
SO_2CH_3	—	100	—
$N(CH_3)_3$	—	100	—

the nitro group during the introduction of the second nitro group). For nitration of monoalkyl and dialkylderivatives of benzene, see, for example, (35). Nitration of poly-cyclic hydrocarbons does not usually lead to the formation of derivatives suitable for identification. When naphthalene is nitrated, the first nitro group enters exclusively into position 1, while the second nitro group enters the second nucleus, at positions 5 and 8, this giving rise to 1,5 and 1,8-dinitronaphthalene.

General Procedure of Nitration of Aromatic Hydrocarbons

The procedure consists in mixing 2−3 mmole of a hydrocarbon with 1.5 ml of concentrated sulfuric acid and adding 1.5 ml of fuming nitric acid. The reaction is brought to its end by gentle warming of the test tube in a water bath. The nitro derivative is isolated by pouring the cooled reaction mixture onto ice. Optimum reaction conditions for actual cases, if not known already, should be determined by changing the concentration of the nitration mixture, the temperature, and the reaction time.

Sometimes it is advantageous to prepare a trinitro derivative [m, p-xylene (36)]. The preparation of mononitroderivatives, which are usually liquid, is generally combined with reduction and the acetylation of the amine formed (37, 38). Nitration is often utilized for the detection and the determination of small amounts of aromatic hydrocarbons, for example, in air. After nitration the nitro derivatives formed can be detected by a sensitive color reaction or polarographically.

Method 1

Reagents: Nitration mixture: equal parts of conc. H_2SO_4 and of fuming nitric acid; 30% NaOH; peroxide-free ether; methyl ethyl ketone; a solution of 16 g NaOH + 12 g NaCl in 100 ml of water.

Procedure: The tested solvent (0.5 ml) is added slowly to 5 ml of ice-cold nitration mixture in a 150-ml Erlenmeyer flask, which is then stoppered and allowed to stand for 1 hr. The reaction mixture is then diluted, under steady cooling, with 50 ml of water, and 15 ml of 30% sodium hydroxide solution is added until the reaction of the mixture is alkaline to litmus. The mixture is then transferred to a separatory funnel and thoroughly extracted with 30−40 ml of water. The ethereal layer is then put into a porcelain dish and the ether is evaporated carefully. The residue is dissolved in 50 ml of butanol. A part of this solution is put into a test tube and mixed with sodium hydroxide solution. The appearance of a red to violet color means a positive reaction.

Method 2

Reagents: Nitric acid ($\varrho = 1.33$); sulfuric acid ($\varrho = 1.84$); acetone; 30% NaOH.

Procedure: Several milligrams of the tested substance are heated with two drops of nitric acid and 15 drops of sulfuric acid on a boiling water bath for 10 min. After cooling, 1 ml of water is added, followed by 10 ml of acetone and 5 ml of alkali hydroxide solution. The reaction is considered positive if the organic layer becomes violet or red (39).

Method 3

Reagents: Nitration mixture: nitric acid ($\varrho = 1.5$) and acetic acid 1 : 1; dimethylformamide; 10% tetramethylammonium hydroxide in water.

Procedure: A drop of chloroform solution of the tested substance, containing 5−100 mg of an aromatic hydrocarbon, or of fluorenone or quinone, is evaporated to dryness in a test tube, and 0.1 ml of the nitration mixture is added. The mixture is then heated for 5 min over a water bath and evaporated to dryness at the same temperature in vacuo. The residue is dissolved in 0.5 ml of dimethylformamide, and after 30 min standing 0.05 ml of tetramethylammonium hydroxide is added. A color characteristic of single hydrocarbons is formed (40).

Nitration of Benzene to *m*-Dinitrobenzene

Reagents: conc. H_2SO_4, conc. HNO_3, ethanol, ice.

Procedure: To the concentrated sulfuric acid (4 ml) in a ground-joint flask 1.3 ml of benzene is added, followed by 4 ml of HNO_3, which is carefully added dropwise under continuous and vigorous stirring. The flask is then connected with a reflux condenser and heated at 45 °C on a water bath for 5 min. The mixture is then poured onto 20 g of finely ground ice, and when this is dissolved the separated *m*-dinitrobenzene is filtered off and washed with 50 ml of water. The yield of the crude preparation is approx. 584 g, mp 74 – 85 °C. The crude product is dissolved in 10 ml of warm ethanol and then diluted with water (dropwise addition) until the solution begins to opalesce. The solution is made clear by gentle heating and it is then allowed to stand for a slow crystallization (yield of *m*-dinitrobenzene, 280 mg; mp 89 – 90 °C).

Nitration of Toluene to 2,4-Dinitrotoluene

Reagents: conc. H_2SO_4, fuming HNO_3, methanol, ice.

Procedure: A 10-ml test tube is filled with 0.15 ml of toluene and 1.5 ml of conc. sulfuric acid. Fuming nitric acid (1.5 ml) is then added dropwise to the mixture, which is then heated at 60 °C in a water bath, with occasional shaking, for 15 min. After cooling, the mixture is transferred into a glass beaker containing 10 g of finely ground ice. When the ice has melted, the nitro compound is filtered off in a filtration tube and washed with 6 ml of water. The product is dried in a stream of dry air. Yield, 168 mg; mp 58 – 62 °C. The crude product, when crystallized from 1.5 ml of methanol, yielded 61 mg of a product melting at 69 °C.

Addition Compounds of Aromatic Hydrocarbons (π-Complexes)

Polynitro compounds with aromatic hydrocarbons give solid derivatives, many of which are suitable for identification purposes. These addition compounds are usually intensely colored and are formed by the mixing of both components in a suitable solvent (benzene, methanol, ethanol, chloroform); they can also crystallize from a melt of both components. The ability of nitro compounds to form the mentioned complexes increases with the number of nitro groups in the molecule. The presence of halogens, amino groups, and hydroxy groups can also increase this ability, while methyl groups have an opposite effect.

The stability of the complex is also influenced by the second component of the addition compound. The number of aromatic nuclei in the molecule has a positive effect. For example, benzene does not form a stable addition compound with picric acid, but naphthalene gives a sufficiently stable compound. Substitution with a methyl, hydroxyl, or amino group has a positive effect on the stability of addition complexes, while the presence of a nitro group decreases the tendency of the second component to form an addition compound.

Although there are different opinions on the nature of forces acting between different components, it seems most probable that the nitro groups of one component induce the polarization of the second component. In relation to this, structural changes which increase the stability of addition compounds are precisely of such a nature as to lead to the increase in the polar character of one component and the polarizability of the other. A further reason for the formulation of molecular compounds on the basis of the influence of electrostatic forces consists in the fact that addition compounds are very easily and rapidly formed and equally easily and rapidly decomposed. The study of crystals of these compounds has also shown that the distance between the components is such as to make a covalent bond very improbable. In contrast to this, in certain cases the properties of molecular compounds do not agree with the formulation, which requires only the influence of electrostatic forces. With dimethyl ester of 4,4′, 6,6′-tetranitrodiphenic acid, indene gives compounds which form at a measurable rate, which points to the possibility that activation energy is necessary for the formation of this compound and also that valence bonds are broken in the course of the reaction. Moreover, the given compound reacts with bromine very slowly, while indene adds bromine rapidly under the same conditions.

For the identification of aromatic hydrocarbons addition compounds with picric acid, styphnic acid, and 2,4, 7-trinitrofluorenone are most commonly used. The last-mentioned reagent is most useful because its addition compounds with aromatic hydrocarbons are more stable. Identification constants of more than 150 addition compounds have been described (41, 42).

A General Method for the Preparation of Addition Compounds of 2,4,7-Trinitrofluorenone with Aromatic Hydrocarbons

To a boiling solution of 2,4,7-trinitrofluorenone (0.1 g) in 12 ml of a mixture of methanol and benzene (5 : 1) 0.3 mmole of the hydrocarbon dissolved in the required amount of the same solvent mixture (4 − 6 ml) is added and allowed to cool. The precipitated addition compound is filtered off (under suction) and washed with 1 ml of methanol, and crystallized from ethanol or an ethanol-benzene mixture.

Preparation of Addition Compounds of Picric Acid with Aromatic Hydrocarbons

The aromatic hydrocarbon (0.5 mmole) is dissolved in hot methanol (2 − 3 ml) and mixed with 1 ml of saturated methanolic picric acid solution.

After cooling, the precipitated product is filtered off and washed with 0.4 ml of methanol.

Addition Compound of Naphthalene with Picric Acid

Reagents: saturated picric acid solution in methanol.

Procedure: Naphthalene (50 mg) is dissolved in 3 ml of hot ethanol (in a 10-ml test tube), and after cooling, it is mixed with 0.8 ml of a saturated alcoholic picric acid solution. After 15 min standing in the refrigerator the addition compound is filtered off in a filtering tube and washed with 0.5 ml of ice-cold methanol. It is then dried for 10 min in a stream of dry air; yield, 50 mg; mp 150 °C (in capillary).

Acyl and Aroylbenzoic Acids

The preparation of aroylbenzoic acid by condensing phthalic anhydride with aromatic hydrocarbons in the presence of $AlCl_3$ (43) is applied only in those cases where the hydrocarbons do not give, on nitration, derivatives suitable for identification purposes:

$$ArH + C_6H_4 \overset{CO}{\underset{CO}{\diagup\diagdown}} O \xrightarrow{AlCl_3} C_6H_4 \overset{COAr}{\underset{COOH}{\diagup\diagdown}}$$

Identification of Benzene as 2,4-Dinitrophenylhydrazone of Acetophenone

Reagents: acetyl chloride, carbon disulfide, dry aluminum chloride, concentrated hydrochloric acid, 5% sodium hydrogen carbonate, 2,4-dinitrophenylhydrazine, sulfuric acid, and anhydrous calcium chloride.

Procedure: Into a 50-ml flask fitted with a calcium chloride tube 5 ml of a 1% acetyl chloride solution in anhydrous carbon disulfide and 0.15 g of $AlCl_3$ are introduced, and after 5 min standing 0.2 ml of benzene dissolved in 2 ml of carbon disulfide is added dropwise with shaking. The mixture is allowed to stand at room temperature until all $AlCl_3$ is dissolved (10 to 15 min). The mixture is then poured carefully onto a mixture of ice and 2 ml conc. HCl; 5 ml of carbon disulfide are added, and the organic layer is separated. This layer is washed gradually with 3 ml of 10% HCl, water, 5% $NaHCO_3$, and again with water until neutral to litmus. It is then dried over $CaCl_2$, and CS_2 is evaporated. The residue in the flask is dissolved in 2 ml of ethanol and mixed with a solution of 0.1 g of 2,4-dinitriphenylhydrazine in 0.5 ml of conc. sulfuric acid, 0.7 ml of water, and 2 ml of ethanol. The crystals are separated by filtration, washed with water, and crystallized from glacial acetic acid (93 mg of a crude product are obtained, mp 245–246 °C; after repeated crystallization the yield was 71 mg, mp 249 to 250 °C).

Oxidation of Aromatic Hydrocarbons

The side chain can be oxidized either with alkaline potassium permanganate solution or with chromic acid. However, a general method of oxidation, similar to the procedure described below for *p*-xylene, gives benzoic acid in all instances of oxidation of monoalkylbenzenes, while the oxidation of the three isomeric xylene gives different dicarboxylic acids. Thus for these cases the oxidation of the side chain represents a suitable identification method.

Oxidation of *p*-Xylene to Terephthalic Acid

Reagents: potassium permanganate, 6 N NaOH, dilute H_2SO_4 (1 : 1), $NaHSO_4$.

Procedure: Potassium permanganate (0.3 g) in 5 ml of water alkalized with two drops of 6 N NaOH is introduced into a 20-ml flask fitted with a reflux condenser, and, after dissolution (heating), 0.1 ml of xylene is added, followed by a boiling stone, and the mixture is refluxed for 3 hr. After cooling, the mixture is acidified with sulfuric acid (testing with Congo red paper), heated to boiling point for a short while, and the excess permanganate is eliminated by addition of several crystals of $NaHSO_3$. If MnO_2 is present, it may be dissolved in the same manner. The terephthalic acid is separated by filtration through a filtering tube, yield 43 mg. To purify the product, it is dissolved in hot 2 N NaOH solution (2 ml), diluted with water (3 ml), filtered through a folded filter paper, and the filtrate is acidified with sulfuric acid to precipitate the free acid. It is filtered in a filtration tube and washed with 5 ml of water (30 mg).

Oxidation of 2,6-Dimethylnaphthalene to 2,6-Naphthalenedicarboxylic Acid

Reagents: hexane, benzene, glacial acetic acid, chromium trioxide, methanol.

Procedure: 2,6-Dimethylnaphthalene (0.1 g) is dissolved in 1 ml of hexane, 0.5 ml of benzene, and 1 ml of glacial acetic acid. To this solution a solution of chromium trioxide in 0.2 ml of water and 1 ml of glacial acetic acid is added dropwise and with shaking, and the reaction mixture is boiled for 30 min. It is then cooled and diluted with 40 ml of water. The precipitate is allowed to settle down and it is then filtered off, washed with hot water, hot 1% sodium hydroxide solution, and again with water. Yield, 50 mg; mp 286.5–288 °C. After crystallization from benzene the yield was 22 mg, mp 286 °C.

Color Reactions of Aromatic Hydrocarbons

Reaction with Aluminum Chloride or Antimony Pentachloride in Chloroform Solution

Aromatic and polycyclic hydrocarbons give characteristic colors with antimony pentachloride or aluminum chloride in chloroform solution. A Friedel-Crafts reaction probably takes place, giving rise to colored triphenylmethane derivatives of the type

$$(C_6H_5)_3C^{\oplus}(AlCl_4)^{\ominus}$$

Reaction with Aluminum Chloride

Reagents: dry chloroform, anhydrous aluminum chloride.

Procedure: Approximately 100 mg of the tested substance are dissolved in 1−2 ml of chloroform in a test tube, and the walls of the test tube are moistened by shaking the tube. Powdered aluminum chloride is then added in such a way that part of it sticks to the moistened sides of the test tube. If an aromatic hydrocarbon is present, a characteristic color is formed, for example, blue if naphthalene is present, purple if phenanthrene is present.

Reaction with Antimony Pentachloride

Reagents: a mixture of antimony pentachloride and dry chloroform or carbon tetrachloride 1 : 2; CCl_4.

Procedure: The tested aromatic hydrocarbon is dissolved in chloroform and a drop of $SbCl_5$ solution is added. Characteristic color or precipitate is formed (44, 45).

Reaction with Sulfuric Acid and Formaldehyde

Unsaturated cyclic hydrocarbons, aromatic hydrocarbons and their derivatives, and polycyclic hydrocarbons give a very sensitive reaction with sulfuric acid and formaldehyde, in which deeply colored resinous substances are formed. In this manner as little as 0.1% of benzene in solvent mixtures can be detected. Saturated hydrocarbons, unsaturated aliphatic hydrocarbons, and cyclic saturated hydrocarbons do not give this reaction.

Reagent: To 2 ml of 30% formaldehyde conc. sulfuric acid is added carefully to make the volume 100 ml.

Procedure: 3 ml of the reagent are added to several milligrams of the tested liquid. A red-brown color represents a proof for small quantities, a brown color for larger quantities of unsaturated cyclic or benzenoid hydrocarbons.

It is important to carry out a parallel control experiment with sulfuric acid in the absence of formaldehyde and to compare the results (46). For a color reaction and a colorimetric determination of polycyclic hydrocarbons by this reaction in the presence of Fe^{3+} see (47).

4. Paper and Thin-Layer Chromatography of Hydrocarbons

As regards aliphatic hydrocarbons, paper chromatography has been used until now only for olefins, and this only after the addition of mercuric acetate (48). Aromatic hydrocarbons of the benzene series could probably be separated chromatographically either after nitration or after acylation according to a Friedel-Crafts procedure. On the other hand, a series of paper-chromatographic methods for the separation of polycondensed aromatics is already available. For a rapid detection of these substances in the presence of more polar ones chromatography in dimethylformamide/hexane can be used. Hydrocarbons have relatively high R_F values and do not separate well. However, they separate well from more polar substances. Identification of polycondensed hydrocarbons in mixtures can be carried out in the system 1-bromonaphthalene/90% acetic acid, or paraffin oil/methanol (49), or on acetylated or alumina-impregnated papers (50—52). Detection is carried out predominantly on the basis of their fluorescence; they can also be detected as characteristically colored spots after spraying the chromatogram with 0.01 M acetonic tetracyanoethylene solution (53) or with 0.5% 2,4,7-trinitrofluorenone solution in acetone or benzene (54). The limit of the detection based on fluorescence is 1 µg, and in the other two cases, tens of micrograms. For thin-layer chromatography, silica gel G as the stationary phase and hexane as the mobile phase are suitable (55).

5. Identification of the Basic Carbon Skeleton

An elegant modern method of identification of the basic carbon skeleton consists in pyrolysis in the presence of hydrogen combined with gas chromatography, which is mainly suitable for aliphatic compounds (56—58). The necessary equipment is commercially available (59).

The classical method of determination of the basic skeleton of polycondensed aromatic hydrocarbons consists in distillation with powdered zinc. During this operation the aromatic compounds are reduced to the parent hydrocarbon, which can be identified on the basis of its melting point, ultraviolet absorption spectra, or, often, by conversion of the resulting hydrocarbons to suitable derivatives—for example, picrates. Hydrocarbons can be also identified by chromatography. The advantage of this method consists in the possibility of working with minute amounts of substances; for example, for paper chromatography an almost imperceptible film of a sublimate would suffice. On the other hand, by-products of the reduction do not represent any inconvenience.

Procedure (60): Into a tube 15 cm long, 7−8 mm in diameter, and sealed on one side, 5 mg of the tested substance is introduced together with approximately ten times as much powdered zinc, an equal amount of anhydrous zinc chloride, and several milligrams of sodium chloride. After mixing the components the tube is inserted 6 cm deep into a small electric oven which heats the tube over 20 min to 300 or 360 °C. After 20 min the oven is switched off and allowed to cool. If the resulting hydrocarbon was volatile, it appears on the cooler parts of the tube in the form of a ring of sublimate. In this case the tube is cut just below the ring, the sublimate is washed out with several drops of benzene into a micro test tube, and the solution obtained is applied directly on the chromatogram.

Paper chromatography of the resulting products of reduction is carried out either on papers impregnated with dimethylformamide and hexane as the mobile phase, or, if necessary, on papers impregnated with 1-bromonaphthalene and 90% acetic acid as the running solvent. The detection is carried out either by observing the chromatogram under an ultraviolet lamp (fluorescence), or by spraying with an acetonic solution of trinitrofluorenone; carbazole and its derivatives can be detected by dipping the chromatogram into an antimony pentachloride solution.

Remarks: Anthraquinone yields anthracene at 300 °C almost quantitatlively. Among its derivatives, positive reaction was given by mono- and disufochloro derivatives, monosulfo and disulfo acids, as well as their potassium and barium salts, chloro- and dichloroanthraquinones, amino-anthraquinones, 1-aminoanthraquinone-2-carboxylic acid, chloro and bromo derivatives of 1-aminoanthraquinone and hydroxyanthraquinones. In all these cases anthracene was detected on the chromatogram as the only, or at least the predominant, spot. Only in those cases in which hydroxy or amino groups were protected by alkyl or acyl groups, as, for example, in 1,5- or 1,8-dimethoxyanthraquinone and acetaminoor benzoylaminoanthraquinones, was no positive reaction obtained. Amino and nitro derivatives of 2-methylanthraquinone, giving unequivocally 2-methylanthracene, derivatives of dibenzopyrenequinone, giving dibenzopyrene, and derivatives of pyrene were also investigated. The latter always gave two spots on chromatograms, of which one corresponded to pyrene and the other to the by-product formed during the sublimation of pyrene under the given conditions. Interesting results were obtained with halo derivatives of carbazole. For example, in the case of 1,3,6,8-tetrabromo-carbazole several spots were detected on the chromatogram: not only small amounts of the initial tetrabromocarbazole and of the resulting carbazole, but also single intermediary steps of dehalogenation, i.e., tri-, di-, and monobromocarbazole.

6. Gas Chromatography of Hydrocarbons

For the separation of olefins and naphthalenes diphenylformamide was used as stationary phase (61), for the separation of the isomers of methyl-cyclohexanes and methylcyclohexenes glycol saturated with AgNO$_3$ (62) was applied, for low-boiling alkanes, olefins, and diolefins monoethyl ether of diethylene glycol (63), dimethylformamide (64), or nitrobenzene (63) were suitable. For the separation of hydrocarbons the following stationary phases are of general value: dibutyl phthalate (67), dioctyl phthalate (68), dinonyl phthalate (66), n-hexadecane (69), silicone oils (68), polyethylene glycol (66), polypropylene glycol (66), tritolyl phosphate (70), 3,5-dinitro-benzoates of polyethylene glycols (71), and water (65).

References

1. Kattwinkel, R.: Brennstoffchem. **8**, 353 (1927).
2. Rank, D. H., Scott, R. W., and Fenske, M. R.: Ind Eng. Chem., Anal. Ed. **14**, 816 (1942).
3. Heigl, J. J., Black, J. F., and Dudenbostel, B. F., Jr.: Anal. Chem. **21**, 554 (1949).
4. Ludzack, F. J., and Whitfield, C. E.: Anal. Chem. **28**, 157 (1956).
5. Janák, J. Chem. Listy **47**, 1184 (1953).
6. Lichtenfels, D. H., Fleck, S. A., and Burow, F. H.: Anal. Chem. **27**, 1510 (1955).
7. Keulemans, A. I. M.: Gas-Chromatographie. Verlag Chemie, Weinheim 1959.
8. Bengen, F., and Schlenk, W., Jr.: Experientia **5**, 200 (1949).
9. Schlenk, W., Jr.: Ann. **565**, 204 (1949); **573**, 142 (1951).
10. Zimmerschied, W. J., et al.: J. Am. Chem. Soc. **71**, 2947 (1949).
11. Matawowski, A.: Chem. Anal. (Warsaw) **12**, 369 (1967).
12. Sharefkin, J. G., and Boghosian, E. M.: Anal. Chem. **33**, 640 (1961).
13. Baeyer, A.: Ann. **245**, 103 (1888).
14. Ilosvay, v. Nagy Ilosva L.: Ber. **32**, 2697 (1899).
15. Tilenschi, S.: Acad. rep. popul. Romane. Filiala Cluj. Studii cercetari stiint. **3**, 99 (1952); C. A. **50**, 11,174 (1956) .
16. Simonsen, J.: The Terpenes. Cambridge Univ. Press 1931.
17. Castells, J., and Meakins, G. D.: Chem. and Ind. **1956**, 248.
18. Johnson, J. R., and McEwen, W. L.: J. Am. Chem. Soc. **48**, 469, (1926).
19. Kharasch, N., et al.: J. Am. Chem. Soc. **71**, 2724 (1949); **75**, 1081 (1953).
20. McMurry, H. L., and Thornton, V.: Anal. Chem. **24**, 318 (1952).
21. Wedgwood, P., and Cooper, R. L.: Analyst **78**, 170 (1953).
22. Waterman, H. I., and Booy, H.: Anal. Chim. Acta **7**, 277 (1952).

23. Schnurmann, R., Maddams, W. F., and Barlow, M. C.: Anal. Chem. **25**, 1010 (1953).
24. Chamberlain, N. F.: Anal. Chem. **31**, 56 (1959).
25. Golumbic, C.: Anal. Chem. **22**, 579 (1950).
26. Chang Y.—Ch., and Wotring, R. D.: Anal. Chem. **31**, 1501 (1959).
27. Conrad, A. L.: Anal. Chem. **20**, 725 (1948).
28. Criddle, D. W., and Le Tourneau, R. L.: Anal. Chem. **23**, 1620 (1951).
29. Zander, M.: Erdöl und Kohle **15**, 362 (1962).
30. Foch Fu-Hsie Yew, Kurland, R. J., and Mair, B. J.: Anal. Chem. **36**, 843 (1964).
31. Weygand, C.: Organisch-chemische Experimentierkunst, p. 409, 2nd Ed., Barth Verlag, Leipzig 1948.
32. Martinsen, H.: Zeit. phys. Chem. **50**, 385 (1905); **59**, 605 (1907).
33. Westheimer, F. H., and Kharasch, M. S.: J. Am. Chem. Soc. **68**, 1871 (1946).
34. Holleman, A. F.: Chem. Revs. **1**, 187 (1925).
35. Franc, J., and Knížek, J.: Collection Czech. Chem. Commun. **24**, 2299 (1959).
36. Cheronis, N. D.: Micro and Semimicro Methods. Vol. 6, Ed.: Weisberger. Technique of Organic Chemistry, p. 318, Interscience, New York 1954.
37. Ipatieff, V. N., and Schmerling, L.: J. Am. Chem. Soc. **59**, 1056 (1937); **60**, 1476 (1938).
38. Newton, A.: J. Am. Chem. Soc. **65**, 2434 (1943).
39. Fabre, R., Truhaut, R., and Péron, M.: Ann. pharm. franc. **8**, 613 (1950).
40. Sawicki, E., and Miller, R. R.: Anal. Chem. **30**, 109 (1958).
41. Orchin, M., et al.: J. Am. Chem. Soc. **68**, 1727 (1946); **69**, 505, 1225 (1947); **70**, 1245 (1948); **71**, 3002 (1949); **73**, 436, 1877 (1951); J. Org. Chem. **18**, 609 (1953).
42. Laskowski, D. E., et al.: Anal. Chem. **25**, 1400 (1953); **26**, 1497 (1954).
43. Lewenz, G. F., and Serian, K. T.: J. Am. Chem. Soc. **75**, 4087 (1953).
44. Hilpert, S., and Wolf, L.: Ber. **46**, 2215 (1913).
45. Ekkert, L.: Pharm. Zentralhalle **72**, 51 (1931).
46. Nastyukov, A.: Zhurn. russ. fiz.-khim. obshch., khim. **36**, 881 (1904), C 1904 II 1042.
47. Rajzman, A.: Analyst **85**, 116 (1960).
48. Huber, W.: Mikrochim. Acta **1960**, 44.
49. Gasparič, J.: Mikrochim. Acta **1958**, 681.
50. Dubois, L., Corkery, A., and Monkman, J. L.: Int. J. Air Poll. **2**, 236 (1960).
51. Bergmann, E. D., and Gruenwald, T.: J. Appl. Chem. **7**, 15 (1957).
52. Spotswood, T. M.: J. Chromatog. **2**, 90 (1959).
53. Tarbell, D. S., and Huang, T.: J. Org. Chem. **24**, 887 (1959).
54. Gordon, H. T., and Huraux, M. J.: Anal. Chem. **31**, 302 (1959).
55. Petrowitz, H. J.: Chem. Ztg. **88**, 235 (1964).
56. Beroza, M., and Coad, R. A.: J. Gas Chromatog. **4**, 199 (1966).
57. Beroza, M., and Sarmiento, R.: Anal. Chem. **35**, 1353 (1963); **38**, 1042 (1966).
58. Beroza, M., and Acree, F., Jr.: J. Assoc. Offic. Agr. Chemists **47**, 1 (1964).
59. National Instrument Laboratories, Inc., Rockville, Md., Bulletin 3—8000.
60. Gasparič, J.: Mikrochim. Acta **1966**, 288.
61. Tenney, H. M.: Anal. Chem. **30**, 2 (1958).
62. Gil-Av, E., Herling, J., and Shabtai, J.: J. Chromatog. **1**, 508 (1958).
63. Podbielniak Information Booklet A-15-E 557.
64. Kaiser, R.: Gas-Chromatographie. Akademische Verlagsgesellschaft, Leipzig 1960.

65. Karger, B. L., and Hartkopf, A.: Anal. Chem. **40**, 215 (1968).
66. Brooks, V. T., and Collins, G. A.: Chem. and Ind. **1956**, 921.
67. James, D. H., and Phillips, C. S. G.: J. Chem. Soc. **1953**, 1600.
68. Scott, R. P. W.: Vapour-Phase Chromatography, Ed.: D. H. Desty. I. Intern. Symp., London 1956, Butterworths Sci. Publ., London 1957.
69. James, A. T., and Martin, A. J. P.: J. Appl. Chem. **6**, 105 (1956).
70. Grant, D. W., and Vaughan, G. A.: Vapour-Phase Chromatography, p. 413, Ed.: D. H. Desty, I. Intern. Symp., London 1956. Butterworths Sci. Publ., London 1957.
71. Petránek, J., and Šlosar, J.: Collection Czech. Chem. Commun. **26**, 2667 (1961).

CHAPTER VIII

HALO COMPOUNDS

If a halogen is detected in an unknown organic substance, the kind of halogen must first be determined (see p. 100) and then the type of bond by which it is connected to the rest of the molecule.

Procedures for the identification of this class of compounds can be divided into methods for the identification of alkyl and cycloalkyl halogenides and methods for the identification of aromatic halo compounds. The type of bond between the halogen and the rest of the molecule is decisive for the choice of the identification procedure. While the procedures for the identification of halo compounds with halogen atom bound to an aliphatic carbon are based on the exchange of a halogen by a suitable residue, the identification of compounds with halogen bound to an aromatic nucleus is generally carried out by additional substitution of the aromatic nucleus.

An elegant method of identification of halo compounds is based on catalytic dehalogenation combined with hydrogenation; the paraffinic (cycloparaffinic) hydrocarbons formed are identified by gas chromatography (1).

1. Determination of the Type of Bond between a Halogen and the Rest of the Molecule

a) Ionic halides. The test is carried out as in inorganic analysis, i.e., by precipitating silver halides, AgX, in dilute nitric acid. A white precipitate indicates a positive reaction. The test can be carried out by paper chromatography in the solvent system n-propanol-ammonia 2 : 1 or acetone-water (4:1) and detection with 5% silver nitrate solution in water and exposing the chromatogram for several minutes to the unfiltered light of a mercury lamp. Halogenides appear as violet-black spots. If chloride is detected in the sample, it is advisable to check whether the sample is free of sodium chloride. A positive reaction for ionic chlorine can, of course, be given by all compounds which are easily hydrolyzable under given conditions, as, for example, lower acyl halides or low-molecular α-halohydrins.

b) If the halogen is not ionized, the solubility of the substance has to be determined as well as the type of bond between the halogen and the rest of the molecule. This can be done by means of several simple tests: with alcoholic silver nitrate, sodium iodide solution in acetone, or by reaction with alcoholic alkali hydroxide solution. The results of these reactions are presented in Table 4 for various types of halo compounds. For a more detailed treatment see (2).

Method A

Reagents: 5% aqueous silver nitrate solution, dilute HNO_3 (1 : 1).

Procedure: The sample (2−5 mg) is dissolved in 2−3 ml of water and acidified slightly with dilute nitric acid, and to this solution the silver nitrate solution is added dropwise, with shaking. In the presence of Cl^-, Br^- and I^- precipitates are formed; white, floculent, and soluble in ammonia in the first case, white but only partly soluble in ammonia in the second case, and yellow in the last case.

Method B

Reagents: 2% $AgNO_3$ solution in 95% alcohol, alcohol, 5% HNO_3.

Procedure: The sample is dissolved in 1−2 ml of alcohol, 2 ml of the reagent are added and the mixture shaken briefly. It is then observed for approx. 5 min to see whether a precipitate is formed in the cold. If not, the mixture is heated to boiling point. If a precipitate separates, its solubility in dilute HNO_3 is tested. Silver halides do not dissolve, while silver salts of organic acids do.

Method C

Reagents: 15 g of sodium iodide are dissolved in 100 ml of acetone. The solution, which turns yellow on standing, should be kept in a brown bottle. If the solution becomes red-brown, it cannot be used any more.

Procedure: To 1 ml of the reagent in a test tube an acetone solution of the sample, containing chlorine or bromine, is added. The mixture is observed to see whether after shaking a precipitate is formed within 3 min at room temperature, or after heating to 50 °C on a water bath.

Method D

Reagents: 5% ethanolic KOH, 5% aqueous $AgNO_3$, dilute HNO_3 (1 : 1).

Procedure: The sample is mixed with 10 ml of the reagent, the mixture is boiled for 10 min, then cooled, acidified with dilute HNO_3, cooled again,

Reactions Used for the Determination of the Nature of Bonds   *Table 4*
between Halogen Atoms and Other Atoms

Reaction	*Halo compound*
2% AgNO$_3$ in ethanol → precipitate at room temperature (procedure B on p. 137)	RCOX, ArSO$_2$X, Ar$_3$CX, Ar$_2$CHX, ROCH$_2$X, ArCH$_2$X, RCH=CHCH$_2$X, RCHBrCH$_2$Br, R$_3$CX, RI, CBr$_4$, RCHXY (Y = COOH, COOR′, CONH$_2$, CN; X = = Br, I)
2% ethanolic AgNO$_3$ → precipitate formed in the cold only slowly, more rapidly on heating	R$_2$CHBr, RCH$_2$Br, RCHBr$_2$, CHBr$_3$, CHBr$_2$CHBr$_2$, R$_2$CHCl, RCH$_2$Cl, RCHCl$_2$
2% ethanolic AgNO$_3$ → substances which do not react even on heating	ArX, RCH=CHX, CCl$_4$, CHCl$_3$, CCl$_3$COOH, CHCl$_2$CHCl$_2$, ArCOCH$_2$Cl, RCHClY, ROCH$_2$CH$_2$X (Y = COOH, COOR′, CONH$_2$, CN; X = Cl, Br)
Acetonic NaI solution → precipitate formed in the cold within 3 min (procedure C on p. 137)	RCOX, RCH=CHCH$_2$X, ArCX$_3$, ArCHX$_2$, ArCH$_2$X, RCOCH$_2$X, RCHXCOOR′, RCHXCONH$_2$, RCHXCN, RCH$_2$Br, ArSO$_2$Cl,* CBr$_4$,* RCHBrCHBrR,* RCHBrCHClR*
Acetonic NaI solution → precipitate formed at 50 °C in 6 min	RCH$_2$Cl, R$_2$CHBr, R$_3$CBr, CHBr$_3$,* CHBr$_2$CHBr$_2$,* RCHClCHClR*
Acetonic NaI solution → no precipitate formed	ArX, RCHCHX, CH$_2$⟨CH$_2$—CH$_2$ / CH$_2$—CH$_2$⟩CHX, RCCl$_3$, CCl$_4$
Filtrate after hydrolysis with alcoholic KOH + AgNO$_3$ → precipitate	Halo compounds: aliphatic, aromatic, and with a halogen atom in the side chain
Filtrate after hydrolysis with alcoholic KOH + AgNO$_3$ → no precipitate	Aromatic compounds with a halogen atom on the nucleus, if not activated as in 2,4-dinitro-l-chlorobenzene

* Free iodine is also formed.

and filtered. Then $AgNO_3$ solution is added to the filtrate to see if a precipitate forms.

2. Identification of Alkyl and Cycloalkyl Halides

The basic reaction utilized for the identification is a nucleophilic substitution:

$$RX + Y \rightarrow RY + X$$

Thiourea, thiophenol, β-naphthol, phthalamide, and the potassium salt of p-hydroxyazobenzene (3) are most commonly used as nucleophilic substitution agents.

In addition, derivatives are described whose preparation is based on the preparation of a Grignard reagent. This reagent, when reacting with cyanates, affords solid derivatives suitable for identification. The procedure can be expressed by the following scheme:

$$RX \xrightarrow{Mg} RMgX \xrightarrow[2. \ H^+,\ H_2O]{1.\ ArNCO} RCONHAr$$

In addition to this, Grignard reagents prepared from alkyl and acyl halogenides can be converted by reaction with dimethylformamide and 2,4-dinitrophenylhydrazine to 2,4-dinitrophenylhydrazones of aldehydes (see p. 218) (4).

This procedure will be utilized only when a simple nucleophilic substitution cannot be applied, for example, the reaction of alkyl halogenide with thiourea.

Reaction of Alkyl and Cycloalkyl Halogenides with Thiourea

The formation of S-alkylthiuronium salts is represented by the equation

$$RX + CS(NH_2)_2 \rightarrow \left[R - S - C {\overset{\displaystyle NH_2}{\underset{\displaystyle NH_2}{\diagup}}} \right]^{\oplus} X^{\ominus}$$

This reaction involves nucleophilic substitution at the carbon atom and can take place in two ways, depending on the reaction conditions and the properties of the reacting component RX.

In bimolecular nucleophilic substitution (S_N2 reaction) a one-step attack of the nucleophilic agent (thiourea) takes place, which extrudes the second nucleophile (halogenide ion) from its link with the carbon atom. Primary and secondary alkyl halides always react in this manner.

In monomolecular substitution (S_N1) the dissociation of the $C-X$ bond in alkyl halide takes place first with the formation of a carbonium ion, followed by its rapid reaction with the nucleophilic agent. Tertiary alkyl halides always react in this way.

$$RX \rightarrow R\oplus + X\ominus$$
$$R\oplus + Y\ominus \rightarrow RY$$

The two reactions differ by the fact that the S_N2 reaction is of first order both with respect to alkyl halide and the nucleophilic agent, while the S_N1 reaction is of first order only with respect to alkyl halide. In other words, in the first case the reaction rate is proportional to the concentration of both reacting species, while in the second case the reaction rate depends only on the concentration of alkyl halide.

The general knowledge of the mechanism and of the kinetics of nucleophilic reactions has served very well for the reaction of thiourea with alkyl halides.

In less-polar solvents the reaction rate of thiourea with alkyl halides drops sharply in the following order: primary > secondary > tertiary; tertiary alkyl halides practically no longer react with thiourea in alcoholic solution (reaction S_N2); however, as soon as the polarity of the medium is increased by the addition of water, tertiary alkyl halides become reactive. The increased polarity of the medium facilitates the S_N1 reaction mechanism.

Reaction of primary and secondary alkyl halides with thiourea is usually carried out in the following manner: thiourea (2 mmole) is boiled with the corresponding alkyl halide (1 mmole) in ethanol (3 ml) or acetone. Reaction times vary appreciably with the nature of the halide. Primary alkyl bromides may be heated only $20-30$ min, and the reaction of primary alkyl iodides requires $10-20$ min; for primary alkyl chlorides the reaction time has to be prolonged to $120-180$ min while adding potassium iodide (0.1 g of KI is first dissolved in a few drops of water). Secondary alkyl bromides require practically the same time for their reaction as primary alkyl chlorides. Secondary alkyl chlorides require $3-10$ hr of boiling.

The reactivity is further influenced by the branching of the carbon chain and by the possible presence of a double bond and position of a second halogen atom. Among the tested monovalent and divalent halo compounds (5), methyl iodide, allyl bromide, benzyl bromide, and methylene iodide were found most reactive, a medium reactivity was observed for pentamethylene bromide, ethyl bromide, n-propyl bromide, n-butyl bromide, and n-amyl bromide, while ethylene bromide and isobutyl bromide reacted most slowly.

Reactions of tertiary alkyl halides with thiourea are carried out in 50 – 90% aqueous alcohol (10 – 15 ml) (6); 5 to 6 mmole of halide and 0.5 g of thiourea are boiled for 2 – 5 hr until the layer of halide disappears.

S-Alkylthiuronium halides are soluble in the reaction medium and are therefore converted to insoluble picrates or 3,5-dinitrobenzoates. While aqueous or alcoholic picric acid solutions are used for the preparation of picrates, 3,5-dinitrobenzoates are precipitated with 3% aqueous sodium 3,5-dinitrobenzoate (7).

Both picrates and 3,5-dinitrobenzoates are purified by crystallization from aqueous alcohol.

Instead of thiourea, N-phenylthiourea can also be used for the reaction with alkyl halogenides, and the S-alkyl-N-phenylthiuronium halogenides formed can be converted to sharply melting picrates, picrolonates or styphnates (8).

General Procedure for the Preparation of S-Alkylthiuronium 3,5-Dinitrobenzoates (Microprocedure)

Alkyl halide (10 – 25 mg) is mixed with 0.5 ml of a 2% acetonic thiourea solution and heated in a sealed glass ampoule in a water bath for various time intervals: iodides, 1 hr; bromides, 2 hr; chlorides (if the mixture was additioned with 10 mg of sodium iodide as catalyst) 3 hr. After cooling, the ampoule is opened and the contents transferred into a small test tube, and the solvent is evaporated on a water bath. The residue is dissolved in two drops of water and mixed by dropwise addition and shaking with 1 ml 2% aqueous sodium 3,5-dinitrobenzoate solution. The separated amo rphous product is dissolved by immersing the test tube into a hot water bath, and is then allowed to crystallize by slow cooling. The product is recrystallized from 0.5 – 1 ml of hot water.

This procedure was utilized for microidentification of the following alkyl halides (the corrected melting points in °C, as determined on a Kofler block, are given in parentheses): methyl iodide (205 – 206), ethyl chloride (188), ethyl bromide (188), n-propyl bromide (176), isopropyl bromide (194), allyl bromide (163), n-butyl chloride (169), n-butyl bromide (169), isobutyl bromide (158), sec-butyl bromide (176), n-amyl bromide (156), n-heptyl bromide (157 – 158), 2-ethylhexyl bromide (134 – 135), n-nonyl bromide (144), n-decyl bromide (145 – 146), n-dodecyl bromide (143), and benzyl chloride (174 – 175).

S-Methylthiuronium 3,5-Dinitrobenzoate

Reagents: thiourea, acetone, 2% aqueous sodium 3,5-dinitrobenzoate solution.

Procedure: A glass tube (10 mm inner diameter, 100 mm long), sealed at one end, is softened in a flame 4–5 cm from the end and is contracted slightly by drawing the ends carefully apart. An ampoule is thus formed which is filled, after cooling, with 10 mg of finely ground thiourea and 0.5 ml of a 3% methyl iodide solution in acetone (correspoding to approx. 15 mg of the halide). The ampoule is sealed and heated for 20 min in a boiling water bath. After cooling, the contents are transferred into a small test tube (3 ml) and the ampoule is rinsed with 0.5 ml of acetone. The solvent is then evaporated on a water bath and the residue dissolved in two drops of water. To this solution 1 ml of the 2% aqueous sodium 3,5-dinitrobenzoate solution is added dropwise with simultaneous shaking, the separated product is dissolved by immersing the ampoule into a hot water bath, and then allowed to cool for crystallization. The crystals are filtered using a filtration tube, washed twice with 0.5 ml of water, and dried. Yield, 35.5 mg; mp, 214–215 °C. Another crystallization from 1 ml of water gives 23 mg of a product, the melting point of which is 216 °C.

S-sec-Butylthiuronium Picrate

Reagents: thiourea, ethanol, picric acid.

Procedure: Thiourea (0.5 g) and sec-butyl bromide (0.5 g) are dissolved in 5 ml of ethanol and refluxed in a 20-ml flask for 2 hr. The reaction mixture is poured into a solution of 0.4 g of picric acid dissolved in a minimum amount of boiling ethanol and allowed to cool. Yield, 252 mg; mp, 167 °C. After crystallization from 20 ml of hot water the product (190 mg) melted at 167 °C.

S-tert-Butylthiuronium Picrate

Reagents: 3% thiourea solution in 50% aqueous ethanol, 5% aqueous sodium picrate, 50% aqueous ethanol.

Procedure: In a 10-ml flask provided with a reflux condenser 0.2 ml of tert-butyl bromide and 5 ml of thiourea solution are heated on a boiling water bath for 2 hr. The 5% picrate solution (10 ml) is then added and the mixture heated until it is transparent, and then allowed to cool slowly. The crystallized picrate is filtered off under suction using a filtration crucible, washed twice with 1 ml of water, and dried. Yield, 0.2 g; mp, 119–129 °C. Crystallization of the product from 2.5 ml of 50% ethanol gave 0.1 g of crystals melting at 157 °C.

3,5-Dinitrobenzoates

Esters of 3,5-dinitrobenzoic acid are used primarily for the identification of alkyl iodides, which are obtained by the cleavage of ethers or esters with hydriodic acid. The reaction is carried out by heating the alkyl iodide with the silver salt of 3,5-dinitrobenzoic acid in benzene or ether (see also p. 201)

$$RI + (NO_2)_2C_6H_3COOAg \rightarrow AgI + (NO_2)_2C_6H_3COOR$$

3. Identification of Fluoro Compounds

For identification of these compounds see (9 – 11).

4. Identification of Aromatic Halo Compounds

In comparison with alkyl halogenides, halides bound to aromatic nuclei are less reactive, and, unless activated by the presence of other substituents (for example, reactive chlorine in 2,4-dinitrochlorobenzene), nucleophilic substitution reactions cannot be used for their identification. For the identification, electrophilic substitution reactions on aromatic nuclei are used predominantly, such as nitration and chlorosulfonation. Only in exceptional and special cases are other procedures used [preparation of Grignard reagent and the conversion to anilides (12); preparation of addition compounds with picric acid – chloronaphthalenes, see p. 127; oxidation of side chains – oxidation of chlorotoluene to the corresponding chlorobenzoic acid, see p. 129].

Nitration of Aromatic Halo Compounds

A halogen atom on the nucleus decreases its reactivity on nitration and directs the substitution to ortho and para positions.

General Method of Mononitration

Concentrated nitric acid (2 ml) is added in several portions and with cooling to a mixture of the aromatic halo compound (0.2 g) and conc. sulfuric acid (2 ml) in a 15-ml test tube. The contents of the tube are then heated to 50 – 55 °C by immersion into a boiling water bath for 15 – 20 min with occasional shaking. After cooling, the mixture is diluted with 10 ml of ice-cold water and the separated nitro derivative is filtered off and washed with 4 – 5 ml of water. If an oily residue is formed, the test tube is put into a refrigerator, or it is allowed to stand at room temperature with occasional scratching of the inner wall of the test tube with a glass rod. The crude nitro derivative is recrystallized from aqueous methanol by first dissolving the product in a minimum amount of hot methanol, diluting the solution with water until a permanent turbidity is formed, and heating again until the solution is clear.

General Method of Dinitration

The procedure is the same as in the case of mononitration, except that fuming nitric acid is used, which is added in small portions, to keep the formation of brown fumes at a minimum. If heating at 45−50 °C for 15−20 min does not cause dinitration, the temperature is increased to 80 °C, and if this temperature is still not sufficient, the nitration is carried out at 100 °C.

2,4-Dinitrochlorobenzene

(Identification of chlorobenzene)

Reagents: conc. H_2SO_4, fuming HNO_3, 50% aqueous ethanol.

Procedure: In a 10-ml test tube a mixture of 0.5 ml of conc. H_2SO_4, 0.5 ml of fuming nitric acid, and 0.2 ml of chlorobenzene is heated over a boiling water bath for 30 min and with occasional shaking. After cooling, the mixture is diluted with 2 ml of water and allowed to cool in a refrigerator. The separated derivative is filtered off in a filtration tube, washed twice with water (2-ml portions), and dried in a stream of warm air. Yield, 180 mg; mp, 44−45 °C. Crystallization from 4 ml of 80% aqueous ethanol gives 114 mg of a product melting at 50 °C.

5. Color Reactions of Halo Compounds

Fujiwara Reaction

When chloroform is heated with pyridine in the presence of alkali hydroxide a red color is formed. The same reaction is given by benzo trichloride, trichloroisobutyl alcohol, and all other substances which, under the conditions of the reaction, set free or change to chloroform, as, for example, chloral, trichloroacetic acid, or sym-tetrachloroethylene. In addition to these compounds, various colors are given by 2,4-dinitrochlorobenzene, trichloroethylene, trichloroethanol, etc. The reaction was utilized for a large number of halo compounds. However, its use is debatable in the case of some of them (13). It is now certain that with carbon tetrachloride the reaction is negative. During the reaction chloroform first reacts with pyridine, upon which the pyridine nucleus is opened, with the formation of a derivative of glutaconic dialdehyde.

Reagents: pyridine, 10% NaOH.

Procedure: Pyridine (4 ml) and the sample are heated in the presence of two drops of alkali in a test tube over a water bath. The formation of a red color means a positive reaction.

For more recent applications see (13a−17).

Other Color Reactions

The following reactions were used for the industrially important benzene hexachloride: with zinc or, better, magnesium, in glacial acetic acid, benzene hexachloride is dechlorinated to benzene, which can then be detected by the Janovsky or Mohler color reaction (p. 356) (18–20). Also, the test proposed for DDT, i.e., nitration and the reaction of the nitro compounds formed with methanolic sodium hydroxide, is recommended (21). Other tests recommended are the reaction of DDT with sulfuric acid in acetic acid (22), the reaction with xanthydrol, alkali, and pyridine (23), and the reaction with benzene according to the Friedel-Crafts procedure (24).

For absorption-tube detection of trichloroethylene in air two tubes were recommended, the first filled with silica gel impregnated with potassium permanganate and orthophosphoric acid (in which chlorine is set free), and the second filled with silica gel impregnated with o-tolidine hydrochloride (for detection of the liberated chlorine) (25).

6. Paper and Thin-Layer Chromatography of Halo Compounds

In principle, alkyl halides can be chromatographed in two ways. Monoalkyl halides, with alkyls from C_1 to C_5, can be converted to corresponding S-alkylthiuronium salts and chromatographed in n-amyl alcohol-acetic acid-water (26).

Procedure: Alkyl halide (1–1.5 mg) is heated in an ampoule on a water bath with 0.8 mg of thiourea and 0.4 ml of acetone for 30 min in the case of iodides, 4 hr in the case of n-alkyl bromides, and 10 hr, with addition of a crystal of sodium iodide, in the case of branched alkyl bromides or n-alkyl chlorides. After cooling and opening of the ampoule, the acetone solution of the S-alkylthiuronium salt formed is applied on the chromatogram in increasing amounts: 6, 12, and 18 μl. The development is carried out in the n-amyl alcohol-acetic acid-water system (45 : 15 : 10). The chromatogram is then air-dried at room temperature and sprayed with a 0.04% ethanolic bromocresol-green solution. S-alkylthiuronium salts appear after a while as blue spots on a yellow background which turns green on standing. For the chromatography of S-alkylphenylthiuronium picrates on thin-layer plates see (27).

Another possibility to chromatograph alkyl halides on paper consists in their conversion to alkyl 3,5-dinitrobenzoates with the silver salt of 3,5-dinitrobenzoic acid according to the equation

$$RI + AgOOCC_6H_3(NO_2)_2 \rightarrow AgI + ROOCC_6H_3(NO_2)_2$$

The chromatography is carried out in systems suitable for these derivatives (see p. 154).

Procedure: A suspension of 50 mg of silver 3,5-dinitrobenzoate, 2 ml of benzene, and a drop of alkyl iodide in a sealed glass ampoule is heated at 100 °C for 120 min. After cooling and opening of the ampoule the benzene solution is applied onto the chromatogram either directly or, if necessary, after concentration.

7. Gas Chromatography of Halo Compounds

Fluorinated compounds were chromatographed on dinonyl phthalate (28), dioctyl phthalate (29), fluorinated esters (30), and silicones (30) as stationary phases.

For the separation of other halo compounds the following stationary phases were used: dibutyl phthalate (31), dinonyl phthalate (32), glycerol (33), paraffin (32), mineral oil (35), silicone oils (32, 34, 36), tritolyl phosphate (37), and water (33).

Experimental conditions are described in the literature quoted.

References

1. Thompson, C. J., Coleman, H. J., Ward, C. C., and Rall, H. T.: Anal. Chem. 34, 154 (1962).
2. Jureček, M.: Organická analysa I, p. 188—195. Nakladatelství ČSAV, Prague 1955.
3. Woolfolk, E. O., Donaldson, E., and Payne, M.: J. Org. Chem. 27, 2653 (1962).
4. Sharefkin, J. G., and Forschirm, A.: Anal. Chem. 35, 1616 (1963).
5. Večeřa, M., and Jureček, M.: Chem. Listy 47, 1342 (1953).
6. Veibel, S., and Lillelund, H.: Bull. Soc. Chim. (5), 5, 1153 (1938).
7. Jureček, M., and Večeřa, M.: Collection Czech. Chem. Commun. 19, 77 (1954).
8. Thomas, J., and Baker, W. A.: J. Pharm. Pharmacol. 12, 460 (1960).
9. Simons, J. H., Black, W. T., and Clark, R. F.: J. Am. Chem. Soc. 75, 5621 (1953).
10. Cheronis, N. D., and Entrikin, J. B.: Semimicro Qualitative Analysis, 2nd Ed., p. 470, Interscience Publishers, New York 1958.
11. Herkes, F. E., and Burton, D. J.: J. Chromatog. 28, 396 (1967).
12. Underwood, H. W., Jr., and Gale, J. C.: J. Am. Chem. Soc. 56, 2117 (1934).
13. Jenšovský, L., and Bardoděj, Z.: Pracovní lékařství 6, 301 (1954).
13a. Seto, T. A., and Schultze, M. O.: Anal. Chem. 28, 1625 (1956).
14. Hildebrecht, C. D.: Anal. Chem. 29, 1037 (1957).
15. Friedman, P. J., and Cooper, J. R.: Anal. Chem. 30, 1674 (1958).
16. Freytag, H.: Zeit. Anal. Chem. 145, 24 (1955).
17. Hunold, G. A., and Schühlein, B.: Z. Anal. Chem. 179, 81 (1961).
18. Schechter, M. S., and Hornstein, I.: Anal. Chem. 24, 544 (1952).
19. Scheibe, E., Zaumseil, I., and Dressler, A.: Pharmazie 13, 613 (1958).
20. Rathenasinkam, E.: Analyst 83, 688 (1958).
21. Schechter, M. S., and Haller, H. L.: J. Am. Chem. Soc. 66, 2129 (1944).

22. Chaikin, S. W.: Ind. Eng. Chem., Anal. Ed. **18**, 272 (1946).
23. Stiff, H. A., Jr., and Castillo, J. C.: Ind. Eng. Chem., Anal. Ed. **18**, 316 (1946).
24. Bailes, E. L., and Payne, M. G.: Ind. Eng. Chem., Anal. Ed. **17**, 438 (1945).
25. Gage, J. C.: Analyst **84**, 509 (1959).
26. Večeřa, M., and Gasparič, J.: Chem. Listy **48**, 1360 (1954); Collection Czech. Chem. Commun. **19**, 1175 (1954).
27. Rink, M., Lenhard, R., and Jaeger, H.: J. Chromatog. **29**, 396 (1967).
28. Evans, D. E. M., and Tatlow, J. C.: J. Chem. Soc. **1955**, 1184.
29. Percival, W. C., Anal. Chem. **29**, 20 (1957).
30. Desty, D. H.: Vapour-Phase Chromatography. I. Intern. Symp. London 1956. Butterworths Sci. Publ., London 1957.
31. Green, S. W.: ibid., p. 388.
32. Pollard, F. H., and Hardy, C. J.: Anal. Chim. Acta **16**, 135 (1957).
33. Desty, D. H.: Vapour-Phase Chromatography. I. Intern. Symp. London 1956, p. 115. Butterworths Sci. Publ., London 1957.
34. Desty, D. H.: ibid., pp. 52, 115, 332.
35. Cuthbertson, F., and Musgrave, W. K. R.: J. Appl. Chem. **7**, 99 (1957).
36. McFadden, W. H.: Anal. Chem. **30**, 479 (1958).
37. James, D. H., and Phillips, C. S. G.: J. Chem. Soc. **1953**, 1600.

CHAPTER IX

ALCOHOLS

The most suitable method for the detection and identification of alcohols is the utilization of the reaction with 3,5-dinitrobenzoyl chloride; the reaction has been very thoroughly studied and it can be used in combination with paper chromatography for simultaneous detection and identification. As they serve for identification purposes, the esters of 3,5-dinitrobenzoic acid were prepared from a maximum number of alcohols. Another advantage consists in the possibility of preparing these derivatives even when the alcohols are available only in the form of their aqueous solutions.

If the 3,5-dinitrobenzoates have melting points which are too high, benzoic acid esters are used for the identification of alcohols, namely glycols. On the other hand, in cases where the 3,5-dinitrobenzoates are low-melting substances, the alcohols are identified in the form of urethanes, preferably naphthylurethanes. Identification procedures are also elaborated for microscale work; a partial disadvantage consists in the fact that the samples for identification must be anhydrous.

Other derivatives are used in special cases either because no other derivatives of the corresponding alcohol are described, or because the commonly used derivatives are not suitable. For example, for the identification of glycols and their monoalkyl ethers pseudosaccharin ethers are suitable, while tert. alcohols are easily converted to S-alkylthiuronium salts via the corresponding alkyl chlorides.

If the preparation of the derivatives by the methods described is not feasible, or if no suitable derivative is known, primary and secondary alcohols can be oxidized to corresponding carbonyl compounds.

If authentic samples are available, the simplest and quickest way of identification is paper, thin-layer, or gas chromatography, or spectroscopy.

Another method for the detection of the alcoholic group is acetylation and the reactions based on the oxidation and the formation of colored complexes. The nitrole reaction or use of the Lucas test makes the differentiation of the type of binding of the hydroxy group possible. The xanthogenate reaction can also be applied for sensitive detection of primary and secondary alcohols. The utilization of chemiluminescence in the identification of some aliphatic alcohols has also been described (1). For a rapid identification of certain glycols the determination of the so-called critical dissolution temperature (2) has been proposed. A color reaction with N-bromo- and N-iodosuccinimide used for the differentiation of primary, secondary, and tertiary alcohols is also known (3).

1. Identification of Alcohols by Their Conversion to Esters

Acid chlorides are most commonly used for the esterification of alcohols. A simple method of esterification consists in a gentle heating of an acid chloride with an excess of an alcohol. The use of an inert solvent as a medium, preferably in combination with a tertiary amide, may be of advantage. Acetylation of low-molecular alcohols with acetyl chloride is seldom used for the preparation of derivatives because the acetates formed are usually liquids. In the case of higher-molecular alcohols (steroids), however, acetylation is the usual identification method. This method is used for the detection of the hydroxy group, by saponification of the isolated acetate and testing the presence of acetic acid in the distillate. Acetylation with acetyl chloride is slower than the reaction with acetic anhydride. The reaction proceeds easily with primary and secondary hydroxyl groups, but acetyl chloride does not react with tertiary alcohols under normal conditions. Acetates of tertiary alcohols can be obtained in good yields on reaction with acetyl chloride in ether solution on addition of metallic magnesium (4).

With pyridine as the solvent acetylation usually takes place even in the cold and the esters can be isolated by acidification of the reaction mixture with dilute sulfuric acid. Since tertiary alcohols do not react in this case, the method can be utilized for the separation of tertiary alcohols.

Benzoylation can also be carried out with benzoyl chloride alone or in the solution of some inert solvent. However, the usual (Schotten-Baumann) procedure is carried out in water in the presence of an alkali (i.e., in the heterogeneous phase) and the reaction is enhanced by vigorous shaking or stirring. The use of a tertiary base (usually pyridine) as solvent is also quite common. For example, the alcohol can be benzoylated by heating it with a slight excess of benzoyl chloride in a mixture of pyridine and chloroform. Pyridine is then separated by extraction with dilute sulfuric acid, and the ester is isolated by evaporating the chloroform. The residues of benzoic acid are eliminated by washing the ester with a solution of sodium hydrogen carbonate, sodium carbonate, or sodium hydroxide. Benzoic acid esters do not represent a universal type of compound for the identification of alcohols, because the majority of benzoates of aliphatic alcohols are liquids; they serve for the identification of polyhydroxy compounds.

A suitable substitution of the aromatic nucleus of benzoyl chloride (nitration, bromination), or the use of chlorides of naphthalene-carboxylic acids, or, especially, anthraquinone-carboxylic acids, results in their esters being solid substances, even when prepared from aliphatic alcohols. The most popular derivatives for the identification of alcohols are, however, the esters of 3,5-dinitrobenzoic acid.

Acetylation (Detection)

Alcohols can be converted with acetyl chloride to corresponding acetates, which can be detected by the hydroxamate test. Since tertiary alcohols with acetyl chloride predominantly give tertiary alkyl chlorides, the reaction is carried out in the presence of dimethylaniline, which binds the liberated hydrogen chloride.

Procedure: Acetyl chloride (0.05 ml), dimethylaniline (0.05 ml), and 0.1 ml of the tested alcohol are shaken occasionally in a micro test tube. After 3−4 min the mixture is diluted with 0.5 ml of ice-cold water, stirred thoroughly (decomposition of acetyl chloride), and the ester allowed to separate. One to two drops of the ester layer are transferred with a micropipette into another test tube, where the hydroxamate test is then carried out: 0.5 ml of a saturated methanolic hydroxylamine hydrochloride solution is added, the mixture is shaken and then mixed with 0.5 ml of a saturated KOH solution in methanol. The mixture is shaken again and heated over a water bath or over a microflame just to the boiling point, and after cooling it is acidified with 2 N HCl and additioned with 1−2 drops of a 1% ferric chloride solution. Pink, red, or violet color is proof of the presence of alcohol in the original sample.

3,5-Dinitrobenzoates (Identification)

The esters are prepared either by reacting excess alcohol with 3,5-dinitrobenzoylchloride, or by acylation in an alkaline medium according to the Schotten-Baumann procedure, or by carrying out the reaction in pyridine diluted with an inert solvent (benzene). Esters are usually isolated by pouring the reaction mixture into water or by extracting it with ether and evaporating the solvent. Unreacted reagents, or the by-products formed (for example, anhydrides if pyridine was used) are separated by washing with dilute acid solution and then with sodium hydrogen carbonate, sodium carbonate, or sodium hydroxide, depending on the nature of the ester. Those esters which do not crystallize well are often purified by dissolution in ether, washing with 50% potassium hydroxide, then with dilute hydrochloric acid (1 : 1), drying of the ethereal layer, and distilling of the solvent. Some esters can be purified by vacuum distillation.

Reaction time and temperature depend on the nature of the alcohol. Primary and secondary alcohols are acylated in ether or benzene in the presence of pyridine after 2 hr standing. Tertiary alcohols react much more slowly, and therefore require higher reaction temperatures. They also incline to the formation of by-products (olefins, tertiary chlorides; these are formed especially in the absence of acceptors — pyridine, for example). It is advantageous to use isopropyl ether as solvent for the preparation of tertiary 3,5-dinitrobenzoates of alcohols, and to add pyridine to the reaction mixture.

The prepared esters are usually purified by crystallization. Aqueous ethanol, light petroleum, cyclohexane, ligroin, and toluene are suitable as solvents.

The preparation of 3,5-dinitrobenzoates is advantageous even when the alcohol is available in the form of a dilute aqueous solution. Derivatives are prepared by shaking an aqueous solution of alcohol with a solution of 3,5-dinitrobenzoylchloride in a mixture of ether and benzene in the presence of pyridine. The aqueous phase should be saturated with potassium carbonate.

3,5-Dinitrobenzoates give crystalline complexes with aromatic amines. 1-Naphthylamine is especially suitable for identification purposes. Complex compounds are prepared by mixing both components in a mixture of alcohol and ether. These complexes crystallize well and serve as a second series of derivatives. They can also be used for the purification of 3,5-dinitrobenzoates which do not crystallize well.

One of the few disadvantages of 3,5-dinitrobenzoates is that the esters of certain higher alcohols are not easily isolated and that their melting points are low and close together. For example, the melting point of 3,5-dinitrobenzoate of n-amyl alcohol is 46 °C, of n-heptyl alcohol 52 °C, and of n-decyl alcohol 56 °C. In such cases it is advantageous to use X-ray diffraction analysis (5) or infrared spectroscopy (6) for identification.

General Method of Preparation of 3,5-Dinitrobenzoates

The alcohol (1.5 — 2 mmole, containing less than 5% of water) is heated in a test tube with 1 mmole of the reagent (a control of its melting point is necessary, to see whether or not it is hydrolyzed; it might also be necessary to recrystallize it from carbon tetrachloride) over a small flame, being careful to keep the reaction mixture liquid. When derivatives of lower alcohols are prepared heating for 3 min suffices; for higher and secondary alcohols the reaction time should be increased three to four times. On cooling, the reaction mixture solidifies. It is then ground with a spatula to a fine powder directly in the test tube. The powder is shaken thoroughly with 2% $NaHCO_3$ solution at 50 °C for 1 min. The solid ester, washed several times with water, is dissolved in hot methanol, and water is added dropwise to the solution to incipient turbidity.

Ethyl 3,5-Dinitrobenzoate

Reagents: redistilled pyridine, 3,5-dinitrobenzoyl chloride, potassium carbonate, benzene, ether free of alcohols, 1% sulfuric acid.

Procedure: To 10 ml of a 0.5% aqueous ethanol (50 mg of ethanol) in a 100-ml flask with a ground glass stopper 0.1 ml of pyridine and 1 ml of benzene are added. The mixture is cooled with ice, and 11 g of potassium carbonate are added gradually, followed by 0.5 g of 3,5-dinitrobenzoyl chloride dissolved in 2 ml of benzene, under constant shaking, which is then continued for three additional minutes. Ether is then added (30 ml) and the shaking is continued for 10 min more. The ethereal layer is separated and the aqueous layer is extracted twice more with ether, first with 20 ml and then with 10 ml. The combined ethereal extracts are freed from pyridine by washing the extract twice with 10 ml of 1% H_2SO_4 and eventually with 10 ml of water. The organic extract is then dried and evaporated. The residue is dissolved in 10 ml of hot ethanol and filtered into a crystallization flask. Water is added dropwise to incipient turbidity, and the solution made transparent by heating and allowed to cool slowly. The precipitated product is filtered off on a filtration crucible and dried. Yield, 216 mg; mp, 91−92 °C. A repeated crystallization from 5 ml alcohol and 1 ml water gave 169 mg of product, mp 91−92 °C.

Methyl 3,5-Dinitrobenzoate

Reagents: 3,5-dinitrobenzoyl chloride, 2% aqueous sodium carbonate solution, ethanol.

Procedure: Methanol (2 ml) is boiled for 5 min with 0.5 g of 3,5-dinitrobenzoyl chloride in a test tube. After the addition to 10 ml of water the mixture is allowed to stand in a refrigerator and the solidified product is filtered off on a sintered-glass filtration funnel and washed with 10 ml sodium carbonate solution. Yield, 0.5 g; mp, 105−106 °C. For further crystallization the crude ester is dissolved in the necessary amount of hot ethanol (approx. 5 ml) and diluted slowly with water until the solution becomes turbid (approx. 1.5 ml). It is then heated again, to dissolve the turbidity, and allowed to crystallize. Yield, 0.4 g; mp, 108 °C.

Benzyl 3,5-Dinitrobenzoate

Reagents: 10% 3,5-dinitrobenzoyl chloride solution in benzene, pyridine, cyclohexane, 50% KOH, HCl (1 : 1).

Procedure: Benzyl alcohol (0.5 ml) is mixed in a test tube with 15 ml of the 3,5-dinitrobenzoyl chloride solution and 3 ml of pyridine, the test tube is stoppered with a rubber stopper, shaken thoroughly, and allowed to stand at room temperature for 2 hr. The solution is then transferred to a separatory funnel and is extracted twice with 5 ml of 50% KOH (after each shaking it is diluted with water), twice with 5 ml of HCl, and twice with 5 ml of water. After distilling off the organic solvent the crude preparation is crystallized first from approx. 80 ml of cyclohexane (yield, 0.9 g; mp, 111−113 °C) and then from 65 ml of cyclohexane (yield, 0.85 g; mp, 113−113.5 °C).

tert-Butyl 3,5-Dinitrobenzoate

Reagents: dibutyl ether (dried over sodium), 3,5-dinitrobenzoyl chloride, pyridine, 1% H_2SO_4, 2% Na_2CO_3, ethanol.

Procedure: 3,5-Dinitrobenzoyl chloride (0.1 g) and 0.006 ml of tert-butanol are mixed with 0.2 ml of pyridine and sealed in an ampoule and heated in a boiling water bath for 30 min. The contents of the ampoule are then poured into 10 ml of water, and the solid phase is separated by centrifugation. The supernatant is carefully poured off and the residue is stirred with 5ml of 1% sulfuric acid and centrifuged again. The solid substance is then dissolved in 1.5 ml of ethanol, water is added to the solution until incipient turbidity, and the separated crystals (the crystallization might be enhanced by scratching with a glass rod) are separated in a filtration tube (with suction) and dried in a stream of dry air. Yield, 45 mg; mp, 142 to 143 °C.

Addition Compound of Methyl 3,5-Dinitrobenzoate with 1-Naphthylamine

Reagent: 10% solution of 1-naphthylamine in 80% aqueous ethanol.

Procedure: Methyl 3,5-dinitrobenzoate (30 mg) is dissolved in 1 ml of ether and mixed with 1 ml of 1-naphthylamine solution. The precipitated addition compound is allowed to stand in a refrigerator for 15 min, it is then filtered off using a filtration tube and washed first with 0.5 ml of ice-cold 80% alcohol, then twice with 0.5 ml of icy water. The product is then dried in a stream of dry air (yield, 50.1 mg; mp, 119 °C). Crystallization from 5 ml of cyclohexane gave 30.7 mg of product, mp 120 °C.

For the preparation of 3,5-dinitrobenzoyl chloride see (7).

Paper and Thin-Layer Chromatography
of 3,5-Dinitrobenzoates (8)

Preparation of Derivatives

a) In a micro test tube (3 ml), 0.05 ml of an alcohol is heated with approximately 50 mg of 3,5-dinitrobenzoyl chloride at $105-110$ °C for $10-15$ min (oil bath). The reaction mixture is mixed with 1 ml of benzene and five drops of a 5% sodium carbonate solution and heated, with thorough stirring with a glass rod. The benzene solution can be used directly for application on the chromatogram.

b) A solution of 0.5 g of 3,5-dinitrobenzoyl chloride in 30 ml benzene is added to 10 ml of a dilute aqueous solution of the alcohol (containing approximately 1.5 mmole of dissolved alcohol) in a separatory funnel. Potassium carbonate (10 g) is then added to the cooled solution with shaking. After 20 min standing at room temperature (in the case of secondary alcohols the mixture is allowed to stand overnight) the benzene solution is extracted gradually with 20% KOH, water, HCl (1 : 1), and again with water. After drying the benzene solution with sodium sulfate, benzene is distilled off and the residue, dissolved in a minimum amount of benzene, is applied onto the paper.

Detection of 3,5-Dinitrobenzoates

This is carried out by spraying the chromatogram first with a solution of stannous chloride (0.7 g of $SnCl_2 . 2 H_2O$, 15 ml of conc. HCl, and 100 ml of water), and after 30 min standing with an acidified 1% alcoholic p-dimethylaminobenzaldehyde solution (95 parts of ethanol and 5 parts of conc. HCl) 3,5-dinitrobenzoates appear as yellow spots on a white background.

Solvent Systems

Aliphatic alcohols $C_1 - C_5$: dimethylformamide or formamide as the stationary phase, hexane or cyclohexane as the mobile phase (see Fig. 27). Higher aliphatic alcohols: paraffin oil as the stationary phase, a mixture of dimethylformamide-methanol-water in various ratios, (8 : 1 : 1, 4 : 1 : 1), (2 : 1 : 1), as the mobile phase. Glycols (as monoesters): formamide as the stationary phase, a mixture of benzene and hexane 1 : 1 as the mobile phase (see Fig. 28). Aliphatic cyclic alcohols: dimethylformamide as the stationary phase, hexane or cyclohexane as the developing solvent. For thin-layer chromatography of 3,5-dinitrobenzoates of alcohols on silica gel G (see 9). The solvent system cyclohexane-carbon tetrachloride-ethyl acetate (10 : 75 : 15) is suitable for the separation of $C_1 - C_5$ aliphatic alcohols.

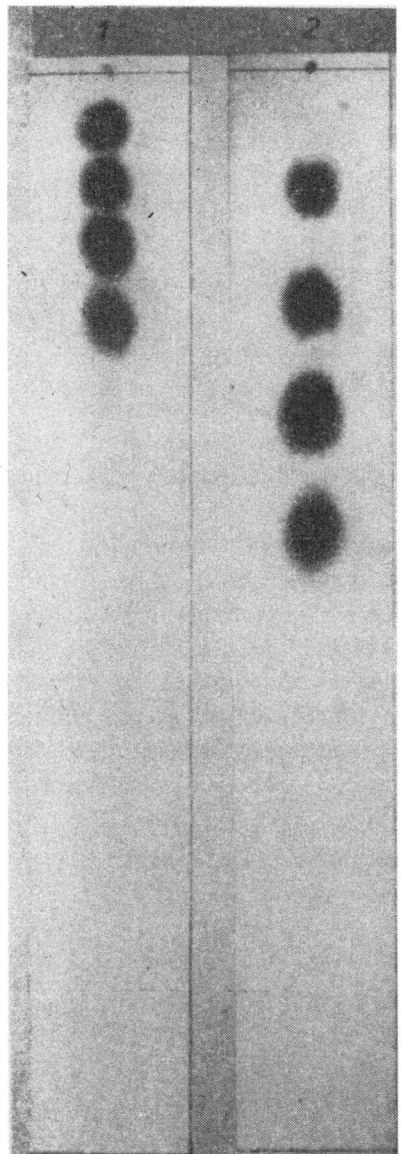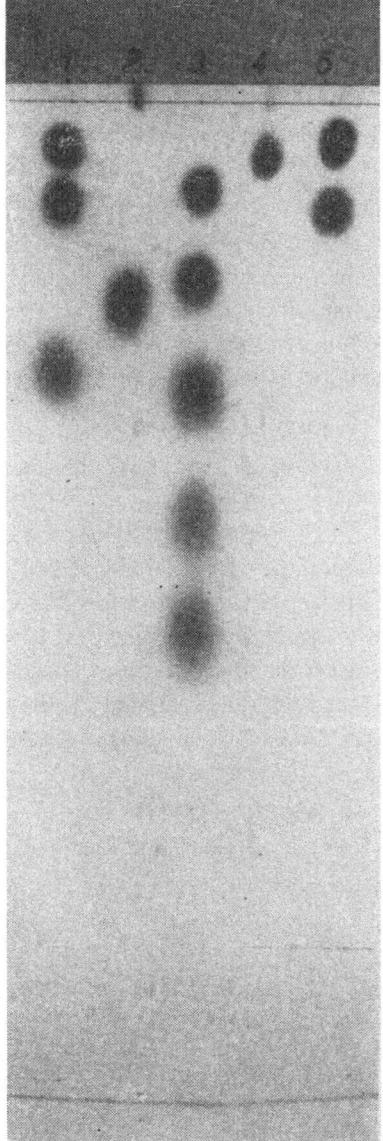

Fig. 27. Chromatogram of 3,5-dinitrobenzoates of $C_1 - C_4$ alcohols run in dimethylforma-mide/hexane. The impregnation of the paper was carried out (1) with 50% dimethyl-formamide in benzene, (2) with 25% dimethylformamide in benzene. The R_F values increase with the increasing chain length.

Fig. 28. Chromatogram of glycol and chlorohydrin ethers in 50% dimethylformamide/hexane: (1) monomethyl-, monoethyl-, and mono-*n*-butyl ether of ethylene glycol, (2) monobutyl ether of diethylene glycol, (3) methyl, ethyl, *n*-propyl, and *n*-butyl alcohol, (4) α-dichlorohydrin, (5) ethylene chlorohydrin and 1,2-propylenechlorohydrin.

Benzoyl Derivatives (Identification)

Benzoylation is suitable mainly for the preparation of derivatives of poly-hydroxy compounds.

General Preparation Procedure

To a 10-ml test tube containing 0.1 g (approximately three drops) of an alcohol, 0.5 ml of benzoyl chloride and 5 ml of 10% NaOH are added; it is shaken vigorously for 5 min, and then allowed to stand for 30 min at room temperature with occasional shaking. Larger lumps of the solid phase are triturated with a glass rod. After the addition of 5 ml of water the suspension is filtered and the product washed twice with 5 ml of water. Benzoates usually crystallize from aqueous ethanol.

Tribenzoyl Glycerol

Reagents: benzoyl chloride, 20% NaOH, 5% acetic acid, 30% ethanol, ether.

Procedure: To 3 ml of a 2% aqueous glycerol solution (60 mg) in a separatory funnel 3 ml of 20% NaOH are added and the mixture cooled with ice. Benzoyl chloride (0.4 ml) is then added and the mixture shaken thoroughly with cooling for 10 min. It is then diluted with 10 ml of water and extracted three times with 10 ml of ether, and the ethereal extract is washed with 5% acetic acid. Ether is then distilled off, the residue is mixed with 5 ml of water, and is filtered off with suction using a filtration tube. Yield, 173 mg; mp, 70 °C. After crystallization from 3 ml of ethanol the yield was 126 mg, mp 70−71 °C.

Remarks Concerning the Mechanism and Kinetics

Esterification can be represented by the general reaction

$$R-C\overset{O}{\underset{X}{\diagdown}} + R'OH \rightarrow HX + R-C\overset{O}{\underset{OR'}{\diagdown}}$$

This reaction takes this course if the C−X bond is weaker than the C−OR' bond. To decide which will be weaker, the residues X and OR' can be compared from the point of view of their affinity to electrons. The more acidic HX is in comparison with R'OH, the easier will the change take place. In the reaction of acid chlorides with alcohols HX is hydrochloric acid and R'OH is an alcohol. Thus the conditions necessary for the reaction are fulfilled, because HCl is a much stronger acid than R'OH.

In the reaction of alcohols with acid chlorides there is a nucleophilic addition of the alcohol to the carbonyl carbon atom, made positive by the presence of the carbonyl oxygen and the chlorine atom:

$$R-C\overset{O}{\underset{\underset{\overset{|}{O}}{\uparrow}\diagdown Cl}{}} \rightarrow R-\overset{O^\ominus}{\underset{\underset{R'\ \ H}{\overset{|}{O}^\oplus}}{\overset{|}{C}}}-Cl \rightarrow HCl + RCOOR'$$

A carbonium ion can also be formed first, which is then combined rapidly with a molecule of alcohol:

$$RCOCl \rightarrow Cl^{\ominus} + R{-}C^{\oplus}{=}O$$

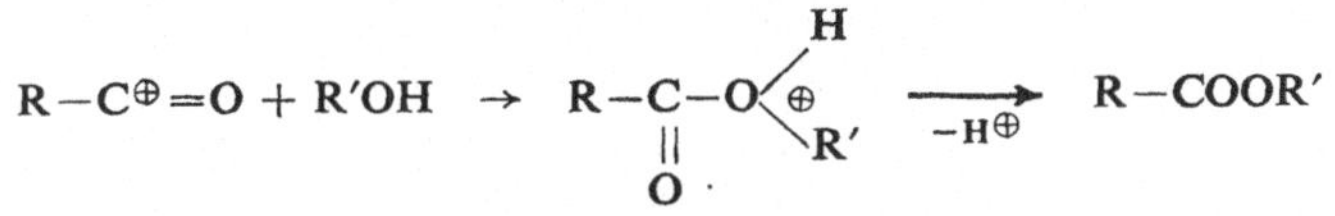

Both types of reaction have been observed.

Acid catalysis has not been observed during the esterification reactions, and it has not yet been decided whether a base catalysis takes place during the reaction. Schotten-Baumann acylation could probably serve as an example of base catalysis. However, the reaction takes place in the heterogeneous phase and its kinetics have not been studied.

The rate of the reaction of acid chlorides with alcohols is strongly influenced by solvents, as can be demonstrated by comparing the relative reaction rates of benzoylation of ethanol in various solvents, carried out under identical conditions (10):

Solvent	Relative reaction rate	Solvent	Relative reaction rate
Diethyl ether	1	*n*-Butyl chloride	6.8
Di-*n*-butyl ether	1.4	Benzene	7
Ethyl acetate	2	Chlorobenzene	7
Acetone	5.1	Hexane	8.6
Anisol	6	Carbon tetrachloride	9.1
Nitrobenzene	6.4	Dimethylaniline	26.5

From this list it is evident that the esterification rate in oxygen-containing solvents is generally lower than in hydrocarbons and their halogenated derivatives. The favorable properties of tertiary amines as solvents for acylation are well known and they are usually explained by the formation of a reactive intermediate from the base and the acylating agent:

$$R_3N + R'COCl \rightarrow R_3NCOR'^{\oplus}Cl^{\ominus}$$

The rate of esterification is further influenced by the structure of the reacting alcohol. A quantitative idea can be obtained by comparing the values measured (11) during the reaction of alcohols with *p*-nitrobenzoyl chloride in ether ($t = 25$ °C, concentration of the reacting substances 1 mole/100 g of ether).

n-Primary alcohols	$k \times 10^4$	n-Primary alcohols	$k \times 10^4$
Methanol	1840	Hexanol	850
Ethanol	850	Heptanol	690
Propanol	660	β-Phenylethanol	400
Butanol	740	β-Phenylpropanol	200
Pentanol	790	Benzyl alcohol	170

Primary alcohols with branched chain	$k \times 10^4$	Primary alcohols with branched chain	$k \times 10^4$
2-Methyl-1-propanol	310	3-Methyl-1-pentanol	770
2-Methyl-1-butanol	360	3-Methyl-1-hexanol	750
2-Methyl-1-pentanol	340	4-Methyl-1-pentanol	680
3-Methyl-1-butanol	730		

Secondary alcohols	$k \times 10^4$	Tertiary alcohols	$k \times 10^4$
2-Propanol	100	tert-Butanol	27
2-Butanol	74	tert-Pentanol	25
2-Pentanol	59	3-Methyl-3-pentanol	14
2-Hexanol	65		
3-Pentanol	36		
4-Heptanol	27		
α-Phenylethanol	5		
α-Phenylpropanol	5		
α-Phenylbutanol	5		

From the data it is evident that primary alcohols are most reactive, and, with the exception of methanol, their reactivities are pretty similar. The reactivity decreases with the branching of the chain at the α-carbon atom, and the presence of the phenyl group also decreases the reactivity, the influence of the aromatic substituent diminishing with the distance from the hydroxy group. Secondary and, especially, tertiary alcohols are much less reactive than primary ones.

Of course, the esterification rate of alcohols is also influenced by the chemical properties of the acylating agent.

In the case of esterification with benzoic acid derivatives, the reaction rate is decreased if the aromatic nucleus is substituted with electron-donating groups, and, conversely, the substituents attracting electrons increase the reactivity of the corresponding chloride, because the carbon atom of the carboxyl group becomes more positive and so more accessible to the nucleophilic attack.

The literature gives information on the relative esterification rates of isopropyl alcohol with benzoic acid derivatives with substituents in the positions ortho and para.

Acyl chloride	Relative esterification rate
p-Methylbenzoyl-	0.64
Benzoyl-	1.00
p-Iodobenzoyl-	1.26
p-Chlorobenzoyl-	1.53
p-Bromobenzoyl-	1.78
o-Nitrobenzoyl-	1.90
p-Nitrobenzoyl-	10.00

Esterification of Alcohols with Other Reagents

Esterification with acid anhydrides is very widespread, especially for preparative use. Acetyl derivatives are prepared by reacting alcohols with acetic anhydride; for identification purposes this method is mainly used with polyhydric alcohols (see p. 311) and with higher-molecular alcohols (sterols, etc.). The alcohol is heated in acetic anhydride, usually in the presence of anhydrous sodium acetate, or also in pyridine, chloroform, or benzene solutions. The isolation is carried out either by diluting the reaction mixture with water, which is followed by the precipitation of a solid derivative, or by extraction with ether. Sometimes it is first necessary to neutralize the acetic acid formed. The reaction of alcohols with acetic anhydride can be catalyzed with a drop of sulfuric acid, but attention must be paid to the possibility of side reactions taking place, as, for example, acetolysis in the case of polysaccharides, acetylation of aromatic rings, or sulfuric acid addition. It is therefore more advantageous to use perchloric acid, which partly prevents side reactions and does not give rise to inorganic esters, as is the case with sulfuric acid.

Phthalic anhydride can also be used for esterification, but acid phthalates are generally low-melting substances which have a special use for the separation of D,L-alcohols into their optical antipodes (12).

For identification purposes substituted phthalic anhydrides were proposed, such as, for example, 3-nitrophthalic anhydride (13, 14) and tetrachlorophthalic anhydride (15).

While 3-nitrophthalic anhydride is more suitable for the identification of larger quantities of alcohols (over 20 mg), tetrachlorophthalic anhydride was used for the identification of tertiary alcohols after their reaction with Grignard reagent (16).

$$R_3COH + C_2H_5MgBr \rightarrow R_3COMgBr + C_2H_6$$

For the identification of alcohols with 2,4-dinitrobenzenesulfenyl chloride, including thin-layer chromatography of 2,4-dinitrobenzenesulfenates of alcohols, see (17).

Carbamic Acid Esters — Urethanes

The esterification of alcohols with isocyanates or azides gives urethanes:

$$RNCO + ROH \rightarrow RNHCOOR$$
$$RCON_3 \rightarrow N_2 + [RCON<]$$
$$[RCON<] \rightarrow RNCO$$
$$RNCO + ROH \rightarrow RNHCOOR$$

The reaction takes place smoothly with primary and secondary alcohols in the majority of cases by simple mixing or gentle warming of both components, if necessary in an inert solvent (light petroleum). Some alcohols react only at their boiling point or at a still higher temperature, in a sealed tube; the reaction can be catalyzed by traces of sodium acetate or potassium carbonate (18, 19).

Tertiary alcohols react either very slowly or do not react at all, and dehydration of alcohols to olefins takes place instead (20).

Phenyl- and α-naphthylisocyanates are used for the preparation of derivatives. The second reagent is more advantageous because it is not so sensitive to traces of water, it is not lachrymatory, and it has a higher boiling point, hence, higher temperature can be used for the preparation of less-reactive alcohols. Melting points of derivatives obtained with its help are higher than the melting points of the corresponding phenylurethanes. m-Nitrophenylisocyanate with alcohols gives derivatives which crystallize well but are not easily purified from the substituted urea formed during the reaction (21).

For micropreparation α-naphthylisocyanate is especially suitable; phenylisocyanate reacts more reluctantly, and urethanes are formed which are more soluble than the corresponding naphthyl derivatives.

The disadvantage in the use of isocyanates for the preparation of derivatives is their sensitivity to humidity, so that anhydrous alcohols are a necessity. The dehydration of higher alcohols is easier, and this reagent is therefore used predominantly for the identification of higher alcohols. In the presence of water in the alcohol a side reaction takes place and a symmetrically substituted urea is formed,

$$RNCO + H_2O \rightarrow RNHCOOH \rightarrow RNH_2 + CO_2$$
$$RNH_2 + RNCO \rightarrow RNHCONHR$$

Urethanes can be separated from the disubstituted urea on the basis of their different solubilities in light petroleum (substituted urea is insoluble). From the point of view of an easy separation of by-products, the reaction with 3-bromopropionylisocyanate is suitable (22). Crystallization of urethanes is carried out from carbon tetrachloride, light petroleum, cyclohexane, or alcohol.

The use of azides of carboxylic acids for the preparation of urethanes has practically the same advantages and disadvantages as the use of iso-cyanates. m-Nitrobenzazide is used mostly, liberating nitrogen on heating at 100 °C (toluene, xylene, and ligroin are used as solvents), forming m-nitrophenylisocyanate as an intermediary product, which reacts imme-diately with alcohols, with the formation of urethanes:

Carboxylic acid azides react with thiols in a similar manner. Esters of m-nitrophenylthiocarbamic acid are formed as reaction products.

Primary and secondary alcohols react with this reagent well; tertiary alcohols require $3-4$ hr of boiling in toluene, if they react at all.

In the presence of water in the alcohol, di-(nitrophenyl) urea is formed as a by-product, which is also formed on the reaction of tertiary alcohols with this reagent if the reaction temperature is too elevated and thus causes their dehydrogenation.

The isolation of derivatives after the reaction of m-nitrobenzazide with alcohols is easy; most alcohols afford derivatives which crystallize immediately after the cooling of the mixture, or after distilling off a part of the solvent. Lower-melting derivatives are purified by crystallization from light petroleum or ligroin, higher-melting derivatives (over 60 °C) are crystallized from ethanol or toluene.

If an ester group is present in the molecule of the alcohol on the α or β carbon, consecutive cyclization reactions take place, for example:

Enols and hydroxy acids (23) also react with the reagents mentioned.

General Preparation Method of α-Naphthylurethanes

In a suitably small tube 2.5 mmole of α-naphthylisocyanate are mixed with 1 mmole of anhydrous alcohol, and the tube is stoppered with a rubber stopper. It is then heated on a water bath at $60-70$ °C for 15 min. After cooling, the solidified urethane (scratching with a glass rod) is triturated in the test tube using a spatula, a minimum amount of light petroleum is added ($0.2-0.5$ ml), and the mixture warmed gently to incipient boiling. The liquid (supernatant), containing soluble impurities and a small part of the derivative, is taken off carefully with a micropipette. Light petroleum (5 ml) is added to the solid residue and heated again to boiling point. The solution is filtered through a cotton wool plug, thus eliminating the substituted urea which is sometimes present in the mixture. After cooling the filtrate, crystallization of the derivative usually sets in. If not, the solution should be concentrated.

α-Naphthylurethane of Cyclohexanol

Reagents: α-naphthylisocyanate, cyclohexane.

Procedure: A small dry tube is filled with 10 mg of cyclohexanol, and 15 mg of α-naphthylisocyanate are added to it. The tube is quickly closed with a rubber stopper and put into a larger test tube (20 ml). It is then heated for 5 min by immersion in a hot water bath (80 °C). After this it is cooled and allowed to stand for 20 min, after which the solidified reaction mixture is triturated with a glass rod. Cyclohexane (1 ml) is added, and the mixture is boiled briefly and then filtered through a filtration tube which has been warmed beforehand and moistened with a few drops of boiling cyclohexane. The derivative separated from the filtrate on cooling is filtered off using a filtration tube and washed with 0.2 ml of cold cyclohexane. Yield, 3 mg; mp, $127-128$ °C.

α-Naphthylurethane of sec-Butanol

Reagents: α-naphthylisocyanate, cyclohexane, calcium chloride.

Procedure: Anhydrous sec-butanol (0.1 ml) and 75 mg of α-naphthylisocyanate are mixed with 5 ml of cyclohexane in a 20-ml flask provided with a reflux condenser fitted with a calcium chloride tube (exclusion of humidity), and the mixture is refluxed on a water bath for 20 min. The hot solution is filtered through a prewarmed filtration tube into a 10-ml test tube. The solution is cooled with ice and the separated urethane is recrystallized from cyclohexane. Yield, 35 mg; mp, 96 °C.

Esters of Dithiocarbonic Acid — Xanthogenate Reaction (Detection and Identification)

Under the influence of carbon disulfide and alkali hydroxides alcohols are transformed to xanthogenates, i.e., monoesters of dithiocarbonic acid,

$$\text{ROH} + \text{CS}_2 + \text{KOH} \rightarrow \text{S}=\text{C}\Big\langle{\overset{\text{OR}}{\text{SK}}} + \text{H}_2\text{O}$$

which can be recognized by their yellow color. The presence of xanthogenates can be detected by converting them to the insoluble copper or nickel salts (24, 25) or to strongly colored complexes with molybdates of the composition $MoO_3(R-O-SC-SH)$. These compounds are formed in a strongly acid medium (mineral acids) and they are soluble in benzene, chloroform, or carbon disulfide. This test is positive both for primary and secondary alcohols. Tertiary alcohols give a positive xanthogenate test (they form yellow xanthogenates), but with the molybdate the test is negative, be cause xanthogenates of tertiary alcohols hydrolyze under the conditions of the reaction. A positive test is also given by monoalkyl ethers of glycols: monoalkyl ethers of diethylene glycol give a red oil instead of a yellow precipitate. Esters which undergo saponification under the conditions of the test behave like alcohols and give a positive reaction. Compounds with a $-CH_2-CO-CH_2-$ grouping in the molecule react with carbon disulfide and alkali hydroxide, giving orange-red compounds which react with molybdates to form chocolate-brown precipitates soluble in chloroform. Hence, in the presence of compounds of this type the xanthogenate test is not reliable for the detection of small amounts of alcohols.

Reagents: carbon disulfide, ground NaOH, 5% aqueous $CuSO_4$, 1% aqueous ammonium molybdate, 2 N H_2SO_4, chloroform.

Procedure: To several drops (10-25 mg) of the tested alcohol in a test tube a few drops of carbon disulfide and 200-300 mg of sodium or potassium hydroxide (preferably ground) are added, and the contents are shaken for 5 min. If the reaction is positive, a pale yellow precipitate is formed.

The test can be continued in either of two ways.

1. After a while 5 ml of water are added, followed by dropwise addition of a 5% copper sulfate solution. The separation of a brown precipitate which rapidly turns yellow (copper alkylxanthogenate) represents a positive reaction.

2. After shaking for 5 min a few drops of a 1% ammonium molybdate solution are added to the reaction mixture, which is then acidified with 2 N H_2SO_4. After dissolution of the solid alkali the mixture is extracted with chloroform. A violet color of the chloroform layer proves the presence of a primary or a secondary alcohol.

For the differentiation of isomeric alcohols their transformation to *p*-bromophenacyl alkyl xanthates was proposed (O-alkyl-S-*p*-bromophenacyl dithiocarbonates) (26).

2. Identification of Alcohols
by Their Conversion to Ethers

The most widespread method of identification of alcohols by their conversion to ethers consists in using the reaction of alcohols with trityl chloride, with the formation of trityl ethers:

$$RCH_2OH + (C_6H_5)_3CCl \rightarrow RCH_2OC(C_6H_5)_3 + HCl$$

The reaction is carried out by heating the components in pyridine, and it takes place easily with the primary alcoholic group, while secondary and tertiary alcohols react either reluctantly or not at all. Trityl ethers are prepared mainly for the identification of glycols and cellosolves [monoalkyl ethers of glycols] (27), and serve principally for the introduction of various groups into the molecule of polyhydroxy compounds. For example, in the preparation of 2-(p-nitrobenzoate) of glycerol, 1,3-bistrityl ether of glycerol is prepared first, which is then esterified with p-nitrobenzoyl chloride, and eventually the trityl groups are split off with hydrobromic acid (28).

For the identification of higher alcohols the reaction with bromomethylphthalimide (29) is also suitable:

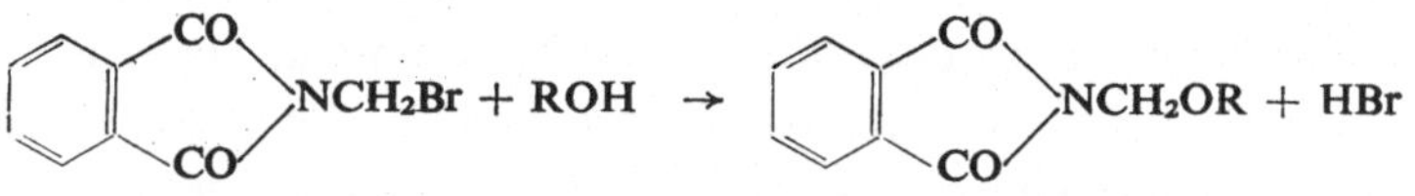

For the identification of glycols, their monoalkyl ethers, and esters of glycolic acid, alkyl pseudosaccharin ethers are suitable, which can be prepared by reacting the hydroxy compound with pseudosaccharin chloride in chloroform in the presence of pyridine, or also without any solvent (30, 31).

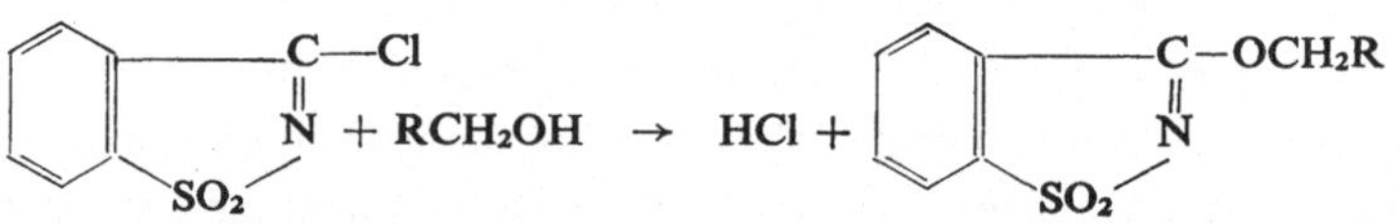

Ethylene Glycol Ditrityl Ether

Reagents: Pyridine, trityl chloride, acetone.

Procedure: To a solution of 3.4 g of trityl chloride in 7 ml of pyridine contained in a ground-glass-joint flask provided with a reflux condenser, 0.35 ml of ethylene glycol are added and the mixture is heated over a water bath for 15 min. After cooling, the reaction mixture is diluted with 30 ml of water and the separated oil crystallized in the refrigerator. After crystallization from acetone the yield was 1.5 mg, mp, 189—190 °C.

3. Oxidation of Alcohols

Oxidation with Chromic Acid

Similarly to potassium permanganate, chromic acid oxidizes primary alcohols to aldehydes and secondary alcohols to ketones. With tertiary alcohols chromic acid esters are formed (32, 33) which are yellow to red.

Reagents: light petroleum or carbon tetrachloride, CrO_3.

Procedure: The tested alcohol is mixed in a test tube with light petroleum or carbon tetrachloride and an excess of CrO_3. If the alcohol contains easily oxidizable primary or secondary hydroxy groups, the solution becomes brown to green. The tertiary alcoholic group gives a wine-red to dark red color if the rest of the molecule does not react with the oxidant. After the separation of chromium trioxide, esters can be isolated. Some esters are liquids, others are solid substances.

Oxidation with Agulhon's Reagent (Detection)

This reaction is based on the formation of a blue color on oxidation of alcohols with potassium dichromate and sulfuric acid. The reaction is also positive with other easily oxidizable organic compounds. However, if nitric acid is used instead of sulfuric acid, the characteristic blue-violet color is given in the cold only by compounds containing the OH group (34). This reaction has been used for detecting small amounts of alcohols in esters, ethers, ketones, and biological and technical liquids (35, 36).

Reagents: 5 ml of conc. nitric acid are mixed with glacial acetic acid to make a volume of 100 ml (solution A); 15% potassium dichromate solution (solution B).

Procedure: To 5 ml of the tested liquid 10 ml of solution A (freshly prepared, if possible) and 0.1 ml of solution B are added and the mixture is shaken thoroughly. If an alcohol is present, a blue color is formed. In the absence of alcohols the mixture remains unchanged for 30 min.

Oxidation with Bromine (Detection)

Bromine reacts with tertiary alcohols according to the equation

$$(CH_3)_3C-OH + Br_2 = (CH_3)_3C-Br + HBr + \frac{1}{2}O_2$$

If this reaction takes place in the presence of carbon disulfide, the liberated oxygen oxidizes CS_2 to sulfuric acid, which can be detected as barium sulfate (37, 38).

Reagents: dry bromine, dry carbon disulfide, 5% aqueous barium nitrate solution.

Procedure: A mixture of the dehydrated sample, dry bromine, and an excess of pure carbon disulfide is allowed to stand in a stoppered flask at room temperature for several hours. Water is then added, and, after shaking briefly, the barium nitrate solution is added. If the tested substance was a tertiary alcohol, a white precipitate of barium sulfate is formed.

Note: Since this reaction takes place in the presence of water, it is important to use anhydrous reaction components and to exclude humidity during the reaction.

Oxidation of Alcohols to Aldehydes and Ketones

Primary alcohols can be identified by oxidizing them to aldehydes with chromic acid or potassium permanganate in the presence of 2 N sulfuric acid. The aldehyde formed can be isolated by distillation and identified in the form of its dimedone derivative (see p. 230). Primary alcohols can be identified in this manner in the presence of secondary and tertiary alcohols, because dimedone condenses only with aldehydes, i.e., with products of a mild oxidation of primary alcohols. Carbonyl compounds can be isolated from an aqueous distillate by precipitating them with 2,4-dinitrophenyl-hydrazine (see p. 218). If paper chromatography cannot be used for identification (see p. 222), it is advisable to oxidize at least 500 mg of alcohol.

Detection of Methanol

a) *Reagents:* 3% $KMnO_4$ in 15% H_3PO_4, 5% oxalic acid solution in dilute H_2SO_4 (1 : 1), Schiff's reagent (p. 217).

Procedure: 2 ml of a dilute aqueous solution of the sample are mixed with 1 ml of $KMnO_4$ solution and, after 10 min standing, with 1 ml of the oxalic acid solution. Then 2 ml of Schiff's reagent are added to the colorless solution. If the sample contains methanol, a red-violet color appears.

b) *Reagents:* 5% H_3PO_4, 5% $KMnO_4$, saturated $NaHSO_3$ solution, chromotropic acid, conc. H_2SO_4.

Procedure: One drop of the tested alcohol is mixed with one drop of 5% phosphoric acid, then another drop of 5% $KMnO_4$, and, after 1 min of standing, with the saturated sodium hydrogen sulfate solution, just the amount necessary for decoloration (if part of the brown precipitate remains undissolved, some dilute H_3PO_4 and $NaHSO_3$ solution is added). To the decolorized solution 10−20 mg of chromotropic acid are added, and then, dropwise, 5 ml of conc. sulfuric acid. The solution is heated in a 60 °C warm water bath for 10 min. The violet color of the solution is proof of the presence of methanol in the sample.

Note: When carrying out the detection of methanol in the presence of ethanol it is important to do so in a sufficient excess of the oxidizing agent, i.e., after the addition of potassium permanganate the tested solution should not further decolorize.

Detection of Ethanol

a) *Reagents:* saturated $K_2Cr_2O_7$ solution, conc. H_2SO_4, piperidine, 3% aqueous sodium nitroprusside solution.

Procedure: 20 ml of aqueous ethanol solution are mixed with 5 ml of saturated potassium dichromate solution and 1 ml of conc. sulfuric acid, and the mixture is distilled until the volume of the distillate is 5 ml. After the addition of a few drops of piperidine and sodium nitroprusside solution a blue color is formed.

b) *Reagents:* 12% hydrogen peroxide, 4% ferrous ammonium sulfate in 1% H_2SO_4, piperidine, 3% aqueous sodium nitroprusside solution.

Procedure: To 5 ml of dilute aqueous ethanol solution four drops of hydrogen peroxide (12%) and 0.5 ml of 4% ferrous ammonium sulfate are added and allowed to stand for 10 min. A few drops of piperidine are then added and the solution is filtered to get rid of the precipitated iron oxides. Three drops of the 3% sodium nitroprusside solution are added to the filtrate to give a blue color (39).

4. Reactions Based on the Conversion of Alcohols to Alkyl Halogenides

Nitrole or Pseudonitrole Test (Detection of the Primary, Secondary, or Tertiary Hydroxyl Group)

This test makes it possible to differentiate primary and secondary OH groups. The analyzed alcohol is converted under the influence of iodine and red phosphorus to the corresponding iodide. In this case hydriodic acid cannot be used, because intramolecular rearrangements might take place. Under the influence of $AgNO_2$, alkyl halides give nitro compouds in addition to esters of nitrous acid, and these are changed in the presence of nitrous acid to erythronitrolanes (from primary alcohols) or to pseudonitroles (from secondary alcohols). Nitro compounds derived from tertiary alcohols do not react with HNO_2.

$$R-CH_2OH \xrightarrow{P + I_2} R-CH_2-I \xrightarrow{AgNO_2} R-CH_2-NO_2 \xrightarrow{HNO_2} R-C\begin{smallmatrix}NOH\\NO_2\end{smallmatrix}$$

$$\begin{smallmatrix}R\\R'\end{smallmatrix}CH-OH \xrightarrow{P + I_2} \begin{smallmatrix}R\\R'\end{smallmatrix}CH-I \xrightarrow{AgNO_2} \begin{smallmatrix}R\\R'\end{smallmatrix}CH-NO_2 \xrightarrow{HNO_2} \begin{smallmatrix}R\\R'\end{smallmatrix}C\begin{smallmatrix}NO\\NO_2\end{smallmatrix}$$

$$\underset{R}{\overset{R}{\underset{R}{>}}}C-OH \xrightarrow{P + I_2} \underset{R}{\overset{R}{\underset{R}{>}}}C-I \xrightarrow{AgNO_2} \underset{R}{\overset{R}{\underset{R}{>}}}C-NO_2$$

Alkaline solutions of erythronitrolanes are yellow-red to red, while pseudonitroles are blue when in a liquid state or in solution, and colorless when in a solid state.

Reagents: $AgNO_2$, dry sand, alkyl iodide prepared by reacting the alcohol with iodine and phosphorus, saturated $NaNO_2$ solution, 40% NaOH, dilute H_2SO_4.

Procedure: 1 g of $AgNO_2$, ground with a small amount of dry sand, and 0.5 g of purified alkyl iodide are introduced into a small distillation flask. In the case of higher-molecular compounds 1 g of alkyl iodide should be used. The nitrole reaction often takes place in the cold with evolution of heat. The reaction mixture is then heated over a free flame and the distillate is shaken with 0.5 ml of sodium nitrite solution in alkali for 1 – 2 min. After the addition of a small amount of water, sulfuric acid is added with caution, upon which the reaction products of primary alcohols produce an orange to dark red color. The solution can be decolorized either by acidifying it more strongly or by alkalizing it. Pseudonitriles can also be recognized by their odor and have lachrymatory properties.

In the presence of a secondary group in the initial alcohol the reaction mixture assumes a blue-green to dark blue color if acidified with dilute sulfuric acid. The addition of alkali hydroxide solution does not change the color. The separated pseudonitroles can be extracted with chloroform, in which they dissolve with a blue color.

This reaction is not suitable for the detection of secondary OH groups in the presence of primary hydroxyls, because small amounts of primary iodide prevent the pseudonitrole reaction. The nitrole reaction, on the other hand, is not disturbed by the presence of secondary hydroxy groups. This is the reason this test is reliable for the detection of primary alcohols, while for secondary alcohols it is of value only in the absence of primary alcohols. Aromatic alcohols (benzyl alcohol) do not give the nitrole reaction. While the nitrole reaction is positive with alcohols up to C_{16}, the pseudonitrole reaction cannot be used for alcohols above C_5.

Lucas Test (Detection of Primary, Secondary, and Tertiary Hydroxy Groups).

This test serves to distinguish primary, secondary, and tertiary alcohols and is based on the varying rates of formation of alkyl chlorides from alcohols under the influence of the reagent.

Tertiary alcohols react with the reagent immediately, with the separation of alkyl chlorides, the formation of which is indicated first by incipient turbidity and after a while by the separation of a layer. Secondary alcohols react in this manner after $2-5$ min; after 10 min a distinct layer is formed. Primary alcohols do not react under these conditions, and if they are lower than C_6, they are soluble in the reagent. Impurities can also be a cause of the formation of turbidity, but they do not produce a layer. Allyl alcohol behaves as a secondary alcohol, because its OH group is activated by the double bond of the neighboring carbon atom.

Reagent: 136 g of anhydrous $ZnCl_2$ are dissolved in 105 g of conc. HCl. The solution should be cooled before use.

Procedure: Approximately 0.1 ml of the alcohol is added to 0.5 ml of the reagent in a micro test tube. The mixture is shaken thoroughly for 1 min and then allowed to stand at $25-30$ °C; the reaction is positive if the solution becomes turbid in consequence of the separation of the insoluble alkyl halide. Tertiary alcohols react immediately, secondary alcohols within $2-5$ min, and primary alcohols only slowly and on heating. If the alcohol is not soluble in the reagent, the determination can still be carried out because, if the reaction takes place, the aqueous phase becomes turbid from the finely separated chloride.

Tertiary alcohols react with HCl alone. The rate of this reaction is such that the corresponding alkyl chloride is formed at room temperature within a few minutes. Primary and secondary alcohols do not react under similar conditions. The following test can be utilized for the differenciation of tertiary alcohols from secondary alcohols: To 1 ml of alcohol in a test tube 6 ml of conc. HCl are added and allowed to stand. After 2 min one can see whether the separation of the alkyl halide has set in (turbidity) and whether a layer is formed.

Identification of Alcohols by Their Conversion to S-Alkylthiuronium Salts

This procedure is suitable predominantly for the preparation of derivatives of tertiary alcohols, because on mere shaking with conc. hydrochloric acid they give tertiary alkyl chlorides which react with thiourea in aqueous-alcoholic solution to give S-tert-alkylthiuronium chlorides. These are then isolated in the form of picrates (see p. 142).

S-tert-Butylthiuronium Picrate

Reagents: 3% thiourea solution in 50% aqueous ethanol, 5% aqueous sodium picrate solution, 50% aqueous ethanol, conc. HCl.

Procedure: In a separatory funnel 1 g of tert-butanol is mixed with 10 ml of conc. HCl and the mixture is shaken for 5 min. The upper layer (tert-butyl chloride) is separated and washed twice with 5 ml of water. It is then dissolved in 25 ml of 3% thiourea solution, and the mixture is transferred into a 100-ml flask provided with a reflux condenser and is heated over a boiling water bath for 2 hr. Sodium picrate solution (10 ml) is then added to the mixture, which is heated until the solution is clear, and is then allowed to cool slowly. The separated picrate is filtered off with suction onto a filtration crucible and washed twice with 1 ml of water. Yield, 0.12 g; mp, 152−153 °C. Crystallization from 2 ml of 50% ethanol gives 88, mg, mp, 155−156 °C.

5. Colored Complexes of Alcohols

Reaction with Ceric Ammonium Nitrate (Detection of Alcohols)

Alcohols form red complexes with ceric salts, for example, with nitrate or perchlorate:

$$ROH + [Ce(NO_3)_6]^{2\ominus} \rightarrow [Ce(RO)(NO_3)_5]^{2\ominus} + HNO_3$$

The reaction is positive for alcohols with a number of carbon atoms lower than 10. For higher alcohols the test is not suitable. Under similar conditions hydroxy acids, hydroxy aldehydes, and other compounds containing alcoholic groups also give the reaction. Phenols give a green-brown precipitate in an aqueous solution, in dioxane a dark red to brown color is formed. Aldehydes, ketones, acids, alkyl halogenides, esters, and other compounds containing only C, H, O, and halogen atoms do not disturb the reaction. Aromatic amines, amine hydrochlorides, and compounds with groups which are easily oxidized to chromophores give colors or precipitates. This is the reason certain aromatic amines give a reaction similar to phenols (40). Compounds which are very rapidly oxidized decolorize the reagent sooner than a positive reaction can be observed. Various colors or precipitates are also given with this reagent by thiophene and some of its derivatives having a hydrogen atom or an acetyl group in the α-position (41).

Lower alcohols give a stronger coloration than higher alcohols, but the color is less stable. With the four isomeric butanols the intensity of the coloring decreases in the following order: iso, secondary, normal, and tertiary. On the other hand, the coloring with tert-butanol is most stable. Perchlorate is a more sensitive reagent, but nitrate gives a more stable color.

Reagents: 40 g of $(NH_4)_2Ce(NO_3)_6$ are dissolved in 100 ml of 2 N HNO_3 under reflux. A 0.5 N solution of $(NH_4)_2Ce(ClO_4)_6$ in 6 N perchloric acid can also be used; dioxane (for alcohols insoluble in water).

Procedure: (a) Water-soluble substances: 0.5 ml of the reagent is diluted with 3 ml of water, the mixture is shaken, and, after the addition of 4—5 drops of the sample, a change in color is observed. If the reaction is positive, the yellow color of the solution turns red.

(b) Water-insoluble substances: 0.5 ml of the reagent is diluted with 3 ml of dioxane. If a precipitate is formed, 3—4 drops of water are added and the mixture is shaken until dissolution of the precipitate is complete. The solution is then added (4—5 drops) to the mixture and color is observed.

Reaction with the Vanadium — Hydroxyquinoline Complex

This test is based on the formation of a bright red color which is given by vanadium-hydroxyquinoline complex with alcohols. It is specific for alcohols and it is also given by ether sof polyalcohols and keto-alcohols. Some other oxygenated organic compounds also give a color under identical conditions, which is different, however, from that given by alcohols: gray (esters, ethyl acetate, diethyl phthalate), yellow (citric, tartaric, lactic dacis, amines), light yellow (glucose, lactose), bluish (phenols), or green (mercaptans).

A negative reaction is given by hydrocarbons, alkyl halides, aldehydes, ketones, ethers, amino acids, nitro compounds, and nitriles. The test is suitable for the detection of alcohols in other solvents.

a) *Reagents:* A solution of ammonium vanadate prepared by shaking 400 mg of this salt with 1 liter of water and filtration; 2.5% 8-hydroxyquinoline in 6% acetic acid.

Procedure: 0.5—1 ml of ammonium vanadate solution is added to 3—5 ml of the tested solvent, followed by the 8-hydroxyquinoline solution. The volume of aqueous solutions of the reagents added should not be larger than that of the organic solvent being analyzed.

b) *Reagents:* 1 ml of sodium vanadate solution containing 1 mg of vanadium is mixed with 1 ml of 2.5% 8-hydroxyquinoline solution in 6% acetic acid and shaken with 30 ml of benzene. The solution is stable for about one day.

Procedure: A drop of the sample in water, benzene, or toluene is mixed in a micro test tube with four drops of the gray-green reagent solution with occasional shaking and is then heated to 60 °C on a water bath. The formation of a red color is usually observable after 2—8 min. When very small amounts of alcohol are to be detected, a blank is necessary.

Detection with Tetrachloroethane Solution

Reagents: To a hot, slightly acid ammonium vanadate solution buffered with ammonium acetate (pH 3−6), a 2% 8-hydroxyquinoline solution in 5% acetic acid is added until precipitation is complete. The black precipitate is isolated by filtration using a sintered-glass filter, washing with hot water, and drying at 120 °C. The product is further purified by boiling it with benzene (elimination of the coprecipitated 8-hydroxyquinoline), filtration, washing with boiling benzene, and drying at 120 °C; 18−19 mg of vanadium 8-hydroxyquinolinate are dissolved in 250 ml of dry tetrachloroethane by heating it to 70−80 °C. The solution formed contains approximately 2×10^{-4} M vanadium 8-hydroxyquinolinate. This solution can also be prepared directly from ammonium vanadate and 8-hydroxyquinoline: 50 mg of 8-hydroxyquinoline are dissolved in 100 ml of tetrachloroethane or *o*-dichlorobenzene, and approx. 50 mg of ammonium vanadate are added. The mixture is heated at 130 °C for 30 min and then cooled and filtered. If necessary, the solution obtained is diluted until the intensity of the color is equal to that of the substance prepared from the solid. The reagent prepared in the described manner is less sensitive to small amounts of alcohol, because the excess of 8-hydroxyquinoline inhibits the change of the color from black to red.

Procedure: To a solution of the reagent (1−2 ml), 2−3 drops (or approx. 0.1 g) of the tested substance are added (corresponding to a molar ratio 5000 : 1 for a molecular weight of about 100). If a change in color does not take place within 10−15 min, the mixture is warmed gently and the solution is observed after its temperature has dropped to room temperature.

A change of color to red means a positive reaction. This reaction is given by all alcohols, with the exception of those which are insoluble in tetrachloroethane (for example, mannitol and simple sugars).

Notes: (a) When alcohols are tested the mixture should not be heated too much, because the red color disappears at higher temperatures. However, in almost all instances (with the exception of some cyclic alcohols) the color reappears after cooling.

b) The majority of substances which do not react with the reagent do not prevent the reaction of alcohols. Only the presence of phenols (and also 8-hydroxyquinoline) and lower fatty acids inhibits the change in color from black to red if the amount of alcohol is very small.

c) The tetrachloroethane used for the preparation of the reagent should be dry. Small amounts of humidity cause the hydrolysis of this solvent accompanied by the formation of hydrogen chloride. The latter causes the formation of a blue complex salt $(C_9H_6ON)_2V(OH)_2Cl$. The blue solution also gives the same reactions, but it is less sensitive to small amounts of alcohol because the presence of the acid inhibits the color change.

d) Similar results can also be obtained if a tetrachloroethane solution of the green anhydride of vanadium hydroxyquinolate is used, prepared by heating the black complex to 250 °C for 30 min or additional heating of its tetrachloroethane solution in the absence of humidity. This green solution is not suitable, however, for the detection of small amounts of thiols or amines. [For the constitution of the vanadium-hydroxyquinoline complexes see (42)].

e) *o*-Dichlorobenzene is also an excellent solvent for vanadium hydroxyquinolinate and it can be used instead of tetrachloroethane for the preparation of the reagent solution. On heating, the complex is dissolved easily with the formation of a black solution, and the color does not change to blue on heating even when the solvent is moist, because *o*-dichlorobenzene does not hydrolyze readily. On heating the solution to boiling point (180 °C) the color changes to light yellow, but turns black again on cooling. This black solution in almost all instances gives the same reaction as the tetrachloroethane solution, with the exception that it does not turn blue so easily in the presence of acids.

f) Up to the 10th – 15th minute a positive reaction is given by various alcohols in the following ratio: 50 moles of primary: 100 moles of secondary: 500 moles of tertiary alcohol per 1 mole of vanadium hydroxyquinolinate. The minimum concentration of the solution suitable for the test is 10^{-4} N.

g) Various substituents influence the rate of the reaction with 8-hydroxyquinolinate to a varying extent. For analogous compounds the following sequence applies:

$$(-NH_2, =NH, \equiv N) > (-CH_2OH, -CH_2SH) > (=CHOH, =CHSH) >$$
$$> (\equiv COH, \equiv CSH)$$

Other Color Reactions of Alcohols

Komarowsky Reaction (Detection of Alcohols)

Dozens of applications of the color reaction of aliphatic alcohols with aromatic aldehydes in sulfuric acid can be found in the literature under the name "Komarowsky reaction."

Reagents: 1% ethanolic solution of an aromatic aldehyde, conc. H_2SO_4.

Procedure: To 5 ml of a 0.1% aqueous solution of the tested substance 0.5 ml of a 1% solution of salicylaldehyde (*p*-dimethylaminobenzaldehyde, vanillin) in ethanol and 10 ml of concentrated sulfuric acid are added and thoroughly stirred. A brown-red color is formed, the shades of which change with the character of the alcohol.

Notes: The reaction is suitable for the detection of alcohols of C_5 and higher. A positive reaction is also given by hydroaromatic alcohols, phenols, and all compounds with a double bond in the chain or in the ring, as well as compounds with a three-membered ring. Some ketones (acetone, methyl ethyl ketone, cyclohexanone) and aldehydes (acetaldehyde, propionaldehyde, isovaleraldehyde) also react positively. As regards the reaction with esters of acids and glycols, the data in the literature are conflicting.

Deniges Reaction (Detection of Alcohols)

Tertiary alcohols give yellow or red precipitates when boiled with mercuric sulfate (43). Primary and secondary alcohols (ethanol, n-propanol, iso-butanol, phenylethanol) give only white suspensions (turbidity) which can separate colorless crystals on standing. After a prolonged contact the reagent gives a precipitate of mercuric sulfate with isopropanol. With tertiary butanol a yellow color appears immediately, and after a few minutes a rich yellow precipitate is formed. The test is based on the reaction of olefins formed by the elimination of water from tertiary alcohols with mercuric sulfate. This is the reason those tertiary alcohols which cannot split off water and change to olefinic hydrocarbons will not react, as, for example, in the case of triphenylcarbinol. A positive reaction is also given by esters of tertiary alcohols which are saponified under the conditions of the reaction.

Reagent: 5 g of mercuric sulfate are dissolved in 100 ml of water and 20 ml of conc. sulfuric acid; it is also possible to mix 5 g of yellow mercuric sulfate with 20 ml of conc. sulfuric acid and 100 ml of water.

Procedure: Several drops of the tested alcohol are boiled in a test tube with 3 ml of the reagent for 2—3 min. The appearance of a yellow color changing to a yellow-to-reddish precipitate shows that the reaction is positive.

6. Sodium Test for Active Hydrogen

This test is based on the reaction of metallic sodium with substances containing substituents with a hydrogen which can be substituted for a metal, i.e., such functional groups as $-OH$, $-SH$, or $=NH$. A positive reaction is also obtained with substances carrying active methyl groups (acetone and acetophenone), methylene groups (acetoacetate), and methine groups (acetylenes and monoalkylacetylenes).

The test is suitable for alcohols of a medium molecular weight, from C_3 to C_8. Lower alcohols can be obtained only with difficulty in an anhydrous state, and traces of humidity cause a positive reaction. Higher-molecular alcohols react with sodium very slowly, so that the test becomes meaningless. Solid or too-viscous substances can be dissolved in an inert solvent (anhydrous ligroin or benzene). It is necessary to keep in mind that metallic sodium, if cut in a humid atmosphere, absorbs water on its surface, so that after putting it into benzene the liberation of hydrogen can be observed. An experienced experimenter is able to differentiate between the rate of development of hydrogen caused by humidity, which ceases after a while, and the rate of development of hydrogen caused by a positive reaction, which increases with time.

Performance of the test. (a) Approximately 0.5 g of metallic sodium is heated with 5 ml of dry toluene until the metal is melted; 5 ml of cold toluene are then added, and the test tube is stoppered immediately and is vigorously shaken until the sodium solidifies. The larger part of toluene is poured off carefully and the remaining sodium powder is kept in a micro test tube under 1 ml of dry toluene.

(b) The reagent can also be prepared in the following manner: A small piece of sodium is melted in a test tube and the capillary end of a glass tube drawn to a thin-walled capillary is immersed into the melt, while the broader end is connected with a pump. The melted sodium is sucked into the capillary, where it solidifies. Short segments are cut from the capillary before use to assure a fresh sodium surface for the test. The latter is carried out by putting a drop of the alcohol on a microscope slide, immersing a piece of the capillary with its freshly cut end into the center of the drop, and observing whether hydrogen evolves.

The test can also be carried out by adding a grain of the sodium "sand" to a small amount of the tested liquid in a micro test tube. To detect the presence of active hydrogen, color reactions with phenylisopropyl potassium of triphenylmethyl sodium can be used instead of that with metallic sodium.

7. Paper and Gas Chromatography of Alcohols and Their Derivatives

In addition to 3,5-dinitrobenzoates, the use of which for identification was described on p. 154, various other derivatives have been proposed for the identification of alcohols.

The chromatography of xanthogenates, which has often been proposed [for a review see (44)], has certain disadvantages, such as, for example, their instability. They are stable in solution only under certain circumstances, but not sufficiently under conditions of chromatography. The reaction of 3-nitrophthalanhydride (45) with alcohols is also not suitable, because it gives two isomeric esters. This can be avoided by using the diphenic acid anhydride, but for chromatograms only slightly sensitive acido-basic

indicators can be used. It seems, therefore, that the use of colored phenyl-
azobenzoates (46) or *p*-nitrophenylazobenzoates (47) is more suitable,
because no special detection procedure is needed in this case.

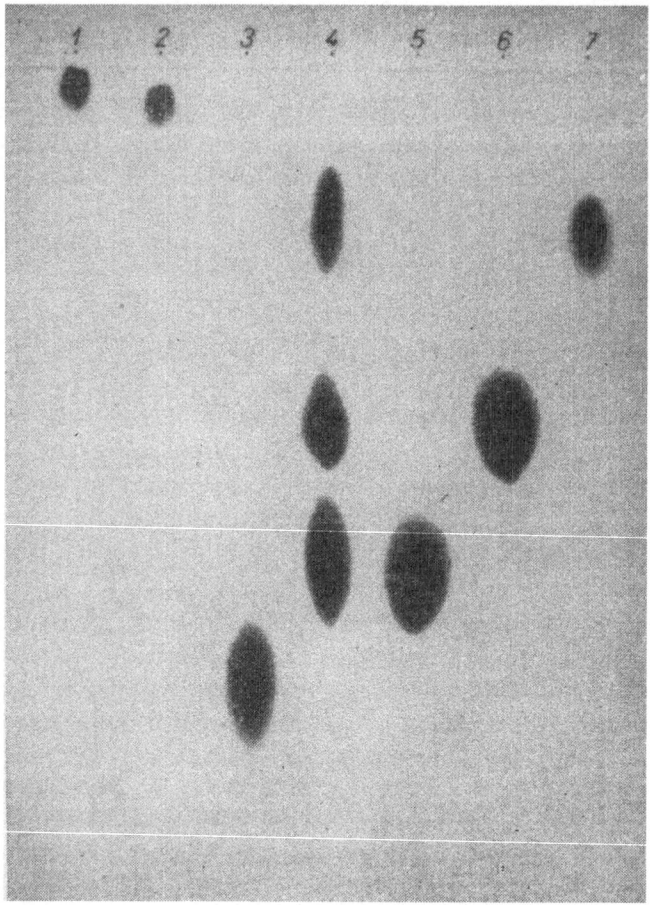

Fig. 29. Chromatogram of polyhydric alcohols in ethyl acetate-ethanol-water (12 : 2 : 1)
(1) Glucose, (2) D-arabitol, (3) 2,3-butylene glycol, (4) mixture of glycerol, ethylene glycol,
and 1,2-propylene glycol, (5) 1,2-propylene glycol, (6) ethylene glycol, (7) glycerol.

For a preparative separation on a chromatographic column (polyethylene, alumina) the conversion of alcohols to 4-dimethylamino 3,5-di-
nitrobenzoates is recommended (48).

Higher alcohols, as well as glycols, can be chromatographed as such
in the ethyl acetate-ethanol-water system (12 : 2 : 1) (49). Detection is
carried out by spraying the chromatogram with ammoniacal silver nitrate

solution (nine parts of 5% $AgNO_3$ + one part of ammonia) and heating to 110 °C in an oven until the background is brown and the polyhydric alcohols appear as black spots on a brown background. The chromatogram can be stabilized by washing it with water and immersing it into an aqueous thiourea solution acidified with sulfuric acid. It is kept there until the background is white (or less dark) and then washed thoroughly in running water (Fig. 29).

To separate alcohols by gas-liquid chromatography silicone oils (50 – 53), tritolylphosphate (54), and mineral oils (55) were used as stationary phases; other substances useful as stationary phases are (56) β, β'-di(propionitrile)ether, tricresyl phosphate, ethyl hexyl sebacate, polyethylene glycol 400, and others, including combinations of various individual substances (57).

8. Various Reactions of Hydroxy Compounds

Oxidation of Hydroxy Compounds with Periodic Acid (58, 59)

At room temperature in aqueous solutions periodic acid oxidizes 1,2-glycols, α-hydroxylaldehydes, α-hydroxyketones, 1,2-diketones, and α-hydroxyacids. The reactivity decreases in the given order. The products formed in the reaction are dependent on the grouping susceptible to oxidation in the original compound, as represented in the following list:

$$-CH(OH)-\boxed{CH_2OH}$$
$$-CO-\boxed{CH_2OH}$$
$$-CH(NH_2)-\boxed{CH_2OH}$$
$$\longrightarrow \quad HCHO$$

$$-CH(OH)-\boxed{CH(OH)}-CH(OH)- \longrightarrow HCOOH$$

$$\boxed{R-CH(OH)}-CH(OH)-$$
$$\boxed{R-CH(OH)}-CO-$$
$$\boxed{R-CH(OH)}-CH(NH_2)-$$
$$\longrightarrow \quad R-CHO$$

$$\boxed{R-CO}-CH_2OH$$
$$\boxed{R-CO}-CH(OH)-$$
$$\boxed{R-CO}-CO-$$
$$\boxed{R-CO}-CHO$$
$$\longrightarrow \quad R-COOH$$

$$R-CO-\boxed{CHO} \longrightarrow HCOOH$$

$$-CH(OH)-\boxed{CH(NHR)-R'} \longrightarrow RNH_2 + R'CHO$$

$$-CH(OH)-\boxed{CH(OH)-CO}-CH_2OH \longrightarrow OHC-COOH$$

$$R-CH(OH)-\boxed{COOH} \longrightarrow CO_2 + H_2O$$

$$-CH(OH)-\boxed{CH(OH)-COOH} \longrightarrow OHC-COOH$$

$$\left.\begin{array}{l} HO-CH_2-\boxed{CO-COOH} \\[4pt] -CH(OH)-\boxed{CO-COOH} \\[4pt] HOOC-\boxed{CO-COOH} \end{array}\right\} \longrightarrow \begin{array}{c} COOH \\ | \\ COOH \end{array}$$

cis-Glycols are cleaved more easily than trans-glycols (60). Periodic acid is reduced to iodic acid,

$$\begin{array}{c} CH_2OH \\ | \\ (CH-OH)_n + (n+1)\,HIO_4 \rightarrow (n+1)\,HIO_3 + 2\,HCHO + nHCOOH + H_2O \\ | \\ CH_2OH \end{array}$$

The detection of polyhydric alcohols is carried out by determining (a) the presence of iodate, or (b) the carbonyl compound formed in the reaction mixture.

Method (a)

Reagents: 0.5 g of H_5IO_6 in 100 ml of water, conc. HNO_3, 3% aqueous $AgNO_3$.

Procedure: Two drops of the reagent are mixed with one drop of HNO_3 in a test tube, and one drop or a small crystal of the tested substance is added to the mixture. This is then shaken for $10-15$ sec and $1-2$ drops of the silver nitrate solution are added. If the reaction is positive, a white precipitate of silver iodate is formed immediately; in the case of a negative reaction a brown turbidity appears, which dissolves on shaking.

It is important to hold to the given procedure, i.e., not to exceed the amount of nitric acid given, because no precipitate would form.

Method (b)

Reagents: 5% KIO_4, N H_2SO_4, saturated SO_2 solution (in water), Schiff's reagent.

Procedure: To one drop of an aqueous or alcoholic solution of a polyhydric alcohol one drop of KIO_4 solution and one drop of N H_2SO_4 are added, and the mixture is allowed to stand for 5 min. The excess periodate is then eliminated by the addition of several drops of SO_2 solution, and after several minutes Schiff's reagent is added to the mixture. If a red or blue color is formed within 30 min, the test may be considered positive.

For the detection of polysaccharides the reaction mixture should be heated to its boiling point and allowed to stand for 5 min.

Oxidation of Hydroxy Compounds with Lead Tetraacetate

Lead tetraacetate oxidizes α-diols in the cold in the same manner as periodic acid (61).

The use of acetic acid as solvent sometimes causes certain difficulties (acetylation); other difficulties arise from the fact that lead tetraacetate sometimes leads to acetylated or methylated products, in consequence of its degradation to radicals:

$$Pb(OCOCH_3)_4 \rightarrow Pb(OCOCH_3)_2 + 2\ CH_3COO\cdot$$
$$CH_3COO\cdot \rightarrow CO_2 + CH_3$$

The course of the oxidation with lead tetraacetate is much dependent on stereochemical differences, for example, on cis- and trans-, and threo- and erythro-isomerism (62).

Reagents: 2 g of red lead (minimum) is dissolved in 100 ml of acetic acid at 80 °C; saturated solution of Na_2SO_3; conc. HCl; Schiff's reagent (p. 217).

Procedure: To several milligrams of the tested substance 1 ml of the red lead solution in acetic acid is added, and after 1 hr standing the mixture is diluted with 5 ml of water. Then Na_2SO_3 solution (0.5 ml) is added to the mixture followed by 2 ml of Schiff's reagent. If a red color appears, the test is positive; the presence of formaldehyde can be checked by the addition of 1 ml of conc. HCl: the color persists.

Dehydration of Glycols

1,2-Diols can be dehydrated by various reagents, and the corresponding aldehydes formed can then be detected either directly in the reaction mixture, or by distilling the mixture and detecting them in the distillate. In cases where the oxidation products can be distilled off they can be transformed with advantage to 2,4-dinitrophenylhydrazones, which may then be identified by paper chromatography or thin-layer chromatography.

Tests with Dilute Acids

Reagents: 100 ml of conc. sulfuric acid are added carefully to 200 ml of water; Schiff's and Legal's reagents (pp. 217 and 234).

Procedure: In a 100-ml flask 50 ml of H_2SO_4 are added to 1 g of the tested substance and the mixture is distilled. After several milliliters have been distilled aldehydes are detected with Schiff's reagent and ketones with Legal's reagent.

Test with Potassium Hydrogen Sulfate

Reagents: $KHSO_4$, saturated 2,4-dinitrophenylhydrazine in 2 N HCl, benzene.

Procedure: A small amount of the tested substance is mixed with 0.5 − 1.0 g of well-ground $KHSO_4$ in a test tube (Fig. 30) and heated over

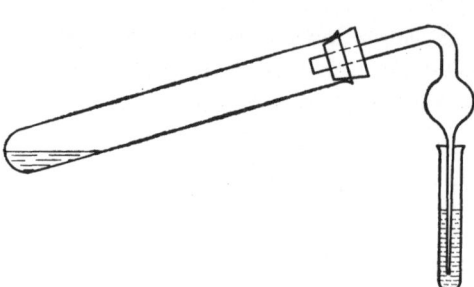

Fig. 30.
Device for the detection
of glycol by dehydration.

a direct flame. The vapors formed by decomposition are distilled into a small tube filled with 1 − 2 ml of the 2,4-dinitrophenylhydrazine solution. The carbonyl compounds escaping from the heated test tube react with the reagent in the small tube, and the solution becomes turbid or a precipitate is formed in it. When the operation is over a few drops of benzene are added to the small tube and the mixture is shaken; 2,4-dinitrophenylhydrazones pass into the benzene layer. This is then applied with a micropipette onto a sheet of chromatographic paper in various concentrations (p. 222).

On the chromatogram ethylene glycol and polyethylene glycols appear as a hydrazone of acetaldehyde; glycerol and polyglycerols as a hydrazone of acrolein; 1,2-propylene glycol and proplylene oxide condensates a hydrazone of propionaldehyde; sugars appear as fural hydrazone; etc.

Instead of using a 2,4-dinitrophenylhydrazine solution for capturing the condensate, pure water can be used and the detection can be carried out by color reactions, but the chromatographic method has the advantages of being applicable for mixtures and of being more reliable.

References

1. Michalski E., Turowska, M.: Chem. Anal. (Warsaw) **5**, 625 (1960).
2. Kartnig, T., and Kren, G.: Sci. Pharm. **31**, 128 (1963).
3. Severin, M.: Bull. Reck. Agron. Gembloux **1**, 298 (1966).
4. Spassow, A.: Ber. **70**, 1926 (1937); **75**, 779 (1942).
5. Garska, K. J., Douthit, R. C., and Yarborough, V. A.: Anal. Chem. **33**, 392 (1961).
6. Douthit, R. C., Garska, K. J., and Yarborough, V. A.: Appl. Spectroscopy **17**, 85 (1963).

7. Saunders, B. C., Stacey G. J., and Wilding, I. G. W.: Biochem. J. 36, 368 (1942).
8. Gasparič, J., and Borecký, J.: J. Chromatog. 5, 466 (1961).
9. Severin, M.: J. Chromatog. 26, 101 (1967).
10. Norris, J. F., and Haines, E. C.: J. Am. Chem. Soc. 57, 1425 (1935).
11. Norris, J. F., Ashdown, A. A., and Cortese, F.: J. Am. Chem. Soc. 47, 837 (1925); 49, 2640 (1927).
12. Pickard, R. H., and Kenyon, J.: J. Chem. Soc. 99, 45 (1911).
13. Nicolet, B. H., and Sacks, J.: J. Am. Chem. Soc. 47, 2348 (1925).
14. Dickinson, G. M., Crosson, L. H., and Copenhaver, J. E.: J. Am. Chem. Soc. 59, 1094 (1937).
15. Teterin, V. K., and Zonis, S. A.: Zhur. Obshch. Khim. 6, 658 (1936).
16. Fessler, W. A., and Shriner, R. L.: J. Am. Chem. Soc. 58, 1384 (1936).
17. Lefebvre, G., Berthelin, J., Maugras, M., Gay, R., and Urion, E.: Bull. Soc. chim. France 1966, 266.
18. Dieckmann, W., Hoppe, J., and Stein, R.: Ber. 37, 4627 (1904).
19. Claisen, L.: Ann. 418, 69 (1919).
20. Neuberg, C., and Kansky, E.: Biochem. Z. 20, 445 (1909).
21. Shriner, R. L., and Cox, R. F. B.: J. Am. Chem. Soc. 53, 1601 (1931).
22. Johnson, H. W., Jr. Day R. J., and Tinti, D. S.: J. Org. Chem. 28, 1416 (1963).
23. Veibel, S., Lillelund, H., and Wangel, J.: Dansk. Tidskr. Farm. 17, 183 (1943).
24. Dubský, J. V.: J. prakt. Chem. (2) 93, 142 (1916); 103, 109 (1921).
25. Lieser, T., and Nagel, W.: Ann. 495, 235 (1932).
26. Berger, J., and Uldall, I.: Acta Chem. Scand. 18, 1353 (1964).
27. Seikel, M. K., and Huntress, E. H.: J. Am. Chem. Soc. 63, 593 (1941).
28. Helferich, B., Moog, L., and Jünger, A.: Ber. 58, 872 (1925).
29. Hopkins, H. H.: J. Am. Chem. Soc. 45, 541 (1923).
30. Meadoe, J. R., and Reid, E. E.: J. Am. Chem. Soc. 65, 457 (1943).
31. Böhme, H., and Opfer, H.: Z. anal. Chem. 139, 255 (1953).
32. Wienhaus, H.: Ber. 47, 322 (1914).
33. Wienhaus, H., and Treibs, W.: Ber, 56, 1648 (1923).
34. Agulhon, H.: Bull. Soc. chim. France (4) 9, 881 (1911).
35. Kellett, E. G.: Analyst 62, 728 (1937).
36. Webb, D. A.: Sci. Proc. Roy. Dublin Soc. 21, 281 (1936); C 1936, II, 2761.
37. Henry, L.: Bull. Acad. Belgique 1906, 424; Rec. trav. chim 26, 116 (1907).
38. Hell, C., and Urech, F.: Ber. 15, 1249 (1882).
39. Doeuvre, J.: Bull. Soc. chim. France (4) 39, 1102 (1926).
40. Rosenthaler, L.: Pharm. Acta Helv. 27, 272 (1952).
41. Hartough, H. D.: Anal. Chem. 20, 860 (1948).
42. Bielig, H. J., and Bayer, E.: Ann. 584, 96 (1953).
43. Deniges, G.: Compt. rend. 126, 1145 (1898); C. 1898 I, 1166.
44. Gasparič, J., and Borecký, J.: J. Chromatog. 4, 138 (1960).
45. Siegel, A., and Schlögl, K.: Mikrochim. Acta 40, 383 (1953).
46. Woolfolk, E. O., Beach, F. E., and McPherson, S. P.: J. Org. Chem. 20, 391 (1955).
47. Hecker, E.: Ber. 88, 1666 (1955).
48. Freytag, W., and Baustian, M.: J. Chromatog. 13, 558 (1964).
49. Borecký, J., and Gasparič, J.: Collection Czech. Chem. Commun. 25, 1287 (1960).
50. Peyrot, P.: Chim. et Industrie 78, 3 (1957).

51. Tenney, H. M.: Proceedings of the Symp. Advances in Gas-Chromatog. Am. Chem. Soc., New York 1957.

52. Cropper, F. R., and Heywood, A.: Nature (London) **174**, 1063 (1954).

53. Ambrose, D., and Purnell, J. H.: Gas-Chromatography; Desty, D. H. (ed). Butterworths Sci. Publ., London 1958.

54. Peyrot, P.: Chim. et Industrie **78**, 3 (1957).

55. Haskin, J. F., Warken, G. W., Pristley, L. J., and Yarborough, V. A.: Proceedings of the Symp. Advances in Gas-Chromatog. Am. Chem. Soc., New York 1957.

56. Bayer, E.: Gas-Chromatographie. Springer, Berlin—Göttingen—Heidelberg 1959.

57. Bober, H.: Beckman Ref. No. 2, p. 17 (1965).

58. Smith, G. F.: Analytical Applications of Periodic Acid and Iodic Acid and Their Salts, 5th Ed. Smith Chemical Comp., Columbus, Ohio, 1950.

59. Fleury, P. F.: Chim. anal. **35**, 197 (1953).

60. McCasland, G. E., and Smith, D. A.: J. Am. Chem. Soc. **73**, 5164 (1951).

61. Criegee, R.: Ber. 64, 260 (1931); Angew. Chem. **53**, 321 (1940).

62. Criegee, R., Büchner, E., and Walther, W.: Ber. **73**, 571 (1940).

CHAPTER X

PHENOLS

Reactions used for the identification of phenols are generally the same as those used for alcohols, i.e., they are based on the reactivity of the hydroxy group, and only partly on the substitution of the aromatic ring. In contrast to alcohols, the hydroxy group of phenols is more acidic. Therefore, the phenols are soluble in alkalies in the form of phenolates, which are practically undissociated even in aqueous solutions. Carbon dioxide can liberate the free phenol. This property can be utilized for the separation of phenols from alcohols and from substances which are not precipitated by carbon dioxide.

Many color reactions based mainly on the reactivity of the aromatic nucleus are used for the detection of phenols.

In addition to chemical reactions, chromatographic as well as spectral (NMR) (1) methods are also suitable for the separation and identification of phenols.

1. Reactions of the Hydroxy Group (Identification)

Similar to alcohols, phenols also react with acid chlorides or anhydrides to give the corresponding esters. The properties of the derivatives formed are dependent on the nature of the starting phenol. Monocyclic phenols can be identified well by converting them to 3,5-dinitrobenzoates, unless the melting points of these derivatives do not exceed 200—220 °C. Acetylation is suitable for dicyclic and polycyclic phenols whose melting points are over 100 °C. For the identification of phenols with a sterically hindered OH group the reaction with benzenesulfonylisocyanate (2) has been recommended,

$$C_6H_5SO_2NCO + ArOH \rightarrow C_6H_5SO_2NHCOOAr$$

The alkylation of phenolates with monochloroacetic acid, leading to the formation of corresponding aryloxyacetic acids, can be used if a larger amount (0.1—0.2 g) of phenol is available. The advantage of these derivatives consists in the possibility of determining their neutralization equivalent, and hence their molecular weight. For microidentification urethanes are suitable.

Reactions of phenols with inorganic salts leading to complexes should be classed with the color reactions of phenolic hydroxyl.

Most phenols are soluble in dilute sodium hydroxide solution (for test see p. 105). However, exceptions are known—the so-called cryptophenols. The hydroxy group of these phenols is situated on the aromatic nucleus between substituents with branched chains or between other substituents easily donating electrons.

Phenols are not soluble in sodium hydrogen carbonate (for the test see p. 105), unless substituted with strongly electrophilic substituents.

Alkaline solutions of phenols darken when exposed to air, because of oxidation and polymerization. Alkaline solutions of o - and p-nitrosubstituted phenols are strongly colored, o-substituted phenols are orange, and p-substituted phenols are yellow.

3,5-Dinitrobenzoyl Derivatives

The reaction of phenols with 3,5-dinitrobenzoyl chloride is expressed by the equation

In view of the lower reactivity of the phenolic group when compared with the alcoholic group, pyridine is usually used as the reaction medium.

General Method of Preparation of 3,5-Dinitrobenzoates

In a small flask fitted with a reflux condenser 1 mmole of phenol and 1 mmole of 3,5-dinitrobenzoyl chloride (for purity control see p. 151) dissolved in pyridine are heated for 60 min. After cooling, 10 ml of 1% sulfuric acid are added, the mixture is filtered, and the product washed with 20 ml of 5% sodium hydroxide and 20 ml of water. Finally, the product is washed with 10 ml of methanol (best done by suspending the product in it and decanting or filtering) and eventually filtered again under suction.

Methanol, ethanol, or cyclohexane are used most commonly for crystallization. If after the addition of sulfuric acid the product separates in the form of an oil, and if it cannot be brought to crystallization, the reaction mixture is extracted with ether, and impurities are washed from the extract several times with water and dilute alkali. The ester is then isolated by drying the ethereal solution and eliminating the solvent by distillation.

3,5-Dinitrobenzoate of o-Cresol

Reagents: 3,5-dinitrobenzoyl chloride, pyridine, 5% H_2SO_4, 5% NaOH.

Procedure: In a 100-ml flask with a reflux condenser 20 ml of pyridine, 1 ml of o-cresol, and 2.3 g of 3,5-dinitrobenzoyl chloride are mixed and refluxed gently for 1 hr. After cooling, the reaction mixture is poured into a beaker containing 400 ml of 5% sulfuric acid. When thoroughly cooled the mixture is filtered, and the product washed with 30 ml of water. The crude product is then suspended in a beaker containing 200 ml of 5% NaOH and filtered off on a filtration crucible. The ester is then washed with 50 ml of water and dried at $80-90$ °C. Yield, 1.3 g; mp, $132-134$ °C. When crystallized from cyclohexane the yield drops to 0.63 g, mp $132-138$ °C. Further crystallization from cyclohexane gives 0.46 g of the derivative, mp $135-138$ °C.

3,5-Dinitrobenzoate of o-Xylenol (3,4-Dimethylphenol)

Reagents: 3,5-dinitrobenzoyl chloride, pyridine, 5% H_2SO_4, 2% NaOH, ethanol.

Procedure: o-Xylenol (1.2 g) and 3,5-dinitrobenzoyl chloride (2.3 g) are added to 20 ml of pyridine in a 100-ml flask and refluxed for 1 hr. After cooling, the solution is poured into 400 ml of 5% sulfuric acid, and immediately after solidification the product is filtered off and washed until neutral. The dry ester is thoroughly ground in a mortar with 200 ml of 2% NaOH and filtered off using a filtration crucible, washed with 100 ml of cold water (yield, 1.3 g; mp, $162-164$ °C), and crystallized from 130 ml of ethanol. Yield, 1.0 g; mp, 165 °C.

Acetates

General Method for the Preparation of Acetates

The acetylation of phenols is carried out by heating them with a ten to twentyfold excess of acetic anhydride, which also serves as a solvent in this case. Anhydrous sodium acetate is used as a catalyst. Acetylation can also be carried out without heating under the catalytic effect of a drop of concentrated sulfuric acid (see p. 159). If the phenol is insoluble in acetic anhydride, pyridine or some other inert solvent such as benzene, chloroform, or ether is added. For weakly reactive phenols acetylation with acetyl chloride in boiling benzene, catalyzed with powdered magnesium, was found suitable (1 − 2 mmole of phenol, 5 ml of benzene, 1 − 2 mmole of acetyl chloride, 0.1 g of powdered magnesium, refluxing for 60 min).

Methanol, ethanol, and their mixture with water were found to be best for the crystallization of acetates.

Pyrogallol Triacetate

Reagents: acetic anhydride, freshly fused sodium acetate, sodium carbonate, anhydrous pyridine.

Procedure: In a 25-ml ground-joint flask fitted with a reflux condenser a mixture of 0.5 g pyrogallol, 5 ml acetic anhydride, and 0.5 g melted sodium acetate is refluxed for 1 hr. After cooling, 50 ml of water were added and allowed to stand for 1 hr before filtering on a filtration crucible. The product is washed with 10 ml of water and dried in a stream of dry air (yield, 0.78 g; mp, 165 °C). After crystallization from 12 ml of ethanol the yield is 0.54 g, mp 166 °C.

Phloroglucinol Triacetate

Reagents: acetic anhydride, fused sodium acetate, ethanol.

Procedure: A 10-ml flask containing a mixture of 2 ml of acetic anhydride, 0.1 g of sodium acetate, and 0.15 g of phloroglucinol is heated under reflux in a water bath until dissolution, enhanced by shaking, is complete. The mixture is then refluxed for an additional hour. After cooling, the reaction mixture is poured gradually and with constant stirring into 10 ml of ice-water. Filtration is carried out using a filtration tube and the product is washed twice with 1 ml water. Yield of the crude product, 86 mg; mp, 105 °C. Crystallization from 2 ml of ethanol gave 51 mg of a pure product, mp 105 − 106 °C.

Aryloxyacetic Acids

Phenols react in alkaline solution with monochloroacetic acid according to the equation

$$ArONa + ClCH_2COONa \rightarrow ArOCH_2COONa + NaCl$$

The aryloxyacetic acid is isolated by precipitation after acidicifation of the reaction mixture. The procedure is suitable if at least 0.1 − 0.2 g of the phenol is available.

General Method of Preparation of Aryloxyacetic Acids

Phenol (2 mmole) is mixed with 4 mmole of chloroacetic acid dissolved in 10 − 20% NaOH and heated for 1 − 2 hr at 90 − 100 °C. Usually, 2 − 3 ml of alkali are taken, but if the reaction components are insoluble, water is added. The crude acid is isolated by acidification of the reaction mixture with dilute hydrochloric acid (test with Congo red) and is purified in the following way: It is first extracted with ether or ethyl acetate, the extract is washed with water and is then shaken with 10% sodium carbonate solution. The aryloxy acid passes into the aqueous layer, from which it is precipitated by acidification. It is purified by crystallization from water.

Phenyloxyacetic Acid

Reagents: monochloroacetic acid, 20% NaOH, HCl (1 : 1), ethyl acetate, 10% Na_2CO_3.

Procedure: Phenol (0.2 g) is shaken in a 10-ml test tube with 2 ml of 20% NaOH and 0.3 g of chloroacetic acid and the mixture heated on a water bath for 1 hr. After cooling, 3 ml of water are added and the mixture is made acid (test with Congo red) with dilute hydrochloric acid (1 : 1). The solution is transferred to a separatory funnel and extracted gradually with 15, 10, and 5 ml of ethyl acetate. The pooled extracts are reextracted with 5 ml water and 10 ml of 10% sodium carbonate. Carbonate-containing solution is acidified with HCl (1 : 1) until distinctly acid (test with Congo red), concentrated to half its volume, allowed to stand for 30 min in a refrigerator, and filtered using a filtration tube. The product is washed with 3 ml of water (yield, 50 mg; mp, 96 − 97 °C) and crystallized from 2.5 ml of hot water. Yield, 20 mg; mp, 98 − 99 °C.

Esters of Carbamic Acids − Urethanes

General Procedure for the Preparation of Urethanes

The general working conditions are analogous to those with alcohols (p. 159). For the preparation of urethanes of reactive phenols it is sufficient to heat both components (molar ratio) for 2 − 5 min at 140 − 150 °C in a thoroughly dry test tube. Less-reactive phenols (usually those substituted on the nucleus with groups which increase the acidity of the phenolic group) − for example, *p*-nitrophenol − require the reaction to take place in pyridine and with the catalysis of a small amount of a tertiary aliphatic amine (a drop of trimethylamine in cyclohexane). The reaction is carried out by 30 − 60 min of refluxing, and for the isolation of the urethane formed the mixture is diluted with 5% sulfuric acid. Urethanes are purified by crystallization (methanol, ethanol, light petroleum, cyclohexane) and/or sublimation.

p-Cresyl-α-naphthylurethane

Reagents: α-naphthylisocyanate, 2% NaOH, cyclohexane.

Procedure: *p*-cresol (2 mg) and α-naphthylisocyanate (4 mg) are heated at 140 − 150 °C for 5 min in a micro test tube by immersing it in an oil bath. When cooled, the solidified melt is ground with a glass rod, and 0.2 ml of 2% NaOH is added. The mixture is heated mildly and stirred thoroughly with the rod. When the solid particles are settled the liquid is separated with a micropipette and the solid phase is washed twice more, each time with 0.2 ml of water. After thorough drying, the crude product is dissolved in 1.5 ml of boiling cyclohexane and transferred immediately to the microsublimation flask. The test tube is washed once more with 0.5 ml of boiling cyclohexane. After evaporation of the solvent the residue is purified by sublimation (yield, 2.7 mg; mp, 142−144 °C).

2. Color Reactions of Phenols (Detection)

Reaction with Ferric Chloride

General Considerations

Solutions of phenols give characteristic colors in reactions with ferric chloride. In these reactions complex compounds are formed, having, in the case of monovalent phenols, the ratio phenol: $FeCl_3 = 1 : 1$. According to recent studies (1,2) the colored complex is formed by gradual expulsion of solvent molecules from the solvated ferric ion by the phenol,

$$ArOH + [Fe(H_2O)_6]^{3\oplus} \rightarrow H_2O + [Fe(H_2O)_5(ArOH)]^{3\oplus} \rightarrow$$
$$\rightarrow [Fe(H_2O)_5(ArO)]^{3\oplus} + H_3O^{\oplus}$$

This agrees well with previous experience that polar solvents are necessary for the reaction, mainly those containing oxygen or nitrogen atoms in the molecule. In nonpolar solvents the reaction takes place only in the presence of pyridine. The latter functions not only as a proton acceptor, but probably also takes part in the formation of the complex, because other bases do not produce the color. The presence of acids, for example, acetic or benzoic acid, prevents the formation of a strong coloring. If these acids are added to the color already formed, the solutions are decolorized. For the same reason p-hydroxybenzoic acid does not give a color reaction with ferric chloride. On the other hand, salicylic acid gives a strong violet color even in the presence of acetic acid. This can be explained by the formation of a chelate-type salt,

formed from phenols with −OH, −COOH, −SO₃H, and −CHO groups in the ortho-position to the hydroxy group. The stability of these salts in different solvents and in the presence of acetic acid depends on the strength of the chelate bond. In this case acetic acid does not change pH substantially, but the described reactions take place by the competitive effect of the acetate ion on the solvated ferric ion. The ability to form a chelate-type ferric complex in a given solvent arises from the ability of this solvent to accept electrons from ferric chloride. Of the solvents which can be used for this test (water, methanol, benzene, ether, dichloro diisopropyl ether), chelation takes place best in methanol and water; we can assume, therefore, that this is caused by their high polarity. In benzene, which is the weakest base, the formation of the chelate bond is minimum, whereas ether is between the two extremes. The strongest chelation was observed in *o*-dihydroxybenzenes, the weakest in phenols where the *o*-position was occupied by strongly electronegative substituents (nitro, carbonyl, carboxyl). These groups can be represented by mesomeric structures with a positive charge on the nitrogen or oxygen atoms, which compete with the ferric ion in their influence on the electron pair, so weakening the iron-oxygen bond. As the nitro group is the most electronegative of all the above-mentioned groups, *o*-nitrophenol is least inclined to form a chelate (the abbreviation "solv." in the formula means water or some other suitable solvent):

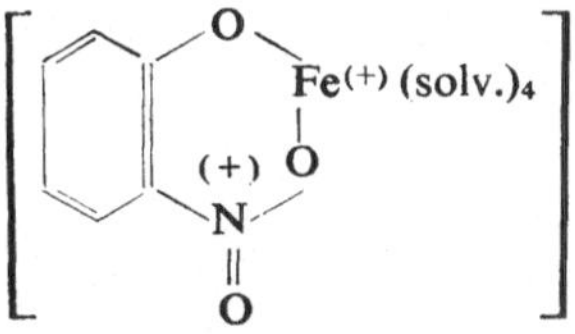

All these facts can be utilized in the application of the color test. Generally, the reaction is carried out in different solvents, sometimes with the addition of pyridine. Phenols can be differentiated according to the strength with which they form chelate bonds in the presence of acetic acid or with a suspension of ferric hydroxide in water or methanol with subsequent acidification. A list of chelate-forming substances is given in Table 5.

A positive reaction is obtained, however, with a series of nonphenolic compounds: aromatic amines, enols or hydroxymethylene compounds, tautomeric β-diketones, β-ketoesters, derivatives of malonic acid esters or of cyanoacetate with a negative substituent at the $-CH_2$ group, some isonitrites, hydroxylamine derivatives, oximes, hydroxamic acids, and certain derivatives of sulfur.

Hydroquinones represent a special case. Dissolved in water they give a blue color which disappears rapidly as a consequence of air oxidation to quinones.

Some Chelate-Forming Substances* *Table 5*

Substance	Ether	di-Cl-diiso-propyl ether	Ben-zene	MeOH	Water
p-Aminobenzamidoxime	−	−	−	+	−
2-Amino-5-nitrophenol	−	−	−	−	+
o-Aminophenol	−	+	−	+	+
p-Aminophenyl-acetamidoxime-HCl	−	−	−	+	−
p-Aminosalicylic acid	+	−	−	+	+
Benzamidoxime	−	−	−	−	−
n-Butyl salicylate	−	−	−	−	−
Catechol	+	+	+	+	+
Diethyl acetylsuccinate	−	−	−	+	−
2,4-Dihydroxybenzoic acid	−	−	−	+	+
2,5-Dihydroxy-1,4-benzoquinone	+	+	−	+	+
4,4′-Dihydroxy-3,3′-di-aminophenyl sulfone	−	−	−	+	+
Ethyl acetoacetate	−	−	−	−	−
Gallic acid	+	+	+	+	+
o-Hydroxyacetophenone	−	−	−	−	−
2-Hydroxy-3-methoxy-benzaldehyde	−	−	−	−	+
1-Hydroxy-2-naphthoic acid	+	+	−	+	+
3-Hydroxy-2-naphthoic acid	+	+	−	+	+ −
8-Hydroxyquinoline	+	+	+	+	+
o-Nitrophenol	−	−	−	−	−
Phenyl salicylate	−	−	−	−	−
5-Phenylsalicylic acid	−	−	−	+	+
n-Propyl gallate	+	+	+	+	+
Pyrogallol	+	+	+	+	+
Salicylaldehyde	−	−	−	+	−
Salicylic acid	−	−	−	+	+
Sulfosalicylic acid	+	−	−	+	+

* Here + indicates that chelate formation takes place, − that chelate formation does not take place.

Detection with Ferric Chloride

Reagents: 1. Anhydrous ferric chloride (10 g) dissolved in water (100 ml). The solution is filtered; its color is orange.

2. Anhydrous ferric chloride (10 g) dissolved in abs. methanol (100 ml). The filtered solution is orange.

3. Anhydrous ferric chloride (0.5 g) is suspended in 100 ml anhydrous benzene. After thorough shaking the suspension is allowed to settle down and is then decanted. The solution is dark brown. It cannot be filtered through filter paper because this takes water from it and the insoluble hydrate precipitates.

4. Anhydrous ferric chloride (5 g) dissolved in peroxide-free ether (100 ml). The filtered solution is dark brown.

5. Anhydrous ferric chloride (10 g) dissolved in 2,2'-dichlorodiisopropyl ether. The filtered solution is dark brown.

Procedure: Approximately 20 mg or two drops of the sample are dissolved or suspended in 1.5 ml of one of the following solvents: water, methanol, benzene, ether, 2,2'-dichlorodiisopropyl ether. One drop of one of the solutions described under 1 − 5 is added and then one drop of pyridine, which causes the formation of the characteristic color of the complex ferric salt (red, violet, blue). Two drops of acetic acid are then added, and if the color, which is different from that in the control experiment, persists, the presence of a chelate bond can be considered as proved (3, 4).

Reaction with Molybdates

In a neutral or slightly acid medium, *o*-dihydroxybenzenes form colored complexes with molybdates (5)

For the reaction either alkaline molybdates (6) or molybdates of organic bases, as, for example, monoethanolamine or ethylamine, are used (7).

Detection with Ammonium Molybdate

Reagents: 14% ammonium molybdate in water; acetic acid.

Procedure: If acetic acid (0.5 ml) and ammonium molybdate solution (1 ml) are added to an aqueous solution of catechol (2 ml), a red brown color is formed.

Detection with Ethylammonium Molybdate

Reagent: An equivalent amount of ammonia is added to 13.2 g of ethylamine hydrochloride, the solution is diluted with water, and 4.4 g of powdered MoO_3 are dissolved in the solution. The mixture is allowed to stand for some time before the clear supernatant is decanted and diluted to a volume of 100 ml. It should be kept in a brown glass bottle. Ammonium formate.

Procedure: Ammonium formate (5 g) is dissolved in methanol (100 ml) and this solution is added to a solution of several mg of catechol in 5 ml methanol. If necessary, it is further diluted with methanol and then additioned with 1 ml of the reagent. A yellow-orange color is formed.

Coupling Reactions

Phenols couple with diazonium salts, giving strongly colored azo dyes. Phenols having a free para position couple at this position; if this is occupied, substtitution takes place at the ortho position. If both these positions are occupied, the coupling does not usually take place, but in the case of certain substituents they can be displaced by the azo-group, such as, for example, when the coupling takes place with *p*-hydroxybenzoic acid, *p*-hydroxybenzenesulfonic acid, 2-hydroxy-1-naphthoic acid, 1-halogeno-derivatives of 2-naphthol, etc.

The most commonly used reagents are diazotized *p*-nitroaniline and diazotized sulfanilic acid, which are prepared shortly before use by diazotization of the corresponding amine. The use of stabilized diazonium salts is very advantageous (i.e., double salts with $ZnCl_2$, fluoroborates, naphthalene disulfonates, anti-diazotates). These salts can be stored for a long time and do not have oxidative effects (free nitrogen oxides) like diazonium salt solutions prepared in the usual manner (oxidation of catechols during the coupling).

In addition to phenols, phenolic esters and ethers can also couple in certain cases, with simultaneous dealkylation. Coupling is also observed with substances containing active methylene groups, such as, for example, with ethyl acetoacetate, aliphatic nitro compounds, acetoacetanilides, compounds carrying a CN, SO, or SO_2 group at the α-position to a methylene group, furthermore some β-dicarbonylic compounds, compounds with a methyl group vicinal (α-position) to a nitrogen atom of an N-heterocyclic compound; in addition, coupling of active diazonium salts with cyclohexanone, acetone, mesitylene, 3,5-benzpyrene, and 1,3-butadiene takes place.

The detection of aromatic mono and polyhydroxy compounds can be carried out on the basis of color reactions of the product of coupling with diazotized sulfanilic acid (influence of pH) (8).

Detection with Diazotized *p*-Nitroaniline

Reagents: *p*-Nitroaniline (0.5 g) is dissolved in hot N HCl (100 ml), cooled with ice to 0 °C, and mixed with a solution (10 ml) of $NaNO_2$ (2.7 g $NaNO_2$ in 100 ml water). After 15 min standing in a dark place at 0 °C this is filtered and used immediately. Sodium acetate.

Procedure: A few milligrams of phenol are dissolved in aqueous sodium acetate solution (5 ml) and mixed with 5 ml of the reagent; a precipitate of a yellow to orange-red color, changing to red after alkalization, is formed.

Detection with Diazotized Sulfanilic Acid

Reagents: 1 ml of a 2% $NaNO_2$ solution, cooled with cold water, is mixed with a precooled 1% solution (5 ml) of sulfanilic acid in 10% hydrochloric acid and allowed to stand for 5 min; N NaOH.

Procedure: Several milligrams of the tested phenol are dissolved in 5 ml of N NaOH and mixed with 1 ml of the reagent. An orange-red to red color is formed.

Detection with 1-Diazo-2-chloro-4-nitrobenzene

Reagents: 0.1 g of 2-chloro-4-nitrobenzenediazonium-1,5-naphthalenedisulfonate dissolved in 100 ml of water and acidified to pH 2 – 3 with a few drops of hydrochloric acid; 10% aqueous sodium acetate solution; 0.5 N KOH.

Procedure: A solution or suspension of the phenol in water is mixed with 1 ml of 2-chloro-4-nitrobenzenediazonium-1,5-naphthalene disulfonate solution, 1 ml of sodium acetate solution, and several milliliters of 0.5 N KOH. After the addition of sodium acetate a yellow to orange precipitate is formed which dissolves in alkali with a red to violet color.

Detection of Resorcinol with 1-Diazo-2-naphthol-4-sulfonic Acid

Reagent: 1% solution of 1-diazo-2-naphthol-4-sulfonic acid in 5% sodium acetate solution.

Procedure: The reagent is added to a few milligrams of resorcinol dissolved in a few milliliters of water, and a red-violet color is formed. With monophenols either the color does not form at all, or only a yellow color appears.

Color Reactions Based on the Formation of Indophenols

The reaction of phenols with aromatic amino compounds in an oxidizing medium gives rise to strongly colored compounds of the quinone anil, indophenol, or indoaniline type of the general formula where R is H in quinone anils, OH in indophenols, and NH_2 in indoanilines. Quinone monoanil is formed, for example, by oxidation of a mixture of a phenol and aniline with sodium hypochlorite,

$$C_6H_5OH + C_6H_5NH_2 \xrightarrow{\text{NaClO}} O=\left\langle\bigcirc\right\rangle=N-\left\langle\bigcirc\right\rangle$$

Reaction with 4-Aminoantipyrine

For practical purposes the most suitable and at the same time the most sensitive reagent for the detection of phenols is 4-aminoantipyrine, which is stable even in an aqueous solution and gives an unambiguous color reaction. The control experiment is colorless and is not sensitive to alkalies (9, 10).

The reaction takes place best in the pH range from 7.5 to 9.5, regardless of the type of buffer used, but it must have a sufficient capacity to prevent a drop in pH during the reaction. For the reaction it is further important that the ratio of reagents be close to that of one mole of aminoantipyrine to one oxidation equivalent of the oxidant, and then a ten to fiftyfold excess of the reagent (phenol as the base of calculation) also be secured. Potassium ferricyanide or alkaline persulfate may be used as oxidants. The former is preferable because of its greater stability in aqueous solution (11).

Reagent: 1% aqueous solution of 4-aminoantipyrine, borate buffer of pH 9−9.5 or a 5% $NaHCO_3$ solution, 3% potassium ferricyanide or 1% persulfate solution.

Procedure: To a few milliliters of an aqueous phenol solution 1 ml of 4-aminoantipyrine solution, 5 ml of borate or hydrogen carbonate buffer, and 1 ml of ferricyanide or persulfate solution are added. A red color is formed.

From the course of the reaction with various phenols the following rules can be formulated:

a) A positive reaction (formation of a red color) is given by substances having at least one phenolic group in the molecule.

b) The reaction does not take place with *p*-substituted phenols; phenols with halogens, carboxy, hydroxy, sulfo, and methoxy groups in the *p*-position are an exception; these groups are displaced during the reaction.

c) The nitro group prevents the reaction if situated at the ortho-position; if the nitro group is in a meta-position, the reaction is not clear-cut.

d) 1-Naphthols give red and 2-naphthols green colors.

e) A positive reaction is also given by β-hydroxypyridine derivatives; among quinolines, 5- and 8-hydroxyquinolines react as 1-naphthol, and 3-, 6-, and 7-hydroxyquinolines as 2-naphthol.

f) Barbituric acid also gives a positive reaction.

g) The pH of the reaction medium should be sufficiently high to prevent the formation of the antipyrine red. The best range is pH 7.5 to 9.5. Sufficient buffering capacity is necessary.

Liebermann Reaction

In sulfuric acid and in the presence of nitrogen oxides, phenols give colored indophenols. In the reaction medium they give a blue or green solution, which after dilution with water turns red, and after alkalization becomes blue. A positive reaction is given by phenols with a free para-position, unless they are substituted with $-OH$, $-NH_2$, $-OCH_3$, $-NO_2$, $-CHO$, $-COOH$, or $-COCH_3$. For example, 3,5-xylenol, *o*-aminophenol, *o*-nitrophenol, *m*-hydroxybenzaldehyde, *m*-hydroxybenzoic acid, and similar compounds do not react.

For different modifications of the reaction of phenols with nitrite in an acid medium see (12−18).

Reagents: concentrated sulfuric acid, sodium nitrite.

Procedure: A few milligrams of a phenol are dissolved in 1 ml conc. H_2SO_4 and a few crystals of sodium nitrite are added to the solution. A blue-green or blue-violet color is formed either immediately or after gentle warming, turning red after dilution with water and changing to blue after alkalization.

Gibbs Reaction

Quinonechloroimines react with phenols with the formation of substituted indophenols, which have a blue or blue-green color in an alkaline medium and a red color in an acid medium.

$$C_6H_5OH + ClN=\!\!\!\left\langle\begin{array}{c}Cl\\ \\ \\Cl\end{array}\right\rangle\!\!\!=O \;\rightarrow\; HO-\!\!\!\left\langle\;\right\rangle\!\!\!-N=\!\!\!\left\langle\begin{array}{c}Cl\\ \\ \\Cl\end{array}\right\rangle\!\!\!=O + HCl$$

The reaction is carried out best in alkaline media at pH 9.0 – 9.6; its sensitivity can be increased by extraction of the blue dye into an organic solvent – for example, butanol.

The reaction is positive with phenols containing a free p-position. However, a negative reaction is observed with salicylic acid, salicylaldehyde, o-nitrophenol, and vanillin. If the para-position is occupied, the reaction does not take place unless the groups at this position are easily displaced. Thus, for example, a positive reaction was observed with the following p-substituted phenols: p-chlorophenol (it gives a color showing the same absorption maximum as free phenol), p-bromophenol, 2,4-dichlorophenol, 4-chloro-6-phenylphenol, and 4,6-dichloro-o-cresol (it gives the same indophenol as 6-chloro-o-cresol). The elimination of a p-hydroxymethyl and an active methylene group also takes place.

2-6-Dichloroquinone-4-chloroimines also react with aromatic amines, with uric acid, theophylline, thiouracil derivatives, glyoxaline-2-thiol derivatives, and p-thiocresol.

Reagent: 1% 2,6-dichloro- or 2,6-dibromoquinone-4-chloroimine in acetone; borate buffer of pH 8 – 10.

Procedure: A small amount of the tested phenol is dissolved or suspended in water and mixed with 5 ml of the phosphate buffer and 1 ml of the reagent solution. A blue color is formed, extractable with n-butanol.

Other Color Reactions

In addition to the color reaction given above phenols give other very important reactions suitable for special cases.

Under the influence of nitrous acid in acetic acid and after subsequent alkalization with ammonia, phenols (and those 2,4,6-trisubstituted) give strong yellow to orange colors (19) as a consequence of the formation of corresponding nitrophenols (20).

If phenols react with nitrites in the presence of mercuric salts (Millon reagent), orange to red colors are formed. Positive reactions are also observed with p-substituted phenols, but not with o,o'-disubstituted phenols (21).

If phenols are reacted with polyhalo compounds in alkalies (22), with aromatic aldehydes, or even with formaldehyde or oxalic acid in an acid medium, di- or triphenylmethane derivatives are formed (23 – 26).

The formation of xanthene derivatives, vividly colored or strongly fluorescing, is utilized for the reaction of phenols with phthalic anhydride and zinc chloride. The strong green fluorescence of the fluoresceins formed is charactetistic of resorcinol derivatives (27); see reactions of dicarboxylic acids on p. 249.

To detect catechol or resorcinol, their color reaction with iodine can be utilized (28, 29).

For p-substituted phenols the reaction with 1-nitroso-2-naphthol is characteristic (30−32).

o-Dihydroxybenzenes give a specific reaction (a violet color) with tartaric acid and ferrous sulfate (33, 34).

Diphenols, especially resorcinol, give a strong color with cadmium or zinc salts (35). Various colors are also obtained when phenols are allowed to react with titanium tetrachloride in sulfuric acid, dimethyl formamide, or methanol (36).

Very characteristic of 1-naphthol is the violet reaction with cupric sulfate and potassium cyanide (37).

Reducing properties of o- and p-dihydroxybenzenes can be utilized for their reaction with ammoniacal silver nitrate solution without heating (38, 39) (see p. 210).

Generally, phenols reduce solutions of phosphomolybdic or phosphotungstic acid, giving rise to the formation of lower oxides of molybdenum or tungsten. This reaction was formerly used to detect the presence of phenols as well as for the colorimetric determination of phenols (40−43). However, it is not specific for phenols, and all reducing organic or inorganic compounds react as well.

Another oxidant very important for phenols is iron (III) ferricyanide, which was found very useful mainly in paper chromatography (see p. 198). This reagent is not specific for phenols either, but its advantage consists in the fact that trisubtituted and sterically hindered phenols, as, for example, 2,4,6-tritert-butylphenol, give a positive reaction with it too. For its application to the colorimetric determination of hydroquinone see (44).

Great possibilities for the detection and identification of phenols are offered by the color reaction with tetracyanoethylene (also in combination with other color reactions) which makes the detection of a p-substituent possible (45).

3. Chromatographic Methods

Paper Chromatography

Virtually all phenols can be chromatographed as such, as their volatility and easy oxidability do not represent serious drawbacks. The most suitable solvent systems for various groups of phenols are given in the following list:

phenol and closest homologs	formamide/hexane
phenols with alkyls up to C_8	dimethylformamide/hexane
phenols with alkyls from C_8 to C_{16}	paraffin oil/methanol or methanol/water 4 : 1
diphenols (dihydroxybenzenes)	formamide/chloroform
triphenols (trihydroxybenzenes)	formamide/chloroform-ethyl acetate 1 : 1 or 1 : 2
naphthols and dihydroxynaphthalenes	formamide/benzene — ethyl acetate 4 : 1
	formamide/chloroform — ethyl acetate 4 : 1
	formamide/carbon tetrachloride

Spraying the chromatogram with a solution of iron(III) ferricyanide represents a general method of detecting phenols: the reagent consists of equal volumes of 1.5% aqueous ferric chloride and 1% potassium ferricyanide. Dark blue spots are formed on a light green background. The color can be fixed by drawing the sheet through dilute hydrochloric acid (1 : 1) and washing it with running water. In addition, detection reactions with various diazonium salts are commonly used, such as, for example, with diazotized sulfanilic acid or diazotized p-nitroaniline. Again, the use of stabilized diazonium salts is recommended. The detection with diazotized sulfanilic acid is carried out as follows: 25 ml of a 0.3% sulfanilic acid solution in 8% HCl are mixed with 1.5 ml of a 5% $NaNO_2$ solution immediately before use. Spraying of the chromatogram with this reagent is followed by a spray with 20% sodium carbonate solution.

However, this method cannot differentiate certain isomeric phenols, such as, for example, p- and m-cresols. In such cases it is useful to carry out the coupling of these phenols with diazotized p-nitroaniline and then to chromatograph the o- and p-azo dyes formed in a formamide/hexane system (46, 47, 49).

Gas Chromatography

In the gas chromatography of phenols the following materials have been used as stationary phases: Apiezon (50, 51), erythritol, mannitol and inositol (50), and glycerol mono-oleate (52). Di-n-octyl sebacate fixed on Embacel as a support is also suitable if the temperature is 178 °C and the gas flow 20 − 30 ml/min. The following relative elution volumes were measured (53): phenol 1.00; *o*-cresol 1.24; 2,6-xylenol 1.43; *m*- and *p*-cresol 1.49; *o*-ethyl-phenol 1.75; 2,4- and 2,5-xylenol 1.84; 2,3-xylenol 2.20; 2-ethylphenol 2.20; 3,5-xylenol 2.26; 3,4-xylenol 2.55.

For the separation of higher tert-butylated phenols silicone oil and 3,5-dinitrobenzoate of polyethylene glycol 400 were found suitable (45). A suitable phase for the separation of polyhydric phenols was proposed by von Rudloff (48).

References

1. Crutchfield, M. M., Irani, R. R., and Yoder, J. T.: J. Am. Oil Chem. Soc. **41** 129 (1964).
2. McFarland, J. W., and Howard, J. B.: J. Org. Chem. **30**, 957 (1965).
3. Soloway, S., and Wilen, S. H.: Anal. Chem. **24**, 979 (1952).
4. Soloway, S., and Rosen, P.: Anal. Chem. **25**, 595 (1953).
5. Haight, G. P., Jr., and Paragamian, V.: Anal. Chem. **32**, 642 (1960).
6. Quastel, J. H.: Analyst **56**, 311 (1931).
7. Janák, J.: Chem. Listy **48**, 1171 (1954).
8. Légrádi, L.: Mikrochim. Ichnoanal. Acta **1965**, 865.
9. Emerson, E.: J. Org. Chem. **8**, 417 (1943); **8**, 433 (1943).
10. Mohler, E. F., and Jacob, L. N.: Anal. Chem. **29**, 1369 (1957).
11. Svobodová, D., and Gasparič, J.: Collection Czech. Chem. Commun. **33**, 42 (1968).
12. Brunner, H., and Krämer, C.: Ber. **15**, 174 (1882); **17**, 1847 (1884).
13. Nietzki, R., Dietze, A., and Mäckler, H.: Ber. **22**, 3020 (1889).
14. Schoutissen, H. A. J.: Rec. Trav. Chim. **40**, 753 (1921); ref. C. A. **16**, 1226 (1922).
15. Eichler, H.: Z. anal. Chem. **96**, 21 (1934).
16. Fischer, O., and Hepp, E.: Ber. **36**, 1807 (1903).
17. Gasparič, J.: Chem. and Ind. **1962**, 43; Z. anal. Chem. **199**, 276 (1964).
18. Ware, A. H.: Analyst **52**, 335 (1927); ref. Zeit. anal. Chem. **78**, 233 (1929).
19. Lykken, L., Treseder, R. S., and Zahn, V.: Ind. Eng. Chem., Anal. Ed. **18**, 103 (1946).
20. Gasparič, J.: Chem. and Ind. **1962**, 43.
21. Inglett, G. E., and Lodge, J. P.: Anal. Chem. **31**, 248 (1959).
22. Francois, M., and Séguin, L.: Bull. Soc. Chim. France **49**, 680 (1931).
23. Kolšek, J., and Perpar, M.: Z. anal. Chem. **167**, 161 (1959).
24. van Urk, H. W.: Pharm. Weekbl. **66**, 101 (1929); ref. C. A., **23**, 1717 (1929).
25. Inglett, G. E., and Lodge, J. P.: Anal. Chem. **31**, 248 (1959).
26. Nicolas, L., and Burel, R.: Chim. Anal. **38**, 316 (1956).
27. Formánek, J., and Knop, J.: Z. anal. Chem. **56**, 273 (1917).
28. Willard, H. H., and Wooten, A. L.: Anal. Chem. **22**, 670 (1950).
29. Janák, J., and Matoušek, L.: Chem. Listy **48**, 1176 (1954).
30. Gerngross, O., Voss, K., and Herfeld, H.: Ber. **66**, 435 (1933).
31. Maciag, A., and Schoental, R.: Mikrochemie **24**, 250 (1938).
32. Mlodecka, J.: Chem. analit. (Warsaw) **4**, 45 (1959).
33. Mráz, V.: Chem. Listy **44**, 259 (1950).

34. Leibnitz, E., Behrens, U., and Czech, H.: J. prakt. Chem. (4) 11, 73 (1960).
35. Vaskevič, D. N., and Gol'dina, C. A.: Zh. Prikl. Khim. 24, 1214 (1951).
36. Sommer, L.: Z. anal. Chem. 171, 410 (1959).
37. Sass, S., Kaufman, J. J., and Kiernan, J.: Anal. Chem. 29, 143 (1957).
38. Kisser, J., and Kondo, Y.: Mikrochemie, Molisch Festschrift 1936, 259.
39. Wildi, B. S.: Science 113, 188 (1951).
40. Folin, O., and Macallum, A. B.: J. Biol. Chem. 11, 265 (1912).
41. Folin, O., and Cioacalteau, V.: J. Biol. Chem. 73, 627 (1927).
42. Vorce, L. R.: Ind. Eng. Chem. 17, 751 (1925).
43. Whettem, S. M. A.: Analyst 74, 185 (1949).
44. Kline, G. M.: Analytical Chemistry of Polymers. Part I, p. 5. Interscience Publ., New York 1959.
45. Smith, B., Persmark, V., and Edman, E.: Acta Chem. Scand. 17, 709 (1963).
46. Franc, J.: Chem. Listy 52, 55 (1958).
47. Gasparič, J., Petránek, J., and Borecký, J.: J. Chromatog. 5, 408 (1961).
48. Rudloff von, E.: J. Gas Chromatography 2, 89 (1964).
49. Latinák, J.: Collection Czech. Chem. Commun. 24, 2939 (1959).
50. Janák, J., and Komers, R.: Collection Czech. Chem. Commun. 24, 1960 (1959).
51. Irvine, L., and Mitchell, T. J.: J. Appl. Chem. 8, 3 (1958).
52. Grant, D. W.: Gas-Chromatography, 2nd Intern. Symp. Amsterdam 1958. Ed. D. E. Desty, Butterworths Sci. Publ., London (1958).
53. Payn, D. S.: Chem. and Ind. 1960, 1090.

CHAPTER XI

ETHERS

1. Aliphatic Ethers

The identification of ethers by classical means, by their conversion to suitable solid derivatives, is a complicated task which cannot be carried out generally.

For the identification of symmetrical aliphatic ethers their pyrolysis at 500 °C has been proposed, during which a carbonyl compound and a hydrocarbon are formed (1):

$$RCH_2OCH_2R \xrightarrow{500\,°C} RCH_3 + RCHO$$
$$R_2CHOCHR_2 \xrightarrow{500\,°C} RCH_2R + RCOR$$

On the pyrolysis of asymmetrical ethers a mixture of hydrocarbons and of carbonyl compounds is formed in an analogous way. The carbonyl compounds formed are then converted to suitable derivatives (see p. 209).

An easier and more convenient method of identification of ethers consists in their cleavage with hydriodic acid, giving rise to alkyl iodide, which can be identified by its transformation to S-alkyl thiuronium 3,5-dinitrobenzoate.

Method of Cleavage of Symmetrical Aliphatic Ethers

The ether (1 mmole) is cleaved by boiling it with hydriodic acid [see, for example, (2)]. The alkyl iodide formed is absorbed in a vessel containing 2 ml of a 2.5% acetonic thiourea solution cooled externally with ice. The absorbing solution is then transferred into an ampoule and processed as in the procedure described on p. 139.

Cleavage of Dibutyl Ether

Reagents: hydriodic acid ($\varrho = 1.7$), 10% NaOH, 5% ethanol, thiourea, picric acid.

Procedure: In a 25-ml flask fitted with a reflux condenser 0.5 ml of dibutyl ether is refluxed with 5 ml of conc. hydriodic acid (boiling stone should be added) for 90 min. After cooling, the mixture is diluted with 10 ml of water, transferred into a 50-ml distillation flask, and neutralized carefully and with cooling with 10% NaOH, until the solution becomes pale yellow. The mixture is then distilled and 20 ml of the distillate are taken in a separatory funnel. After complete separation of the phases the lower, aqueous layer is released from the funnel and the upper layer, alkyl iodide, is diluted with 5 ml of ethanol and transferred into a 25-ml flask containing 0.5 g of thiourea. The mixture is refluxed for 30 min. Picric acid (1 g) is added to the mixture, which is heated again until the picric acid dissolves, and water is added dropwise to the solution to incipient precipitation (turbidity).

The mixture is allowed to stand for crystallization. The separated picrate is filtered off with suction onto a filtration crucible and washed twice with 2 ml of 50% ethanol. Yield, 1.5 g; mp, 177−178 °C.

Detection and Identification of O-Alkyl Groups by Paper Chromatography (3)

(Cleavage with Hydriodic Acid in a Stream of Nitrogen)

The sample (2−3 mg), some phenol on the tip of a spatula, a small amount of red phosphorus, and 2 ml of hydriodic acid are introduced into the reaction flask of the usual apparatus for O-alkyl group determination (2, 4) and the cleavage is carried out in the usual manner in a nitrogen stream (4). The alkyl iodides formed are carried by the stream of nitrogen into an ampoule cooled with ice and containing a suspension of 3−4 mg of silver 3,5-dinitrobenzoate in 1 ml of benzene (Fig. 31). When the absorption of the alkyl iodides is terminated the ampoule is sealed and heated to 100 °C for 2 hr. After cooling, the ampoule is opened and the solution is spotted directly onto the chromatographic paper in several concentrations. If necesary, the solution can be concentrated beforehand. Samples of authentic 3,5-dinitrobenzoic acid esters are applied on the same sheet of chromatographic paper. Chromatography is carried out in the system dimethylformamide/hexane as described on p. 154. For thin-layer chromatography on silica gel G the system cyclohexane-carbon tetrachloride-ethyl acetate (10 : 75 : 15) is used.

The procedure described serves generally for the identification of an alkyl group bound to oxygen, and it was tested on the following substances: ethyl n-butyl ether, di-n-amyl ether, phenacet in, vanillin, n-butyl α-hydroxyisobutyrate, codeine, butyl metacrylate (polymer), anisole, and ethyl acetate.

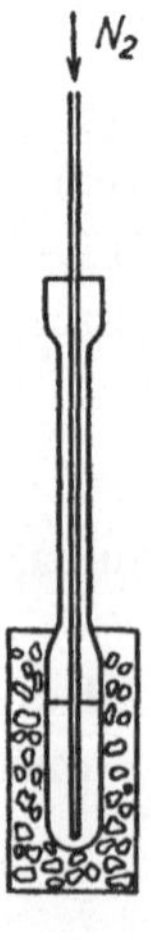

Fig. 31. Ampoule for the absorption of alkyl iodides, cooled with ice and filled with a suspension of silver 3,5-dinitrobenzoate in benzene.

If chromatography is used for identification purposes, the amount of sample necessary for the analysis drops to 1 − 3 mg. The disadvantage of this method is that isomeric alkyls cannot be differentiated by it.

Cleavage of Ethers with Hydriodic Acid in an Ampoule
(Identification of Symmetrical and Asymmetrical Ethers)

Hydriodic acid (1 ml) and the analyzed ether (0.2 ml) are introduced into a glass ampoule, which is then sealed and heated in a drying oven to 130 °C for 30 min. It is recommended to shake the ampoule after the first 15 min of heating. The ampoule is then cooled with ice and opened. Its contents are transferred into a small separatory funnel containing 10 ml of water. The ampoule is rinsed with 5 ml of benzene; this is also poured into the funnel, which is then shaken vigorously. When the layers are well separated the benzene layer is taken and washed with 10 ml of 5% sodium hydrogen carbonate and finally with water. The benzene solution is dried with anhydrous magnesium sulfate and filtered into a test tube made of thick glass; 1.2 g of silver 3,5-dinitrobenzoate is added to it, and the tube is sealed and heated in a drying oven at 100 °C for 30 min. After cooling, the test tube is opened and the contents are filtered to eliminate the silver iodide formed and the excess silver 3,5-dinitrobenzoate. The filtrate and the solutions of authentic 3,5-dinitrobenzoates (as standards) are then applied directly onto a strip of chromatographic paper in various concentrations (if necessary, the filtrate can be concentrated). Chromatography is carried out in 25% dimethylformamide/hexane (see p. 154). For thin-layer chromatography on silica gel G the system cyclohexane-carbon tetrachloride-ethyl acetate (10 : 75 : 15) is used.

Cleavage of Ethers with 3,5-Dinitrobenzoic Acid Anhydride and Stannic Chloride (5)

(Identification of Symmetrical and Asymmetrical Ethers)

The ether (0.05 ml), 3,5-dinitrobenzoic acid anhydride (0.05 g) [for the preparation see (5)], and 1 ml of a 0.5 M $SnCl_4$ solution in benzene (prepared from 6 ml of $SnCl_4$ and benzene to make a total volume 100 ml) are introduced into an ampoule which is then sealed and heated at 120 °C for 30 min. After cooling, the ampoule is opened and the benzenic solution is spotted directly onto a strip of chromatographic paper, where solutions of autenthic samples of esters of 3,5-dinitrobenzoic acid are also applied (see p. 154). Chromatography is carried out in dimethylformamide or formamide/ /hexane. For thin-layer chromatography on silica gel G the system cyclohexane-carbon tetrachloride-ethyl acetate (10 : 75 : 15) is used.

2. Aromatic Ethers

For the identification of aromatic ethers use is made either of the reactivity of the aromatic nucleus (bromination, chlorosulfonation), or else they can be cleaved with hydriodic acid (see p. 201). In the latter case, however, only the alkyl group bound to oxygen is identified. The oxidation of side chains (compare p. 129) has only a limited use; the alkoxy group on the aromatic nucleus increases the stability of addition compounds (see p. 126) with picric acid, which can also be used for identification.

General Method of Preparation of Bromo Derivatives of Aromatic Ethers

Bromo derivatives of aromatic ethers are prepared by dropping bromine into a solution of the ether in glacial acetic acid, ethanol, or chloroform. The isolation is carried out by diluting the mixture with water or by distilling off the solvent. The degree of bromination is dependent predominantly on the properties and the position of the functional groups present on the aromatic nucleus. The amount of bromine used is dependent on the degree of bromination.

Among more common aromatic ethers, monobromoderivatives are formed from o-cresyl methyl ether, ethyl 1-naphthyl ether, ethyl 2-naphthyl ether, and methyl 2-naphthyl ether; dibromoderivatives from veratrol, resorcinol dimethyl ether, diphenyl ether, isoeugenol methyl ether, methyl 1-naphthyl ether, dibenzyl ether, ethyl 2-naphthyl ether, and hydroquinone dimethyl ether; and tribromoderivatives from anethol, eugenol methyl ether, isosafrol.

General Method for the Preparation of Addition Compounds of Aromatic Ethers with Picric Acid

The ether (5 mmole) is dissolved in 5 ml of boiling chloroform and mixed with 5 mmole of picric acid dissolved in 3 ml of boiling chloroform. On cooling the solution, crystals are formed which are filtered off and dried between two pieces of filter paper. Melting point is determined immediately, because these compounds decompose during recrystallization.

Addition Compound of Ethoxynaphthalene with Picric Acid

Reagents: picric acid, chloroform.

Procedure: Ethoxynaphthalene (100 mg) and picric acid (130 mg) are thoroughly mixed in a micro test tube with a glass rod and the mixture is melted by cautious heating. After cooling, the melt is triturated with 0.5 ml of benzene (glass rod) and crystallized from it. Yield, 195 mg, mp, 103 – 104 °C.

3. Vinyl Ethers

While saturated ethers are stable in the presence of acids, vinyl ethers decompose in the presence of cold and dilute acids:

$$\begin{array}{c} R_1 \\ \diagdown \\ R_2 \diagup \end{array} C=C-OR_4 + H_2O \quad \xrightarrow{H^+} \quad \begin{array}{c} R_1 \\ \diagdown \\ R_2 \diagup \end{array} CH-COR_3 + R_4-OH$$
$$\qquad\qquad |$$
$$\qquad\qquad R_3$$

For the decomposition to take place, the acid reaction of an aqueous solution of semicarbazide hydrochloride suffices; under these conditions the formed aldehyde (ketone) gives a semicarbazone.

General Method of Identification of Vinyl Ethers

The vinyl ether (0.5 g) is mixed in a test tube with 5 ml of water and 0.5 g of semicarbazide hydrochloride. Ethanol is added to this mixture until dissolution is complete. Anhydrous sodium acetate (0.4 g) is then added to the mixture and the test tube is heated over a boiling water bath for 30 min. Semicarbazone separates out after cooling (see p. 229).

4. Methylenedioxy Compounds

These compounds occur mainly in natural substances (alkaloids). Chemically, they are formals of o-dihydroxybenzene derivatives, and as such they are split more easily than phenol ethers, but less easily than other cyclic acetals.

$$\text{(benzodioxole)} \; CH_2 + H_2O \quad \xrightarrow{H^+} \quad \text{(catechol)} \begin{array}{c} OH \\ OH \end{array} + HCHO$$

General Method of Cleavage of Methylenedioxy Compounds

The cleavage is carried out by heating the substance (10 – 20 mg) with 10 ml of 50% H_2SO_4 at 80 °C in a small distillation apparatus. A stream of an inert gas is introduced into the apparatus, which carries the distilling vapors with it into a container with an acid 2,4-dinitrophenylhydrazine solution. The precipitated 2,4-dinitrophenylhydrazone is isolated and identified in the usual manner (see p. 221).

However, the formation of formaldehyde on cleaving methylenedioxy groups with acids is not specific for this group. In addition to formaldehyde derivatives, formaldehyde itself is formed by the action of acids on an activated hydroxymethyl group, both free and esterified (6).

Formaldehyde is also formed from certain compounds easily cleaved in acid medium, such as, for example, compounds with a $-NHCH_2NH-$

group, compounds with a primary carbinol group on the aromatic nucleus (benzyl alcohol does not give this reaction, but substituted benzyl alcohols do), and sugars (which under the influence of acids give 5-hydromethyl-furfurol).

A color reaction of the methylenedioxy group is based on the well-known detection of formaldehyde (p. 215) with chromotropic and sulfuric acids (the mixture is heated over a boiling water bath for 30 min) (7, 8).

5. Epoxides

1,2-Epoxides react with 2,4-dinitrothiophenol in a weak alkaline medium according to the equation

$$R-\overset{\displaystyle \diagdown_{\displaystyle O}\diagup}{CH-CH_2} + C_6H_3(NO_2)_2SH \;\to\; R-\underset{\displaystyle \overset{|}{OH}}{CH}-CH_2-S-C_6H_3(NO_2)_2$$

The reaction yields sulfides which crystallize well.

General Method for the Preparation of Sulfides

The epoxide (1 mmole) is dissolved in 5 ml of ethanol and the solution is mixed with 1.1 mmole (10 ml) of an ethanolic solution of 2,4-dinitro-thiophenol. Then 20 ml of saturated $NaHCO_3$ are added to the mixture and it is allowed to stand for 30 − 60 min (9).

Epoxides with larger rings are much less reactive than 1,2-epoxides. Nevertheless, 1,3- and 1,4-epoxides can be easily cleaved with 0.5 N HBr in acetic anhydride during a few minutes.

Tetrahydrofuran gives 4-bromobutanol by the reaction

$$\begin{array}{c} CH_2-CH_2 \\ |\qquad\quad | \\ CH_2\quad CH_2 \\ \diagdown_{\displaystyle O}\diagup \end{array} \xrightarrow[\text{Ac}_2\text{O}]{\text{HBr}} BrCH_2CH_2CH_2CH_2OH$$

However, when asymmetrical epoxides are cleaved, two isomers can be formed

$$R-\underset{\displaystyle \diagdown_{\displaystyle \qquad O\qquad}\diagup}{CH-CH_2-CH_2} \xrightarrow[\text{Ac}_2\text{O}]{\text{HBr}} R-\underset{\displaystyle \overset{|}{Br}}{CH}-CH_2-CH_2-CH_2OH +$$

$$+ R-\underset{\displaystyle \overset{|}{OH}}{CH}-CH_2-CH_2-CH_2Br$$

A color reaction with pyridine has been described for the detection of epoxides (10). 1,2-Epoxides can be detected in a manner similar to 1,2-diols, on reacting them with periodic acid, on the basis of the formation of the colorless precipitate of silver iodate (see p. 177). This reaction is also given by α-hydroxy aldehydes, α-hydroxy ketones, and α-hydroxy acids.

Epoxides can also be identified by paper chromatography after their conversion to corresponding ethers by the reaction with picric acid (11).

Glycidyl ethers were identified in the form of derivatives (2,4-dinitrophenylhydrazones) of corresponding substituted acetaldehydes (alkoxy or acyloxyacetaldehydes) obtained from parent ethers by hydration and oxidation with HIO_4 (12).

6. Gas Chromatography of Ethers

For the separation of simple symmetrical and mixed aliphatic ethers various stationary phases can be used; for example, ethyl hexyl sebacate, chlorinated oils (Convachlor 12), polytrifluorovinyl chloride (Fluorolube S), β, β'-iminobis-(propionitrile), tricresyl phosphate, diphenylformamide, β, β'-bis-(propionitrile) ether, and polypropylene glycol (13).

For the separation of epoxides by gas chromatography see (14).

References

1. Sah, P. P. T.: Rec. Trav. Chim. 58, 758 (1939).
2. Jureček, M.: Organická analysa, p. 466. Přírodovědecké nakladatelství, Prague 1950.
3. Večeřa, M., Gasparič, J., and Spěvák, A.: Chem. Listy 51, 1554 (1957); Collection Czech. Chem. Comm. 23, 768 (1958).
4. Jureček, M.: Organická analysa II, p. 356. Nakladatelství ČSAV, Prague 1957.
5. Jureček, M., Hubík, M., and Večeřa, M.: Collection Czech. Chem. Commun. 25, 1458 (1960).
6. Freudenberg, K., et al.: Ber. 80, 149 (1947); 85, 78 (1952).
7. Bricker, C. E., and Johnson, H. R.: Ind. Eng. Chem., Anal. Ed. 17, 400 (1945).
8. Beroza, M.: Anal. Chem. 26, 1970 (1954).
9. Davies, W., and Savige, W. E.: J. Chem. Soc. 1951, 774.
10. Lohmann, H.: J. prakt. Chem. 153, 57 (1939); Angew. Chem. 52, 407 (1939).
11. Schäfer, W., Nuck, W., and Jahn, H.: J. prakt. Chem. (4) 11, 11 (1960).
12. Ulbrich, V., and Makeš, J.: J. Chromatog. 15, 371 (1964).
13. Kaiser, R.: Gas-Chromatographie, p. 207. Akademische Verlagsgesellschaft, Leipzig 1960.
14. Ulbrich, V., and Dufka, O.: Chem. Průmysl 10 (35), 549 (1960).

CHAPTER XII

CARBONYL COMPOUNDS

When carbonyl compounds are to be detected and identified use is particularly made of condensation and addition reactions. In the case of aldehydes oxidation is also used. The high reactivity of these compounds is due to the character of the carbonyl group. Generally, aldehydes are more reactive than ketones. In the case of aromatic carbonyl compounds the substituents on the nucleus also play an important role, primarily those in the *o*-position, while *m*- and *p*-substituents are without an appreciable influence on the reactivity of the carbonyl group.

The derivatives of aldehydes and asymmetrical ketones in which the structural feature $> C=N-$ is formed can occur in two geometric isomers (cis-trans; I, II) and this must be kept in mind during the preparation of derivatives of this type,

$$
\begin{array}{cc}
\text{R}-\text{C}-\text{R}' & \text{R}-\text{C}-\text{R}' \\
\| & \| \\
\text{N}-\text{X} & \text{X}-\text{N} \\
\text{(I)} & \text{(II)}
\end{array}
$$

Generally, however, the stability of such isomers is lower than in compounds with a $C=C$ bond.

Lists of more important color reactions and of derivatives of carbonyl compounds are given in Tables 6 and 7.

1. Oxidation-Reduction Reactions (Detection and Identification)

Aldehydes are generally easily oxidized to carboxylic acids. Their reducing property is utilized for their detection (Tollens', Fehling's, Benedict's, Nessler's, and Feder's reagents). The acid formed can be isolated and serve for identification as such, or it can be converted to a suitable derivative. Of course, the reaction is not specific. For example, benzoic acid is not only obtained by oxidizing benzaldehyde, but also on oxidation of acetophenone, phenacyl bromide, benzil, benzyl alcohol, toluene, cinnamic acid, ethyl-benzene, etc.

Color Reactions of Aldehydes and Ketones*   *Table 6*

Reaction or reagent	Alde-hydes	Ketones	Reactive group	Page discussed
Tollens	+	−	$-CHO$, $-COCO-$	210
Fehling	+	−	$-CHO$	210
Nessler	+	−	$-CHO$	211
with fluorene	+	−	$-CHO$	213
with carbazole	+	−	$-CHO$	213
with phenols	+	−	$-CHO$	213
with hydroxyaldehydes	+	+	$-CHO$ CH_3COCH_2-	215
with aromatic amines	+	−	$-CHO$	215
with m-diamines	+	−	$-CHO$	216
Schiff	+	−	$-CHO$	217
with hydrazines	+	+	$-CHO$, $-CO-$	218
according to Angeli-Rimini	+	−	$-CHO$	224
with hydroxylamine	+	+	$-CHO$, $-CO-$	224
with o-nitrobenzaldehyde	−	+	CH_3CO-	233
Janovsky	−	+	$R-CH_2-CO-$	234
Legal	+	+	$-CHO$, $-CO-$	234
iodoform	−	+	CH_3CO-	235
with p-phenylenediamine	+	−	$-CHO$	239
with o-diamines	−	−	$-COCO-$ $-COCH_2CO-$	239

* Here + indicates a positive reaction, − a negative reaction.

Derivatives of Carbonyl Compounds*   *Table 7*

Reaction or reagent	Aldehydes	Ketones	Page discussed
Oxidation	+	−	211
Arylhydrazines	+	+	218
Hydroxylamine	+	+	224
Semicarbazide	+	+	228
Dimedone	+	−	230

* Here + indicates that derivatives are formed, − that derivatives are not formed.

Reaction with Tollens' Reagent

Tollens' reagent is a solution of silver oxide in ammonia, from which reducing aldehydes set free metallic silver in the cold:

$$2Ag(NH_3)_2OH + R-CHO \rightarrow 2\,Ag + R-COONH_4 + 3\,NH_3 + H_2O$$

A positive reaction is given, in addition to aliphatic and aromatic aldehydes, by other oxidizable substances, such as, for example, sugars, aromatic amines, aminophenols, polyphenols, and formic acid. With the exception of ketols and diketones, ketones do not react. Organic compounds containing $C=S$ and $S-H$ groups interfere with the reaction, because they precipitate silver sulfide. The sensitivity of the reaction under various conditions was studied by Karaoglanov (1).

Reagents: Ammonia is added dropwise to a mixture of equal volumes of 10% aqueous $AgNO_3$ solution (solution I) and 2 N NaOH (solution II) until the initially precipitated silver oxide is just dissolved. An excess of ammonia should be avoided, because it diminishes the sensitivity of the reagent. The reagent solutions should be preserved separately and mixed immediately before use.

Procedure: Several milligrams of aldehyde are mixed with 1 ml of water and 2 ml of the reagent, and the mixture is allowed to stand for several minutes in the cold. Metallic silver precipitates, either in the form of a black-gray precipitate or by forming a shiny mirror on the sides of the test tube. The formation of the mirror can be sometimes enhanced by gently heating the test tube, which should be cleaned from grease before use by washing it with alkali. Dilute solutions of aldehydes react only after prolonged standing.

Note: Solutions I and II and ammonia are mixed together immediately before use. After the reaction the mixture should not be allowed to stand for any length of time, because Berthelot's silver fulminate can be formed in it, sometimes leading to spontaneous explosions.

Reaction with Fehling's Reagent

On boiling, aliphatic aldehydes reduce the dark blue Fehling's reagent, precipitating a red copper oxide. Ketones and aromatic aldehydes do not react, if they do not contain other reducing groups. A positive reaction is also given by aldoses and ketoses (which isomerize to aldoses in alkaline medium), by α-ketoalcohols, formic acid, etc.

Reagents: Equal volumes of solutions I and II are mixed shortly before use. Solution I: 25 g of $CuSO_4 \cdot 5\,H_2O$ are dissolved in 500 ml of water. Solution II: 175 g of sodium potassium tartarate and 50 g of NaOH are dissolved in 500 ml of water.

Procedure: 50 mg of sample, or 2 − 3 ml of the solution being tested, are added to 2 ml of the reagent and boiled briefly. In the case of easily volatile substances the mixture is allowed to stand in a closed test tube for a certain time before heating it. In the case of a positive reaction the dark blue solution becomes turbid on heating, precipitating a red copper oxide on boiling.

Instead of Fehling's reagent, the less-alkaline Benedict's reagent may also be used (2).

Reaction with Nessler's Reagent

Nessler's reagent is an alkaline solution of mercuric iodide. Under the influence of aldehydes a red-brown precipitate is formed first which turns gray after a short while from the separated mercury. The reagent gives a similar precipitate with ammonium salts. With ketones a bright yellow precipitate is formed. The same is true of guanidine. The reaction is suitable for the detection of aldehydes in methanol and ether.

Reagent: 6 g of mercuric chloride are dissolved in 50 ml of warm water (80 °C) and precipitated by the addition of a solution of 7.5 g of KI in 50 ml of water. The precipitate of HgI_2 is washed by a fourfold decantation with 20-ml portions of water. Then KI (5 g) is added to the precipitate; the mixture is dissolved in a small volume of water, is transferred into a 100-ml flask, and is mixed with 65 ml of 30% NaOH and diluted with water to make the volume of the solution 100 ml. It is then shaken and allowed to stand for 24 hr, after which the clear supernatant is decanted from the formed precipitate into a brown bottle.

Procedure: Equal parts of the reagent and the tested aqueous solution of aldehyde are mixed and the formation of a gray turbidity or precipitate is observed.

Feder's reagent is similar to Nessler's reagent (3).

Identification of Aldehydes by Their Oxidation to Acids

This method can be used for identification when the acid formed can be isolated easily and when it is solid and has a sharp melting point. This is the case mainly with aromatic aldehydes; for the oxidation potassium permanganate, hydrogen peroxide, silver oxide, and other oxidants can be used. In the case of a mixture of aldehydes or if the acids are too soluble (aliphatic acids), the resulting acids can be identified by chromatography on paper (4) (see also p. 259).

Oxidation of Benzaldehyde with Potassium Permanganate

Reagents: 5% NaOH, saturated aqueous potassium permanganate solution, 20% H_2SO_4, Na_2SO_3, 2 N H_2SO_4, 50% aqueous acetone.

Procedure: Benzaldehyde (1 g) suspended in 15 ml of water is mixed with 0.5 ml of 5% NaOH, and to this solution saturated aqueous potassium permanganate solution is added dropwise with stirring (shaking) until the red-violet color persists. The mixture is then acidified with 25 ml of 2 N H_2SO_4, and Na_2SO_3 solution is added slowly just to decolorize the permanganate and dissolve the precipitated MnO_2. After cooling in a refrigerator the separated acid is filtered off and dried. Yield, 1.1 g; mp, 122−123 °C. After crystalization from 5 ml of 50% aqueous acetone the yield is 1 g, mp 123 °C.

Oxidation of Benzaldehyde with Hydrogen Peroxide on a Macroscale

Reagents: 5% NaOH, 3% hydrogen peroxide, 20% H_2SO_4, chloroform, ethanol.

Procedure: In a 500-ml flask 1 g of benzaldehyde is mixed with 20 ml of 5 % NaOH and 30 ml of 3% hydrogen peroxide and the mixture is heated at 70 °C for 15 min with occasional shaking. A few milliliters of ethanol are then added and the mixture heated at 70 °C for another 10 min. After acidification with sulfuric acid (test with Congo paper) and cooling in a refrigerator the separated acid is filtered off onto a filtration crucible. Yield, 0.7 g; mp, 122 °C.

Oxidation of Benzaldehyde with Hydrogen Peroxide on a Microscale

Reagents: 5% NaOH, 3% hydrogen peroxide, 20% H_2SO_4, chloroform.

Procedure: Benzaldehyde (50 mg), 5 % NaOH (5 ml), and 3% hydrogen peroxide (1 ml) are heated at 70 °C in a 20-ml test tube for 10 min. Additional 3% peroxide (1 ml) is added to the mixture under stirring and the mixture is heated over a water bath at 70 °C for another 10 min. After acidification with sulfuric acid the mixture is extracted twice with 3-ml portions of chloroform, the extract is evaporated, and the crude product purified by sublimation. The medium fraction is used for determination of the melting point, which is 122 °C.

2. Condensation Reactions (Detection and Identification)

Carbonyl componuds condense with many reagents with the elimination of a molecule of water and the formation of colored products. Color reactions serving for the detection of carbonyl compounds can be carried out in the presence of condensing agents or in their absence. If these agents are used (especially sulfuric acid), it should be remembered that many compounds can either set free aldehydes under the conditons of the reaction or can be converted to aldehyde (for example, glycols).

Reaction with Fluorene (Ditz's Reagent)

The reaction is carried out in the presence of concentrated sulfuric acid, yielding diphenylmethane derivatives (5).

Reagents: 1% fluorene solution in aldehyde-free ethanol; conc. sulfuric acid.

Procedure: To a mixture of the tested aldehyde and 1 ml of the reagent in a test tube 1 ml of conc. sulfuric acid is added with caution (along the sides of the test tube). If a colored ring is formed on the boundary of the phases, the reaction is positive. In view of the fact that in certain cases the color is formed slowly, it should be observed after 15 min standing. The reaction is as sensitive as the reaction with Schiff's reagent (p. 217). A similar reaction is also given under similar conditions by some hydrocarbons other than fluorene.

Reaction with Carbazole

Carbazole dissolves in conc. sulfuric acid with a sulfurlike color. In the presence of aldehydes characteristic colors are formed. For example, with traces of formaldehyde the color is bright blue; in the presence of a larger concentration of formaldehyde a precipitate is formed. The presence of nitric acid and other oxidants interferes with the reaction, as they produce a similar color. In this reaction di- or triphenylmethane derivatives are probably formed (6).

The reaction is also given by certain α-hydroxy acids, which are transformed under the influence of sulfuric acid to aldehydes, further hexoses, and lactic acid.

Reagent: 1% solution of carbazole in concentrated sulfuric acid.

Procedure: The reagent (1 ml) is added carefully along the sides of the test tube to the tested aldehyde solution (1 ml) in ethanol or acetic acid. A colored stripe is formed on the phase boundary.

Reactions with Phenols

In this reaction, usually carried out in an acid medium (H_2SO_4, HCl), di- or triarylmethane derivatives are formed which are intensely colored (see p. 47). Pomeroy and Pollard (7) investigated 11 phenolic compounds and their sensitivity with 22 different aldehydes. The reactions with resorcinol, β-naphthol, and phloroglucinol are most sensitive.

For a specific detection of formaldehyde, chromotropic acid (1,8-dihydroxynaphthalene-3,6—disulfonic acid; III) is used.

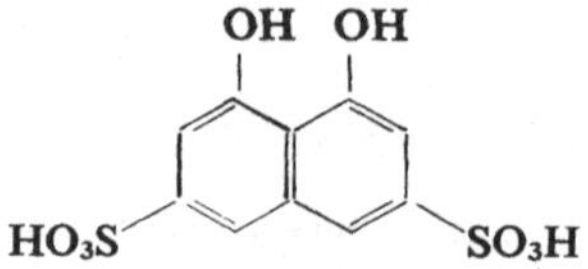

OH OH

HO₃S   SO₃H

(III)

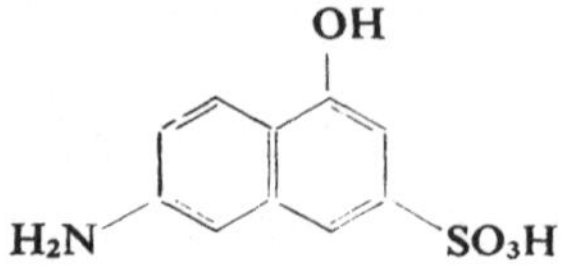

OH

H₂N   SO₃H

(IV)

H₂N   NH₂

O

⊕
C

HO₃S   H   SO₃⊖

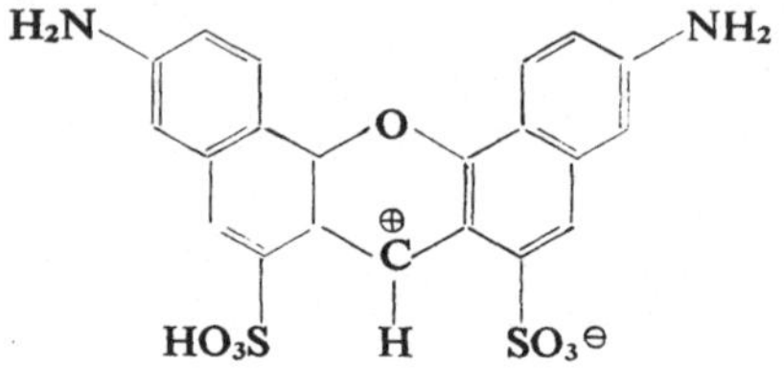

(V)

The reaction also takes place with substances which liberate formaldehyde with sulfuric acid and under the conditions of the reaction (for example, hexymethylenetetramine and dimethylalurea). Similarly, formaldehyde gives a blue color specifically with the I-acid (IV) [giving rise to 3', 3"-diamino-3,4,5,6-dibenzoxanthylium-1,8-disulfonic acid (V) (8)].

Detection of Aldehydes with Resorcinol

Reagents: 1% alcoholic resorcinol solution, conc. H_2SO_4.

Procedure: To a mixture of a few drops of an alcoholic solution of the tested substance and of the reagent 1 ml of conc. sulfuric acid is carefully added. Characteristic colors are formed.

Detection of Aldehydes with Phloroglucinol

Reagent: 1% phloroglucinol solution in concentrated HCl.

Procedure: Equal volumes of the reagent and the aqueous or alcoholic solution of the tested aldehyde are mixed in a test tube and the formation of characteristic color is observed.

Detection of Aldehydes with 1,4-Dihydroxynaphthalene

(Raudnitz's Reagent)

Reagents: 1,4-dihydroxynaphthalene, glacial acetic acid (free of aldehydes).

Procedure: A few milligrams of 1,4-dihydroxynaphthalene are dissolved in 1 ml of acetic acid and mixed with a few milligrams of the tested substance and two drops of concentrated hydrochloric acid. In the presence of an aliphatic or aromatic aldehyde intense red or red-violet color is formed either in the cold or after brief heating.

Detection of Formaldehyde with Chromotropic Acid

Reagent: chromotropic acid, conc. H_2SO_4.

Procedure: To a few drops of a formaldehyde solution 2 ml of sulfuric acid and several crystals of chromotropic acid are added. The mixture assumes a yellow color which changes to pink-violet after heating at 60 °C for 6 min. When small amounts of formaldehyde have to be detected, it is advisable to carry out a control experiment simultaneously.

Reactions with Hydroxy Aldehydes

Vanillin reacts in alkaline media with methyl ethyl ketones, with the formation of a characteristic yellow, orange, or red color (9). It is supposed that alkali salts of corresponding vanillal or divanillal derivatives are formed in this reaction. In addition to ketones, certain aldehydes also react, especially aliphatic ones (with the exception of formaldehyde) if stable in an alkaline medium. Among aromatic aldehydes benzaldehyde (in alcohol) reacts well, while substituted benzaldehydes do not react clearly. Purely aromatic ketones are unreactive.

The reaction of methyl ketones with vanillin or salicylaldehyde can be carried out in an acid medium (H_2SO_4, HCl), but its reaction mechanism is different (10). Recently 2-hydroxy-l-naphthaldehyde (11) was proposed for a very sensitive detection of ketones of general structure RCH_2COCH_2R'.

Detection of Methyl Ketones

Reagents: 1% solution of vanillin in water, 40% sodium hydroxide (for ketones) or 20% sodium hydroxide (for aldehydes).

Procedure: The solutions of vanillin (1 ml) and sodium hydroxide (2 ml) are added to the tested substance (1 − 2 mg) in a test tube, which is then immersed in a boiling water bath, and the color formation is observed.

For water-insoluble substances alcohol is used as solvent.

Reactions with Aromatic Amines

Primary aromatic amines condense in acid media with aliphatic and aromatic aldehydes to give the so-called Schiff's bases:

$$R-CHO + R'NH_2 \rightarrow R-CH(OH)NH-R' \rightarrow R-CH=N-R' + H_2O \xrightarrow{HCl}$$
$$R-CH=\overset{\oplus}{N}H-R' Cl^{\ominus}$$

To carry out this reaction, benzidine, *o*-dianisidine, and N,N-dimethyl-*p*-phenylenediamine are most commonly used (12). Yellow to red colors are formed, and in certain cases colored precipitates are formed. Ketones do not interfere if they are not present in excess.

Reagent: saturated *o*-dianisidine solution in glacial acetic acid.

Procedure: A few milligrams of the tested substance are dissolved in 1 ml of acetic acid mixed with 5 ml of the reagent. Characteristic colors are developed, sometimes changing after heating.

The reaction with benzidine is carried out in the same manner; a 5% benzidine solution in glacial acetic acid is used as the reagent. In this case, too, characteristic colors or precipitates are formed.

Reactions with m-Diamines

Condensation of aromatic *m*-diamines with aldehydes of the general formula $R-CH_2-CHO$ leads to the formation of colored and fluorescing heterocyclic compounds,

$$2 \text{ RCH}_2\text{CHO} \rightarrow \text{RCH}_2\text{CH}=\text{C(R)}-\text{CHO}$$

This condensation can be carried out with aldehydes which have a methylene group in the α-position to the carbonyl group. This feature enables them to be distinguished from aldehydes of a different group. For example, formaldehyde, chloral, isobutyraldehyde, acrolein, glyoxylic acid, benzaldehyde, etc., give a negative reaction. However, certain α, β-unsaturated aldehydes (crotonaldehyde, fural, cinnamaldehyde) give an intense yellow color.

Reagent: *m*-Phenylenediamine hydrochloride (2g) is dissolved in 50 ml of water, and to this solution 50 ml of H_3PO_4 ($\varrho = 1.62$) are added. When kept in darkness the reagent is satisfactorily stable. If it turns pale yellow or becomes fluorescent, it can be purified by filtration with charcoal.

Procedure: To an aldehyde solution in water or in alcohol (blank!) an equal volume of the reagent is added and the formation of a yellow color and green-yellow fluorescence are observed.

Instead of *m*-phenylenediamine, the more sensitive 3,5-diaminobenzoic acid can also be used (13); the reaction also takes place with *m*-toluylenediamine and N,N-dibenzyl-*m*-phenylenediamine (14).

Reaction with Schiff's Reagent

An aqueous solution of fuchsin (*p*-rosaniline hydrochloride) decolorized with sodium hydrogen sulfite or with sodium sulfite becomes red-violet to blue if aldehydes are added to it. The reappearance of this color is not caused by simple oxidation to fuchsin, but represents a specific reaction of the aldehydic group — CHO. However, unsaturated aldehydes and certain aromatic aldehydes (vanillin, *p*-aminobenzaldehyde, *p*-dimethylaminobenzaldehyde) and glyoxal do not react. According to more recent views (15), both amino groups of the fuchsinleukosulfonic acid (VI) react with the

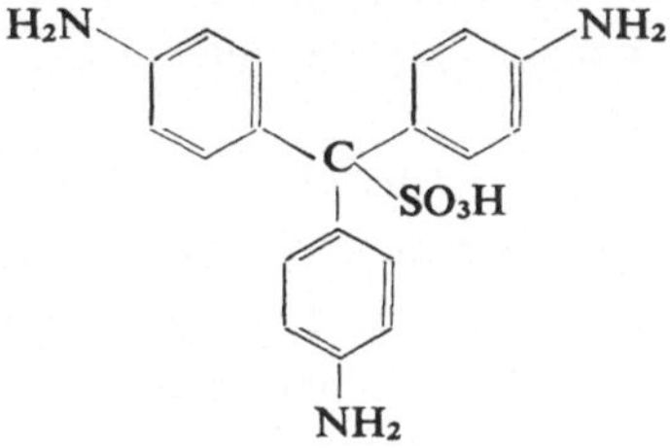

(VI)

aldehyde and the sulfurous acid in a manner similar to that in Mannich's reaction. Upon elimination of sulfurous acid from the central carbon atom, a colored mesomeric cation (VII) is formed.

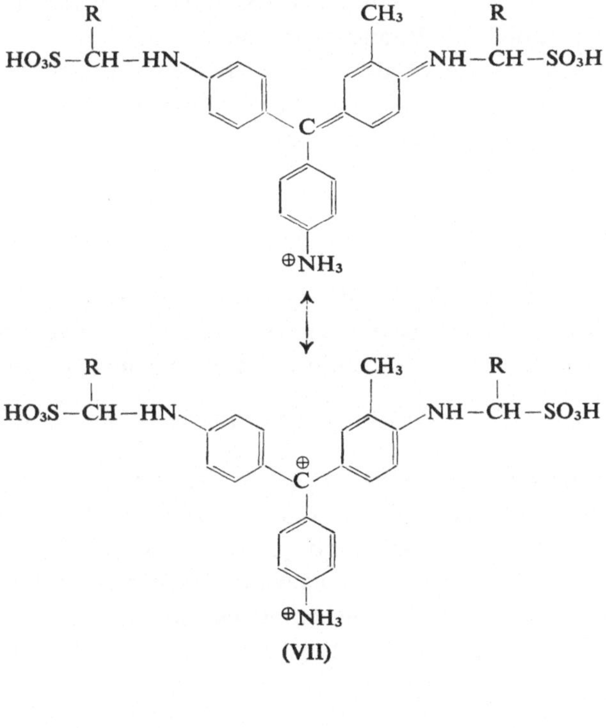

(VII)

Colors formed by aldehydes must be distinguished from the pink color given by numerous substances lacking the −CHO group but capable of binding sulfur dioxide, such as, for example, alkali hydroxides, organic bases, and alkali salts of weak acids. It is therefore necessary to control the pH of the tested solutions and adjust it to 2−3 by addition of an acid or phosphate buffer. In a strongly acid medium formaldehyde alone reacts with the reagent; this feature can be utilized for the detection of formaldehyde in the presence of other aldehydes. The reagent also turns pink on heating. The fact should be kept in mind when vapors escaping during the reaction or distillation are tested. The pink color appearing in such exceptional cases. is of a different hue, lacking the blue shade characteristic of a positive reaction with aldehydes.

Reagent: Fuchsin (0.2 g) is dissolved in 120 ml of warm water, and, after cooling, is mixed with 2 g of anhydrous sodium sulfite dissolved in 20 ml of water, 2 ml of conc. HCl, and enough water to make the volume 200 ml. The reagent is kept in a well-stoppered brown bottle, preferably in darkness. Old reagents loose their sensitivity as a consequence of the escape of sulfur dioxide. If a freshly prepared reagent is brownish, it should be filtered with charcoal.

Procedure: 2 ml of the reagent are added to one drop of the tested substance (or 50 mg of finely ground material) and the change in color of the mixture is observed. In the presence of aldehydes a red to blue color appears. Water-insoluble substances are tested in alcoholic solution. A simultaneous test with alcohol alone should be performed for comparison.

Detection of formaldehyde: 1.2 ml of concentrated sulfuric acid followed by 5 ml of Schiff's reagent are added to 5 ml of a dilute formaldehyde solution. A blue-violet color appears.

Reaction with Arylhydrazines

Carbonyl compounds condense with arylhydrazines with the formation of crystalline arylhydrazones, mostly not very soluble. The reaction is used both for the detection and the identification of carbonyl compounds.

$$R-CHO + Ar-NHNH_2 \rightarrow R-CH=NNH-Ar + H_2O$$

The utilization of phenylhydrazine for the preparation of derivatives is limited because aliphatic aldehydes and ketones give low-melting derivatives, sometimes even liquids. Freshly prepared phenylhydrazones crystallize well, and have sharp melting points within a broad range, but on storing they often darken and their melting points become unsharp. Usually, they can be purified by recrystallization. It is best to dry the recrystallized derivative quickly and to determine its melting point immediately, as phenylhydrazones decompose slowly when exposed to the air for any length of time. The preparations should be kept in darkness with the exclusion of air. Phenylhydrazones of carbonyl compounds which have in the β- or γ-position a hydroxyl or other oxygen-containing functional group cyclize easily. Cyclization is sometimes caused by the presence of a double or triple bond vicinal to the carbonyl group and an amino group in the β-position. Hydrazine reacts with acyl cyanides with the formation of a substituted acylhydrazine (16); some lactones and acid anhydrides also react with phenylhydrazine. For the condensation of carbonyl compounds with phenylhydrazine, the optimum pH of the solution is ~ 5; it is usually adjusted by the addition of alkali acetates. The separation of phenylhydrazones from the solution can be enhanced by the addition of sodium chloride. Many carbonyl compounds react with phenylhydrazine even at room temperature; practically all of them react on heating on a water bath. Less reactive are carbonyl compounds of more complicated natural substances. Hydrazones can occur in two stereoisomeric forms, but not s ofrequently as oximes. The trans form is more common, because of the presence of the large organic residue in the molecule. The original aldehyde or ketone is regenerated by hydrolytic cleavage of phenylhydrazones with hydrochloric acid or by an exchange reaction with formaldehyde or benzaldehyde. For the preparation of phenylhydrazones, ethanol, acetic acid, dioxane, or pyridine is used as solvent.

With 2,4-dinitrophenylhydrazine, aldehydes and ketones give well-crystallized derivatives, which are little soluble in the cold. They can therefore be used with advantage for the isolation of carbonyl compounds from dilute aqueous solutions. Disadvantages in using certain 2,4-dinitrophenylhydrazones arise from the fact that their melting points are within a narrow range (2,4-dinitrophenylhydrazones of: acrylaldehyde 165°; formaldehyde 166°C; acetaldehyde 168 °C; n-caprylaldehyde 104 °C; n-heptaldehyde 108 °C; n-octaldehyde 106 °C; n-undecylaldehyde 104 °C), and their melting points are too high (2,4-dinitrophenylhydrazones of: m-nitrobenzaldehyde 286 °C; vanillin 271 °C; p-nitrobenzaldehyde 320 °C; glyoxal 328 °C).

Most carbonyl compounds react well with 2,4-dinitrophenylhydrazine. In certain cases, however, the reaction cannot be used for identification purposes. The preparation of derivatives of chloral, bromal, methyl vinyl ketone, β-chloropropiophenone. and dimethylpyrone was not successful. The use of concentrated sulfuric acid as the reaction medium during the condensation makes the preparation possible even of derivatives of those carbonyl compounds which react only with difficulty under other conditions (pulegone, fenchone, ionone, p-bromophenacyl bromide). However, the presence of sulfuric acid can cause undesired side reactions. It is also difficult to eliminate the traces of sulfuric acid from the preparation. The literature contains a series of contradictory identification constants, especially melting points, which could lead to erroneous conclusions. It is advisable, therefore, to crystallize the derivatives from two different solvents and also to apply other identification constants, as, for example, mixture melting points, refractive indices of melts, absorption spectra (17), optical crystallographic data (18), X-ray spectra (19) and infrared spectra (20, 21), or paper and thin-layer chromatography (p. 222) (22, 23).

A source of discrepancies is the possibility of formation of geometrical isomers (24 – 29). For example, fural gives two isomers melting at 207 °C and 229 °C, with a mixture melting point of 185 °C (30). It is further important to keep in mind the occurrence of polymorphic modifications, which were observed in the case of formaldehyde, acetaldehyde, propionaldehyde, isobutyraldehyde, butyraldehyde, furfuraldehyde, acetone, methyl ethyl ketone, and methyl n-amyl ketone (31). The formation of mixed crystals was also observed (32). The preparation can be accompanied by rearrangements and side reactions. The rearrangement, accompanied by the formation of a pyrazoline ring, explains the formation of two forms of acetone 2,4-dinitrophenylhydrazone. Carbonyl compounds with other functional groups in the molecule can undergo other reactions (generally cyclization) when reacting with 2,4-dinitrophylhydrazine, and their course is dependent on experimental conditions.

For the crystallization of 2,4-dinitrophenylhydrazones, ethanol, dioxane, toluene, xylene, and ethyl acetate can be utilized. The use of nitrobenzene, pyridine, and glacial acetic acid is recommended only in special cases, for poorly soluble hydrazones. During crystallization from alcohol, esterification of keto acids may take place. The use of ethyl acetate is advantageous because the reagent is soluble in it. Glacial acetic acid is not recommended, because it may react with the unreacted reagent. α-Hydroxy-ketones of the acyloin and benzoin type react very slowly with the reagent. Generally, the presence of the hydroxy group in a carbonyl compound slows down the reaction, while substitution with halogens, methoxy groups, alkyl groups, and double bonds increases the rate of the reaction.

Among newly proposed reagents for the identification of carbonyl compounds, let us mention 3,5-dinitrobenzoylhydrazine (33).

For the detection of carbonyl compounds benzeneazophenylhydrazine-sulfonicacid (34) can be used. Aliphatic aldehydes give a red color, aromatic aldehydes a blue one. With ketones the reaction is less sensitive. Among aldehydes, chloral does not react. The reaction serves for the detection of aldehydes in the presence of esters, alcohols, phenols, amides, amines, and quinones. It leads to the formation of colored benzeneazophenylhydrazones (VIII):

(VIII)

In the analysis of mixtures, or during the identification of small amounts of 2,4-dinitrophenylhydrazones, paper chromatography can be used successfully (35). For preparative separation of 2,4-dinitrophenylhydrazones column chromatography on alumina or calcium sulfate is suitable (36).

Detection of Carbonyl Compounds as Arylhydrazones
Carrying Out the Detection with 2,4-Dinitrophenylhydrazine

Reagent: Saturated solution of 2,4-dinitrophenylhydrazine in 2N HCl.

Procedure: The tested substance (1 – 5 mg) is dissolved in a minimum amount of water and mixed with 1 ml of the reagent. In the presence of carbonyl compounds a turbidity or a yellow to red-yellow precipitate is formed. Some carbonyl compounds (see below) react very slowly and the precipitate is formed only on prolonged standing. The precipitate can be isolated and dissolved in methanol, and the color reaction can be carried out with a methanolic solution of alkali hydroxide. After its addition a red-violet to red-brown color appears. This color reaction is used for the detection of 2,4-dinitrophenylhydrazones on chromatograms.

Detection with Benzeneazophenylhydrazinesulfonic Acid

Reagents: 0.018 g benzeneazophenylhydrazinesulfonic acid (37) dissolved in 100 ml of water; conc. HCl; ethanol; chloroform.

Procedure: The reagent (0.2 ml) and four drops of conc. HCl are added to a drop of the tested solution in a test tube which is then immersed for 30 sec into a boiling water bath and then allowed to cool. A few drops of alcohol are added, followed by 0.5 ml of chloroform and eventually by five drops of HCl. The mixture is thoroughly shaken. In the presence of aliphatic aldehydes the chloroform layer becomes red; in the presence of aromatic aldehydes it assumes a blue color.

Identification of Carbonyl Compounds
in the Form of Arylhydrazones

Benzaldehyde Phenylhydrazone

Reagents: phenylhydrazine (freshly distilled), 50% acetic acid, ethanol.

Procedure: In a 100-ml flask containing 2 ml of phenylhydrazine and 20 ml of water, 50% acetic acid is added dropwise until the dissolution of phenylhydrazine is complete. Benzaldehyde (1 g) is then added followed by ethanol (approximately 50 ml) and the mixture is heated on a water bath until dissolution is complete. The heating is continued for another 10 min on a boiling water bath and the mixture is then allowed to cool. The separated product is filtered off, washed three times with 2-ml portions of 50% acetic acid, and dried in a stream of dry air. Yield, 1.4 g; mp, 155 – 156 °C. Crystallization from benzene (9 ml) gives 1.1 g of the product, mp, 158 – 159 °C.

Acetone 2,4-Dinitrophenylhydrazone (Macroscale)

Reagents: 2,4-dinitrophenylhydrazine (10 g) is dissolved in 130 ml of 80% H_3PO_4 (heating on a water bath) and mixed with 80 ml of 95% ethanol. After cooling, the solution is filtered; ethyl acetate.

Procedure: Acetone (0.5 ml) is dissolved in 10 ml of water and mixed with 50 ml of the reagent. After 3 hr standing the product is filtered off using a filtration crucible. Yield, 2,5 g; mp, 122 – 123 °C. Crystallization from 10 ml of ethyl acetate gave 2.3 g of the product, mp, 124 – 125 °C.

Acetone 2,4-Dinitrophenylhydrazone (Microscale)

Reagents: 2,4-dinitrophenylhydrazine, methanol, conc. HCl, 20% aqueous acetone, ethyl acetate.

Procedure: 2,4-dinitrophenylhydrazine (25 mg), methanol (1 ml), and HCl (0.1 ml) are heated in a 5-ml test tube in a water bath until the dissolution is complete. After cooling, three drops of the 20% aqueous acetone are added and the mixture allowed to stand for 30 min. The separated product is filtered off under suction on a filtration tube. Yield, 25 mg; mp, 125 – 126 °C. Crystallization from 0.5 ml of ethyl acetate gave 14 mg of a pure product, mp 126 °C.

2,4-Dinitrophenylhydrazone of *p*-Bromophenacyl Bromide

Reagents: 2,4-dinitrophenylhydrazine, conc. H_2SO_4, ethanol, ethyl acetate.

Procedure: A solution of 2,4-dinitrophenylhydrazine (1 g) in conc. sulfuric acid (5 ml) is mixed with external cooling with 7.5 ml of water and 25 ml of ethanol. To this solution 1.2 g of *p*-bromophenacyl bromide dissolved in 50 ml of ethanol is added and the mixture allowed to stand for 2 hr. The separated product is filtered off on a filtration crucible; yield, 1.6 g. The melting point is unsharp. After one crystallization from 120 ml ethyl acetate the yield is 0.7 g and the mp 217 – 218 °C; after the second crystallization the yield is 0.6 g, mp 218 °C.

Paper and Thin-Layer Chromatography of 2,4-Dinitrophenylhydrazones

For paper chromatography of 2,4-dinitrophenylhydrazones of aliphatic and alicyclic aldehydes and ketones the solvent systems dimethylformamide/hexane and monomethylformamide/hexane are most suitable. It is best to spot on the paper 5 – 60 µg of 2,4-dinitrophenylhydrazones dissolved in benzene, chloroform, alcohol, or dimethylformamide. Detection is carried out by spraying the chromatogram with 1% alcoholic sodium hydroxide. Derivatives of aldehydes appear as red-brown, ketone derivatives as black-brown spots. The sensitivity of this detection is 1 – 5 µg.

Most of the derivatives of aromatic aldehydes and ketones form streaks on the paper under these conditions. The separation of aliphatic carbonyl compounds is represented in Fig. 32.

Thin-layer chromatography can also be used for the separation and identification of aldehydes and ketones in the form of their 2,4-dinitrophenyl-hydrazones (38 – 40). For simple aliphatic compounds silicagel G and the mobile phase cyclohexane-carbon tetrachloride- ethylacetate (10 : 75 : 15) can be used.

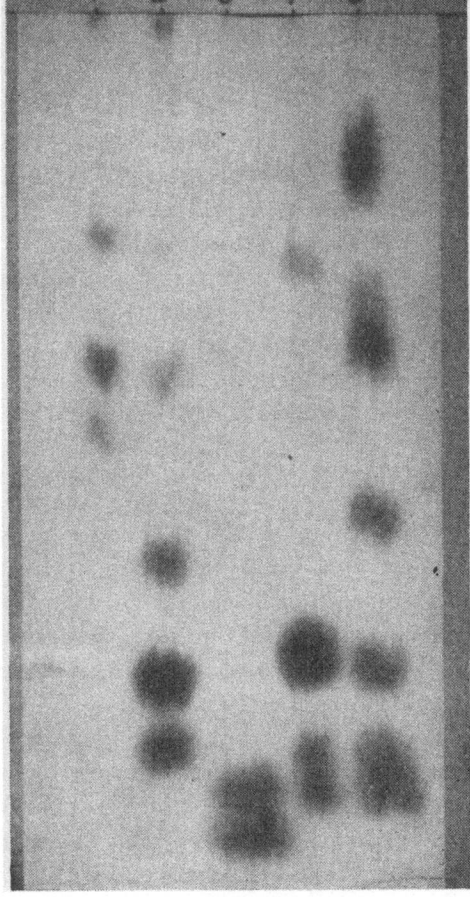

Fig. 32.
Chromatogram of 2,4-dinitrophenyl-hydrazones of carbonyl compounds in 25% dimethylformamide/cyclohexane: (1) formaldehyde, acetaldehyde, and acrolein, (2) formaldehyde, acetalde-hyde, n-butyraldehyde, and n-valer-aldehyde, (3) n-hexylaldehyde and n-dodecylaldehyde, (4) diacetyl (mono-hydrazone), cyclopentanone, cyclo-hexanone, mixture of methylcyclohe-xanones, (5) fural (trans-derivative), fural (cis-derivative), acetaldehyde, acetone, methyl ethyl ketone, diethyl ketone, and methyl isobutyl ketone. Spots near the start are due to 2,4- dinitrophenylhydrazine.

Reaction with Benzenesulfohydroxamic Acid
(According to Angeli-Rimini)

Benzenesulfohydroxamic acid is cleaved with alkali to benzenesulfinic acid and nitroxyl, which reacts with the aldehyde present to give hydroxamic acids:

$$C_6H_5 - SO(OH) = NOH \rightarrow C_6H_5SO_2H + NOH$$

$$R-C\underset{H}{\overset{O}{\diagup}} + NOH \rightarrow R-C\underset{OH}{\overset{NOH}{\diagup}}$$

Hydroxamic acid gives a strong color with ferric chloride. This reaction is suitable for the detection of aldehydes in the presence of ketones. Aromatic p-hydroxyaldehydes, p-aminobenzaldehyde, vanillin, p-dimethylaminobenzaldehyde, salicylaldehyde, and aliphatic aldehydes with a hydroxy group in the γ-position do not react. A positive reaction is observed with benzyl ketones in strongly alkaline media. It is necessary, therefore, to keep the mixture distinctly but not excessively alkaline. This reaction can be utilized for the separation of aldehydes from their mixture with ketones by converting the acid to its insoluble copper salt. Paper chromatography can be used for the identification of aldehydes in mixtures (41).

Reagents: benzenesulfohydroxamic acid (42), 10% aqueous $FeCl_3$, 2 N alcoholic KOH, dilute HCl.

Procedure: Several milligrams of benzenesulfohydroxamic acid are dissolved in 0.5 ml of methanol and mixed with several milligrams of aldehyde and 0.5 ml of 2 N alcoholic KOH. The mixture is heated to the boiling point, and after cooling it is acidified with dilute hydrochloric acid. After the addition of one drop of 10% aqueous ferric chloride solution a wine-red color is produced.

Reaction with Hydroxylamine

The important reaction of the carbonyl group with hydroxylamine has been studied to a wide extent, both its kinetics and its mechanism. The free electron pair on the nitrogen of hydroxylamine is added to the sextet of the carbonyl-group carbon atom, and as the reaction is catalyzed by acids, carbonium ion formed on addition of a proton to the carbonyl oxygen plays a role during the reaction. The addition of hydroxylamine follows, and the reaction is terminated by the splitting off of the proton and a molecule of water. Although acids catalyze the oximation reaction, they also have a reverse influence, because protons are also added to hydroxylamine and form an unreactive ammonium cation. Hence, the optimum reaction course is dependent of a proper choice of pH.

In a strongly alkaline medium the oximation rate increases again, which can be explained by the formation of an anion after the expulsion of a proton from hydroxylamine by a base,

$$NH_2OH + base \rightleftharpoons base\text{-}H + \left[\overline{|O} - \underset{\underset{H}{|}}{\overset{\overset{H}{|}}{N}} | \rightleftharpoons HO - \underset{\underset{}{|}}{\overset{\overset{H}{|}}{N}} | \right]^{\ominus}$$

The anion is a strong nucleophile, and it even reacts with the nonactivated (unprotonated) molecule of the carbonyl compound.

Other conditions being equal, the reactivity of hydroxylamine is higher than that of phenylhydrazine. A comparison of the rates of both reactions with certain ketones is given in Table 8. Aldehydes are oximated at a higher rate than ketones. A number of aromatic ketones cannot be oximated, primarily because of the presence of voluminous groups in the neighborhood of the carbonyl. For example, the preparation of oximes of acetomesitylene, propiomesitylene, and acetoisodurene was unsuccessful. Benzophenone is also oximated with difficulty.

Table 8

Carbonyl compounds	Amount reacted after 1 hr, in %	
	Hydroxylamine	*Phenylhydrazine*
Cyclohexanone	92	40
Acetone	82	66
Methyl ethyl ketone	75	52
Methyl n-propyl ketone	72	38
Diethyl ketone	38	11
Methyl isopropyl ketone	33	15
Acetophenone	8	4

The reaction can be enhanced by heating, but the possibility of subsequent reactions of the formed oximes also increases; in particular, in the case of ketoximes, the so-called Beckmann rearrangement may take place (43). Carbonyl compounds reacting slowly in an acid medium give oximes in an alkaline medium (camphor).

Unsaturated carbonyl compounds can add hydroxylamine to the double bond (44). Benzalacetonphenone reacts with hydroxylamine in an alkaline medium with the formation of a series of reaction products (45).

Oximes of some carbonyl compounds can undergo further cyclization reactions (46). β-Diketones give, for example, heterocyclic compounds. The oxime of propargylaldehyde cannot be isolated because a ring formation takes place and an oxazine is formed. The reaction of hydroxylamine with aromatic α-aminoalkyl ketones is similar to osazone formation (47). Chloral reacts with two molecules of hydroxylamine (48). Carboxylic acids or lactonic oxygen can also react with hydroxylamine (49).

α-Diketones give both monoximes and dioximes (glyoximes) with this reagent. Glyoximes and their alkaline salts are colorless, but with the salts of nickel, divalent iron, and palladium they form intensely colored inner complexes. Exceptions are benzil, camphoquinone, and phenanthrenequinone. Compounds with the groups $-COCH_2OH$, $-CHOH-CH_2-$, and $-CH_2CO-$ can be detected in this way after their oxidation with selenium dioxide (50).

Some oximes are unstable and they can be crystallized only in the presence of hydrochloric acid and an excess of hydroxylamine hydrochloride (51); even pure oximes can sometimes decompose spontaneously. On the oximation of aldehydes with hydroxylamine hydrochloride in alcohol, higher aldehydes may get acetalized, or else the acid formed may cause the polymerization of aldehydes (52).

In contrast to hydrazones, oximes are amphoteric, and hence also soluble in alkali. They can be recuperated from alkaline solutions by saturation with carbos dioxide. Oximes of carbonyl compounds with a lower number of carbons can be distilled under reduced pressure without decomposition.

The detection of carbonyl compounds is based on their reaction with hydroxylamine hydrochloride, during which oxime is formed and hydrogen chloride liberated. The course of the reaction can be followed by the addition of an indicator. The reaction is not given by higher ketones (benzophenone) and carbonyl compounds reacting slowly with the reagent. The test is negative with sugars and quinones.

Reagent: Hydroxylamine hydrochloride (50 mg) is dissolved in 10 ml of ethanol and mixed with 2−3 drops of a universal indicator. The pH of the solution is then adjusted to 3.5−4 by careful addition of 0.1 N sodium hydroxide.

Procedure: A drop of the indicator is added to the tested substance (25−50 mg) dissolved or suspended in 0.2 ml of methanol. Into a second test tube 0.2 ml of the reagent is pipetted. The pH of the sample is then adjusted to match the pH of the reagent by dropwise addition of 0.1 N acid or alkali, and the two solutions are mixed. After 1 min the change in color is observed. If the indicator becomes red, the reaction is positive.

Detection of Diketones

Reagents: Hydroxylamine hydrochloride (1 g) and sodium acetate (1 g) are dissolved in 2 ml of water; 5 % nickel acetate solution.

Procedure: One drop of solution of α-diketone and several drops of hydroxylamine hydrochloride are mixed in a test tube and heated on a water bath for several minutes. One drop of the solution obtained is then transferred on a piece of filter paper and a drop of the nickel acetate solution is spotted in its immediate vicinity. Either immediately or after exposure to ammonia vapors a yellow or red color is produced on the boundary of the two drops.

Identification of Carbonyl Compounds in the Form of Oximes

Oxime of *p*-Nitrobenzaldehyde (Macroscale Procedure for Oximation in Pyridine)

Reagents: hydroxylamine hydrochloride, pyridine, ethanol.

Procedure: In a 50-ml flask a mixture of 1 g of *p*-nitrobenzaldehyde, 2 g of hydroxylamine hydrochloride, 5 ml of pyridine, and 10 ml of ethanol are refluxed for 1 hr. The solvent is then distilled off under reduced pressure and the residue mixed with 5 ml of water. The oxime is obtained by filtration under suction using a filtration crucible. Yield, 0.8 g; mp, 126–127 °C. Crystallization from 10 ml of ethanol yielded 0.7 g of the product, mp 129–130 °C.

Oxime of *o*-Nitrobenzaldehyde (Macroscale Procedure for Oximation with Free Hydroxylamine)

Reagents: hydroxylamine hydrochloride, N NaOH, 50% ethanol.

Procedure: Hydroxylamine hydrochloride (2 g) dissolved in 10 ml of water is mixed with 20 ml of N NaOH and 1 g of *o*-nitrobenzaldehyde, the mixture is heated gently in a water bath, and ethanol is then added until the carbonyl compound is dissolved. After 15 min heating over a boiling water bath the mixture is cooled with ice and the crystallization is enhanced by scratching with a glass rod. The separated oxime is isolated by filtration (filtration crucible). Yield, 1.1 g; mp, 90–91 °C; after crystallization from 3.5 ml of 50 % ethanol the yield is 1.0 g, mp, 92–93 °C.

Oxime of Camphor (Macroscale Procedure for Oximation in a Strongly Alkaline Medium)

Reagents: hydroxylamine hydrochloride, NaOH, ethanol.

Procedure: Hydroxylamine hydrochloride (2 g) and camphor (1 g) in 25 ml of ethanol are dissolved in a flask fitted with a reflux condenser, and 5 g of solid NaOH are added. The mixture is heated on a boiling water bath for 2 hr. The reaction mixture is then poured into 120 ml of water, stirred for several minutes, and filtered, if necessary. The filtrate is neutralized with hydrochloric acid (test with Congo paper), and after 2 hr standing in a refrigerator the separated oxime is filtered off under suction. Yield, 0.7 g; mp, 113−114 °C. Crystallization from 5 ml of ethanol yielded 0.6 g of the product, mp 114−115 °C.

Reaction with Semicarbazide

Semicarbazide is suitable primarily for the identification of lower-molecular-weight, water-soluble ketones. The preparation of semicarbazones of lower aldehydes presents certain difficulties. For example, formaldehyde does not give a crystalline product with semicarbazide even after 10 days standing at room temperature, and if the reaction mixture does not contain sodium acetate, an amorphous condensation product is formed. Acetaldehyde reacts slowly with semicarbazide, and the semicarbazone formed is so soluble that it can be isolated only with difficulty. Aldehydes $C_2 - C_{12}$ yield derivatives with melting points too close to allow reliable identification (semicarbazone of: formaldehyde 169 °C; acetaldehyde 162 °C; acrolein 171 °C; butyraldehyde 106 °C; isovaleraldehyde 107 °C; n-hexylaldehyde 106 °C; n-heptylaldehyde 109 °C; n-octylaldehyde 101 °C; n-decylaldehyde 102 °C). Complications may also be met with during the preparation by prolonged standing or heating of the reaction solution. By-products are formed, especially acetylsemicarbazones and hydrazodicarbonamide: $NH_2CONHNHCONH_2$. This compound can be formed during the preparation of semicarbazones of slowly reacting ketones and it can be a source of error. The mentioned diamide melts at 245−250 °C and is poorly soluble in alcohol, but well soluble in hot water, in contrast to other semicarbazones whose melting points are in the same range. It was found that the melting points of semicarbazones are strongly dependent on the rate of heating; differences of up to 10 °C were found in melting points, depending on whether they vere determined in a normal manner in a capillary or by rapid heating on a Maquenne block. On faster heating, higher melting points are observed which are also easier to reproduce. In the preparation of semicarbazones of α-diketones cyclization takes place with the formation of substituted pyrazoles.

Reactions of carbonyl compounds with semicarbazide, similar to the majority of condensation reactions of this type, are catalyzed by acids. On the other hand, semicarbazide also has a free-electron pair in its molecule capable or accepting a proton, and this makes it unsuitable as a nucleophilic agent.

The addition of acid therefore increases the reactivity of the carbonyl group, but at the same time the concentration of the reagent available for the reaction decreases. At a certain pH the reaction is fastest and its rate drops in more acid or more basic media.

Identification of Carbonyl Compounds
in the Form of Semicarbazones

Semicarbazone of Acetone (A Macroscale Procedure for Carbonyl Compounds Soluble in Water)

Reagents: semicarbazide hydrochloride, sodium acetate.

Procedure: In a 50-ml flask 1 ml of acetone, 1.5 g of sodium acetate, and 1 g of semicarbazide hydrochloride in 10 ml of water are dissolved and refluxed on a boiling water bath for 1 hr. After cooling, the mixture is allowed to stand in a refrigerator, and the derivative is filtered off using a filtration crucible and washed with 3 ml of water. Yield, 0.6 g; mp, 176 – 177 °C. Crystallization from 3 ml of water gave 0.3 g of the product, mp, 178 – 179 °C. A second crystallization from 3 ml of water gave 0.2 g of the product, mp 179 °C.

Those semicarbazones which do not crystallize well can be obtained by scratching the test tube with a glass rod. Crystallization can also be carried out from aqueous ethanol.

Semicarbazone of Benzaldehyde (A Macroscale Procedure for Carbonyl Compounds Insoluble in Water)

Reagents: semicarbazide hydrochloride, sodium acetate, ethanol, 50% ethanol.

Procedure: Water is added dropwise to a solution of benzaldehyde (1 ml) in 10 ml of ethanol until turbidity is produced. The turbidity is eliminated by the addition of several drops of ethanol and the solution is mixed with 1 g of semicarbazide hydrochloride and 1.5 g of sodium acetate. Further treatment is similar to that described in the preceding procedure. Yield, 1.3 g; mp, 216 °C. One crystallization from 65 ml of 50% ethanol gave 1.2 g of a product melting at 219 °C; the second crystallization from 50 ml of 50% ethanol gave 1.1 g of the product, mp 219 °C.

Semicarbazone of Methyl Ethyl Ketone (Semimicroscale Procedure)

Reagents: semicarbazide hydrochloride, sodium acetate, 50% aqueous ethanol.

Procedure: Into a test tube of ∼ 8 ml volume, drawn to a capillary on one end (so that it can be sealed at the appropriate moment), 50 mg of semicarbazide hydrochloride, 80 mg of sodium acetate, and 1 ml of 50% ethanol are introduced and the mixture is dissolved by gentle heating. After cooling, the ketone (50 mg; 0.05 ml) is added, and the tube is sealed and immersed into an 80 °C water bath for 30 min. After cooling, the tube is opened and the contents are concentrated to half their volume over a water bath. After standing in a refrigerator for 30 min and after centrifugation, the supernatant is sucked off with a pipette. The separated crystals are washed twice with two drops of cold water and crystallized from 0.5 ml of water. The product is filtered under suction using a filtration tube and dried. Yield, 12 mg; mp, 141 – 142 °C. After recrystallization from 0.2 ml of water the yield is 5 mg, mp 142 °C.

Reaction with Dimedone

5,5-Dimethylcyclohexane-1,3-dione (IX) (dimethyldihydroresorcinol, dimedone, dimethol, methone) gives, with aldehydes, good crystalline derivatives

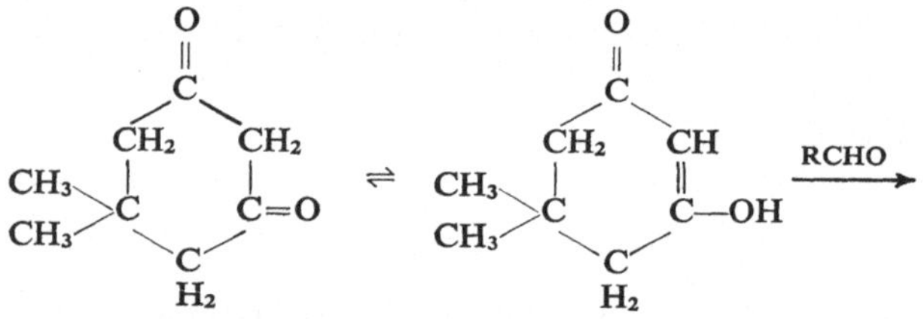

(IX)

the melting points of which range over a wide temperature interval; in particular, the lower-molecular-weight members of the series differ appreciably in their melting points. The melting points of formaldehyde and acetaldehyde derivatives are 191 °C and 141 °C, respectively. Aldimethones can be easily converted to anhydrides (1,5-oxides) – octahydroxanthenes – and a second series of derivatives thus obtained. As the reagent does not react with ketones, it can be used with advantage for the detection and identification of aldehydes in mixtures with ketones.

The reaction of this reagent with aldehydes occurs in two steps; in addition, in acetic acid, water is eliminated and the anhydride (XII) is formed.

The intermediary product (X) is substantiated not only by the fact

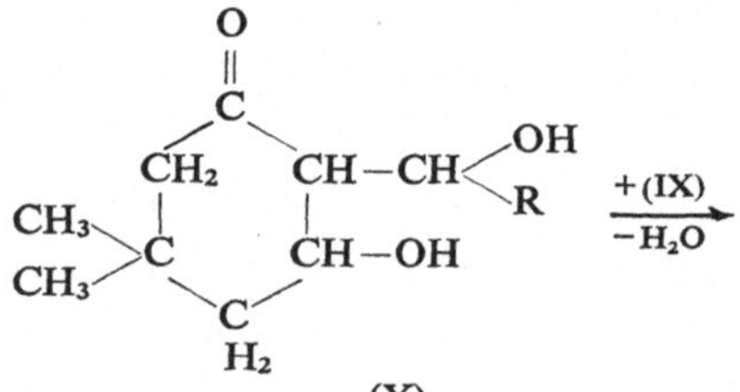

(X)

that a trimolecular reaction in which substance (XI) would be formed directly from two molecules of dimedone is improbable, but also because a substance of the type X was prepared from trichloroacetaldehyde and dimedone.

The reaction is pH-dependent; the reaction with acetaldehyde takes place best and quantitatively at pH 4.

The procedure is especially suitable for the identification of small amounts of aldehydes from very dilute solutions. The reagent is easily accessible and the condensation can be carried out under various conditions; aqueous solutions of both components are mixed together and either allowed to stand in the cold for several days, or they can be refluxed for several hours. However, a prolonged boiling can sometimes lead to dehydration of the product, i.e., to the formation of the anhydride. Aldehydes insoluble in water can be condensed in alcoholic media. This solvent is generally more suitable for the preparation of derivatives of aldehydes with a longer carbon-atom chain, because in water, gummy products are formed.

β-Hydroxy aldehydes are dehydrated during the reaction and give rise to derivatives of the corresponding α,β-unsaturated aldehyde (for example, with dimedone, aldol gives a derivative of crotonaldehyde). The use of aqueous solutions is advantageous, because the reagent is more soluble in water than the condensation products, so that an easy separation is possible. In addition to crystallization, the products can be purified by sublimation. Sublimation can sometimes be utilized for the separation of a mixture of aldehydes.

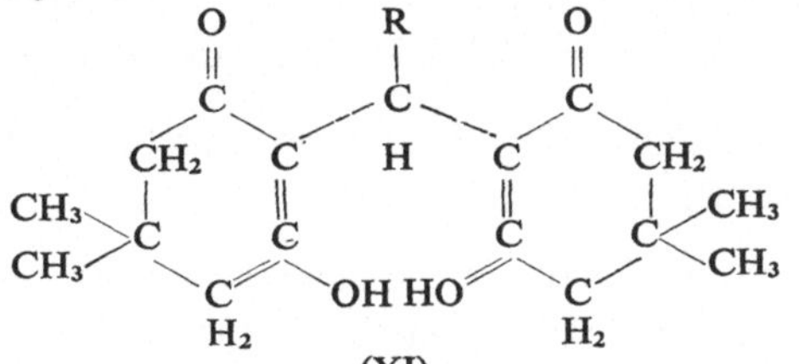

(XI)

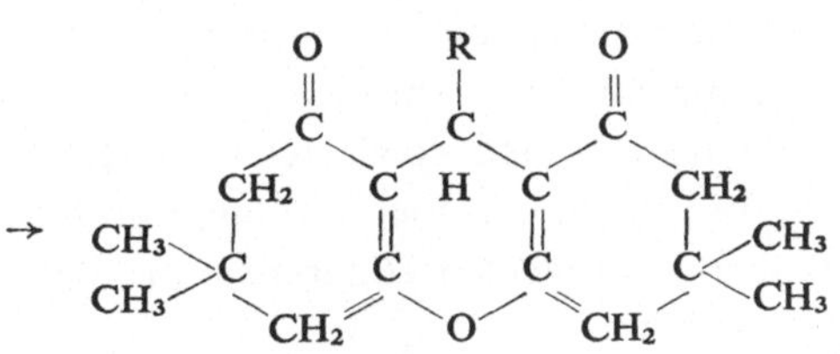

(XII)

Octahydroxanthenes (XII) are prepared either directly by carrying out the reaction in glacial acetic acid, or by dehydrating the products prepared in water by heating them with dilute sulfuric or hydrochloric acid. In addition to melting points, optical and crystallographic constants can also be used for identification purposes (54).

Identification of Carbonyl Compounds in the Form of Dimedone Derivatives

Methone of Formaldehyde (Microscale Procedure)

Reagents: 0.2% aqueous dimedone solution, 30% aqueous methanol.

Procedure: To 10 ml of the reagent in a test tube 1 ml of 0.2% formaldehyde (2 mg) is added and the tube stoppered with a rubber stopper. After standing overnight the separated derivative is filtered on a filtration tube and washed twice with 0.3 ml of 30% methanol. Mp 191 °C; after sublimation, mp 193 °C.

Methone of Crotonaldehyde (Semimicroscale Procedure)

Reagents: dimedone, 50% aqueous methanol, 80% aqueous ethanol, 30% aqueous methanol, piperidine.

Procedure: In a 10-ml flask fitted with a reflux condenser a mixture of 300 mg of dimedone, 3 ml of 50% methanol, 50 mg of crotonaldehyde, and a drop of piperidine is heated on a boiling water bath for 10 min. The hot solution is diluted with water until a persistent turbidity is formed and then allowed to cool in a refrigerator. The separated oily mixture solidifies after longer standing and scratching with a glass rod. The derivative is filtered off on a filtration tube and washed twice with 0.5 ml of 30% aqueous methanol. Yield, 59 mg; mp, 181 °C. Crystallization from 1 ml of 80% aqueous ethanol gives 43 mg and raises the mp to 185 − 186 °C.

Formation of the Anhydride of Crotonaldehyde Methone (Semimicroscale Procedure)

Reagents: conc. HCl, 80% aqueous methanol, 30% aqueous methanol.

Procedure: Crotonaldehyde methone (25 mg) dissolved in 1 ml of 80% methanol is mixed with one drop of concentrated HCl and heated in a small flask fitted with a reflux condenser for 5 min. Water is added to the warm solution until turbidity is formed, and after 30 min of standing in a refrigerator the anhydride formed is collected on a filtration tube and washed twice with 0.5 ml of 30% methanol. Yield, 22 mg; mp, 195 °C. Crystallization from 1 ml of 80% methanol yielded 16 mg of a product, the melting point of which was 201 − 202 °C.

Methone of Crotonaldehyde (Macroscale Procedure)

Reagents: dimedone, 95% ethanol.

Procedure: About 1 g of crotonaldehyde is added to 20 ml of ethanol in a 50-ml flask, followed by 4 g of dimedone. The mixture is refluxed for 1 hr and then cooled in a refrigerator. The separated product is collected on a filtration crucible (suction). Yield, 2.6 g; mp, ~ 169 °C. Crystallization from 15 ml of ethanol and water added to incipient precipitation gave 1 g of the product, mp, 189 °C.

Dehydration of Crotonaldehyde Methone
(Macroscale Procedure)

Reagents: conc. HCl, 30% aqueous methanol, 80% aqueous methanol.

Procedure: In a 20-ml flask 250 mg of crotonaldehyde methone dissolved in 10 ml of 80% methanol is mixed with one drop of conc. HCl and boiled for 20 min. Water is then added until the solution becomes turbid, and the mixture is allowed to stand in a refrigerator for 1 hr. The separated products are then filtered and washed twice with 0.5 ml of 30% methanol. Yield, 170 mg; mp, 197 °C. After crystallization from 80% methanol the yield is 100 mg, mp, 199 – 202 °C.

Reaction of Methyl Ketones
with *o*-Nitrobenzaldehyde (Conversion to Indigo)

Methyl ketones $RCOCH_3$ react with *o*-nitrobenzaldehyde in alkaline media with the formation of indigo,

Among aldehydes, acetaldehyde is quite reactive. This reaction makes it possible to distinguish methyl ketones from methylcarbinols (both classes of substances give a positive iodoform reaction). As the reaction is carried out in an alkaline medium, certain substances are hydrolyzed and their products then give a positive test. Such substances are, for example, geminal dihalo derivatives (1,1-dichloroethane, 2,2-dibromopropane, 2,2-dibromobutane) and certain methyl derivatives (oximes, addition compounds with hydrogen sulfite).

Reagents: *o*-Nitrobenzaldehyde (5 g) is dissolved in 100 ml of ethanol and the solution is kept in a brown bottle for a maximum of one month; 10% NaOH.

Procedure: The sample ($\sim$ 50 mg) is dissolved with stirring and heating to 50 °C in 2 – 3 ml of the reagent, and 1 ml of 10% NaOH is added to the solution. In the presence of methyl ketones the color of the solution changes through yellow and green to blue. With some ketones a blue precipitate is formed. Sometimes it is necessary to heat the mixture again after addition of alkali. After cooling, the blue color is removed with chloroform.

Reaction of Methyl Ketones with m-Dinitro Compounds

This reaction is based on the formation of a red to red-violet color taking place in the reaction of *m*-dinitrobenzene with methyl ketones in an alkaline medium.

More recently a color reaction with 2,2'-dinitrodiphenyl in dimethylformamide was proposed (55) for the detection of dialkyl and diaryl ketones of the general formula $RCOCH_2 - R'$. The mechanism of this reaction is not clear as yet.

Reagents: 1% alcoholic *m*-dinitrobenzene, 5 N NaOH.

Procedure: To a dilute ethanolic solution of the ketone (2 ml) 4 ml of 1% ethanolic *m*-dinitrobenzene solution and 4 ml of 5 N NaOH are added. A red color is produced.

3. Other Color Reactions (Detection)

Reaction with Sodium Nitroprusside (Legal Reaction)

This reaction is based on the formation of a color in a mixture of an aqueous sodium nitroprusside solution with carbonyl compounds in an alkaline medium. A positive test is obtained with aliphatic and certain aromatic aldehydes and aromatic and aliphatic ketones. The literature data on this reaction are conflicting and no general rule can be expressed about which carbonyl compounds give the reaction and which do not. For example, the color is not produced with formaldehyde, glyoxal, benzaldehyde, o-hydroxybenzaldehyde, chloral, vanillin, benzophenone, naphthyl phenyl ketone, trihydroxybenzophenone, benzil, acetophenones substituted on the aromatic nucleus with hydroxyl, etc. The reaction can therefore be used for the differentiation of certain types of aldehydes and ketones, as, for example, acetaldehyde from formaldehyde, and acetophenone from benzophenone or hydroxyacetophenone, etc. Further, various colors formed in alkaline media or after the subsequent acidification of the reaction mixture can also be used for differentiation. Sodium hydroxide can be replaced by ammonia, piperidine

(56), aliphatic amines (57), ethanolamines, or piperazine (58). The colors produced differ for different adehydes and ketones. This can also be utilized for differentiation. For example, in the presence of piperidine, acetaldehyde gives a blue color, propionaldehyde an orange one. The reaction is not specific and is also obtained with sulfur compounds.

Detection of Ketones

Reagents: 1% sodium nitroprusside solution, 5% NaOH, acetic acid.

Procedure: To several ml of an aqueous or alcoholic ketone solution 1 ml of the reagent is added and carefully alkalized with a few drops of 5% NaOH. After shaking the mixture, a red or red-violet color is produced which turns blue after addition of $10-15$ drops of acetic acid.

Detection of Acetaldehyde

a) *Reagents:* saturated aqueous solution of piperazine hydrate, 4% aqueous sodium nitroprusside solution.

Procedure: To an aqueous solution of acetaldehyde (5 ml) cooled with ice are added 1.5 ml of piperazine hydrate solution and 0.5 ml of freshly prepared 4% sodium nitroprusside solution. A red color is produced.

b) *Reagent:* Sodium nitroprusside (0.2 g) is dissolved in a mixture of 8 ml of water and 2 g of piperidine. To carry out the reactions, a strip of filter paper impregnated with this solution is employed.

Procedure: The reagent paper is hung in a test tube with the tested solution and the solution is mildly warmed. In the presence of acetaldehyde the paper strip assumes a blue color.

Reaction with Hypoiodite [Iodoform Reaction (59, 60)]

Compounds containing CH_3CO-, CH_2ICO-, and CHI_2CO- groups linked with a hydrogen or carbon atom (the latter should not carry a strongly activated hydrogen atom, nor groups causing a strong sterical hindrance) react with an alkaline hypoiodite solution to give iodoform,

$$\overset{O}{\overset{\|}{R-C}}-CH_3 + 3\,NaIO \;\rightarrow\; RCOONa + CHI_3 + 2\,NaOH$$

In this reaction substitution with iodine first takes place in the methyl group of the ketone present in the solution with the formation of triiodomethyl ketone. The mechanism is probably the following:

$$R-CO-CH_3 + OH^\ominus \;\rightarrow\; R-CO-CH_2^\ominus + H_2O \text{ (slow)}$$
$$R-CO-CH_2^\ominus + HIO \;\rightarrow\; R-CO-CH_2I + OH^\ominus \text{ (fast)}$$

etc.

The strong inductive effect of iodine in $R-CO-CH_2I$ facilitates the subsequent fast iodination to $R-CO-CI_3$, which is then rapidly hydrolyzed:

$$R-CO-CI_3 + OH^\ominus \rightarrow RCOO^\ominus + CHI_3$$

As the carbanion is formed by a slower reaction, its rate of formation is dependent on the concentration of the base. The reagent, and hence the hydroxyl ions too, should always be in excess. The reaction can be accompanied by disproportionation,

$$3\,NaIO \rightarrow 2\,NaI + NaIO_3$$

which decreases the concentration of the hydroxyl ions. This reaction can take place more rapidly when the ketone is halogenated slowly.

The reaction is also observed with those compounds which under the given conditions yield compounds with a positively reacting group (for example, secondary alcohols are oxidized to ketones).

A review of compounds on which the reaction was tested by various authors is given in Table 9 [see also (61)].

From present knowledge it can be inferred that a positive test with compounds containing only C, H, and one carbonylic oxygen atom means the presence of the CH_3CO- group, and, in a case where the only oxygen atom belongs to a hydroxy group, a CH_3CHOH- group. A negative test with carbonyl compounds containing only C, H, and one O atom is indicative either of the absence of a CH_3CO- group in the molecule or of the fact that it is deactivated electronically or sterically.

Recently, cyanogen iodide was proposed as being a reagent which reacts more slowly and more selectively (60).

Semimicromethod

Reagents: 10% NaOH, 10% solution of iodine in 20% potassium iodide solution, dioxane.

Procedure: To the sample (100 mg) dissolved in 1 ml of water (insoluble substances should be dissolved in methanol or dioxane) are added 3 ml of 10% NaOH and then, dropwise, a 10% iodine solution in 20% aqueous potassium iodide until it is in excess. The test tube is then immersed in water at 60 °C. Additional iodine solution is added until its color persists for 2 min followed by dropwise addition of alkali until the brown color fades. The test tube is taken out of the water bath and 10 ml of water are added to it. Iodoform separates as a yellow solid of characteristic odor, the melting point of which is 120 °C. If the separated iodoform is not lemon-yellow, it is separated by filtration, suspended in 2−3 ml of dioxane, and shaken with 1 ml of 10% NaOH.

Results of the Iodoform Reaction with Various Types of Organic Compounds

Table 9

Positive	Negative
Alcohols	
Ethanol,* isopropanol, sec-butanol, sec-amyl alcohol, methyl-*n*-amylcarbinol, 2-octanol, methyl-isopropylcarbinol, 2,3-butanediol, methylbenzylcarbinol	Methanol, allyl alcohol, trimethylene glycol, mannitol, isobutanol
Aliphatic ketones	
Acetone, methyl ethyl ketone, methyl propyl ketone, methyl isobutyl ketone, methyl *n*-amyl ketone, 2-heptanone, 2-octanone, methyl isohexyl ketone, 4-methyl-2-heptanone, methyl cyclohexyl ketone, methyl-γ-phenoxypropyl ketone, benzylacetone, benzohydrylacetone, 2--phenyl ethyl methyl ketone	
Mixed ketones	
Acetophenone, methyl-*p*-tolyl ketone p-chloroacetophenone, *p*-bromoacetophenone, methyl *p*-anisyl ketone, 2,4-dimethoxyacetophenone, 2-methyl-4-methoxyacetophenone, 5-methyl-2-methoxyacetophenone, acetocumene, 2,5-trimethylacetophenone, *o*-, *m*-, *p*-hydroxyacetophenone, 2,4-dihydroxyacetophenone,* 3-methoxy-4-hydroxyacetophenone, *o*-nitroacetophenone, m-nitroacetophenone,* *p*-nitroacetophenone, *o*-, *m*-, *p*-aminoacetophenone, 2-aceto-1-naphthoxyacetic acid, 2-aceto-4-bromo-1-naphthoxyacetic acid	α-Chloroacetophenone, propiophenone, acetomesitylene, 3,5-dinitroacetomesitylene, 2,4,6-tribromoacetophenone, 3-amino-2,4,6-tribromoacetophenone, 1-aceto--2-naphthoxyacetic acid, 2-methoxy-1--acetonaphthone
Unsaturated ketones	
Mesityl oxide, benzylacetone, 2-methyl--1-phenyl-1-buten-3-one, furfuralacetone	

Table 9 (continued)

Positive	Negative
Diketones	
Acetylacetone, acetonylacetone, benzoyl-acetone, p-bromobenzoyl-acetone, diben-zoylmethane, 1,3-diketohydrindene, 2,6-dimethyl-4-acetylacetophenone	ω-Acetylacetomesitylene, ω-benzoylaceto-mesitylene, di(β-isoduryloyl) methane)
Various	
Acetoxime, diacetylmonoxime, α-phenyl-ethylamine, p-benzoquinone, 2-hydroxy--1,4-naphthoquinone, 1,4-naphthoquino-ne,* compound of the following type	2-Pentene, 1,1-diphenyl-1-propene, 1--chloro-2,3-dihydroxypropane, propioni-trile, isoeugenol, phenylacetylene, rham-nose, acetophenone oxime (gives a posi-tive test after prolonged heating), anethol

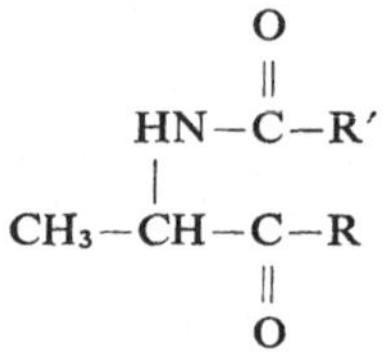

$$
\begin{array}{c}
\quad\quad\quad\ \ O \\
\quad\quad\quad\ \ \| \\
\quad\ \ HN-C-R' \\
\quad\quad\ | \\
CH_3-CH-C-R \\
\quad\quad\quad\ \| \\
\quad\quad\quad\ O
\end{array}
$$

$R = C_2H_5, \ CH_2OCH_3$

$R' = C_2H_5, \ CH_3$

Data in the literature concerning the reaction of resorcinol, hydroquinone, and phloro-glucinol are not consistent.

---

* Slow reaction.

Micromethod

Reagents: 0.1 N aqueous iodine solution in a minimum amount of potassium iodide; 2 N sodium hydroxide.

Procedure: A small amount of the sample (not necessarily dissolved in water) is mixed with 3 ml of iodine solution. Then 2 N sodium hydroxide is added until the color of iodine disappears. A yellow precipitate of char-acteristic odor indicates the formation of iodoform. If the precipitate is not produced immediately, the mixture is allowed to stand in the presence of excess iodine until the color is stable.

Reaction of *p*-Phenylenediamine
with Hydrogen Peroxide, Catalyzed with Aldehydes

In a neutral or acid medium *p*-phenylenediamine is autooxidized to compound (XIII); this reaction is catalyzed by the presence of aldehydes.

(XIII)

This reaction can be used for the detection of aldehydes in the presence of ketones and for the differentiation of aliphatic aldehydes from aromatic ones. In a neutral medium aldehydes produce a black color or a precipitate; other transient colors are sometimes formed which, in the case of aromatic aldehydes, last a bit longer. In an acid medium aliphatic aldehydes behave similarly as in a neutral medium, while in the presence of aromatic aldehydes a yellow color is produced first; sometimes a precipitate is formed which persists for a certain time. Some other derivatives also react similarly to aldehydes, such as, for example, cyanohydrins, aldehyde-ammonia adducts, sodium bisulfite addition compounds of aldehydes, and oximes. For modification with detection tubes containing the reagent see (62).

Reagents: 2% aqueous *p*-phenylenediamine solution, 2 N acetic acid, 3 % hydrogen peroxide.

Procedure: One drop of a 2% *p*-phenylenediamine solution is put on a spot-test plate, followed by two drops of 2 N acetic acid, two drops of 3 % hydrogen peroxide, and one drop of the tested solution. Depending on the quantity of the aldehyde present, the color is produced either immediately or after standing for a while. A blank is performed simultaneously in the absence of acetic acid, because certain aldehydes react more rapidly in acid and others in neutral media.

Reactions with *o*-Diamines

o-Dicarbonyl compounds (for example, α-diketones, α-aldehydoketones, glyoxylic acid, benzoin, *o*-quinones, α-keto acids) condense with *o*-phenylenediamine (63) with the formation of corresponding quinoxalines:

The compounds formed are predominantly poorly soluble, they crystallize well, and are colored or fluorescent. As reagents, *o*-phenylenediamine, *o*-toluylenediamine, *o,o'*-diaminobenzidine, etc., are used. β-Diketones also react, but they give heptazines (64) which are red in acid medium. d,1-1,2-Dianilino-1,2-diphenylethane reacts rapidly and specifically with aldehydes (not with ketones) according to the following equation to give substituted d,1-1,3,4,5-tetraphenylimidazolines (65):

$$C_6H_5-CH-NHC_6H_5 \\ \underset{C_6H_5-CH-NHC_6H_5}{|} + RCHO \rightarrow$$

$$\begin{matrix} & & C_6H_5 \\ & & | \\ & & N \\ C_6H_5-CH & & \\ | & & CHR + H_2O \\ C_6H_5-CH & & \\ & & N \\ & & | \\ & & C_6H_5 \end{matrix}$$

Detection of α-Diketones

Reagents: 2.5% aqueous O,O'-diaminobenzidine hydrochloride solu tion, hydrochloric acid.

Procedure: Dilute aqueous solution of diacetyl (10 ml) is mixed in a test tube with 0.5 ml of O,O'-diaminobenzidine hydrochloride solution and 0.5 ml of hydrochloric acid solution. A yellow-orange color is produced.

Detection of β-Diketones

Reagent: *o*-Phenylenediamine (0.4 g) is dissolved in a mixture of 4.35 g of potassium hydrogen phosphate, 2 ml of 5 N phosphoric acid, 50 ml of 2.4 N sulfuric acid, and water to the mark (100 ml).

Procedures: Several milliliters of the dilute aqueous β-diketone solution are placed in a test tube and 2 ml of the reagent are added. The mixture is allowed to stand to produce a red color. If the tested β-diketone is so poorly soluble in water as to make the reaction negative, a small amount of alcohol should be added to the reaction mixture. The color is gradually formed as the diketone passes into the solution.

4. Chromatographic Methods

Paper Chromatography

Among aldehydes and ketones, only those which are not too volatile or too oxidizable can be chromatographed on paper directly. For example, vanillin and its derivatives can be chromatographed well in the systems formamide/ /hexane or formamide/benzene, and detection can be carried out on spraying with a solution of 2,4-dinitrophenylhydrazine in 2 N HCl, or using other color reactions. In other cases, however, these substances must be converted to suitable derivatives, for example ,2,4-dinitrophenylhydrazones (see p. 222) or hydroxamic acids (see p. 224). Derivatives with Girard's reagent can be also chromatographed (see p. 243).

Gas Chromatography

Volatile aldehydes and ketones can be identified by gas chromatography, for example, on the phases: ethyl hexyl sebacate, silicone oil, tricresyl phosphate, diphenylformamide, polypropylene glycol, and Apiezon-L (66, 67).

5. Regeneration and Isolation of Carbonyl Compounds

During the identification procedure of organic compounds it is often necessary to regenerate the compound from its derivative. Usually, the compound is first transformed to a solid derivative which is purified by crystallization, and then the original compound must be set free in order to be able to measure its physical constants.

Aldehydes and certain ketones form crystalline addition compounds with sodium bisulfite which are poorly soluble in excess saturated solutions of sodium hydrogen sulfite. If these addition compounds are boiled with acids or alkalies, one can regenerate the corresponding carbonyl compound.

This method is suitable for the purification of aldehydes and ketones as well as for their isolation from a mixture with other compounds. Aldehydes or ketones can be expelled from their addition compounds with bisulfite by formaldehyde. Among aldehydes, formaldehyde reacts most easily with bisulfite; with the increasing chain length of other aldehydes, their addition ability decreases. With certain aldehydes (phenyldimethylacetaldehyde, diphenylethylacetaldehyde) the reaction does not take place. Among ketones, methyl ketones of the type CH_3CO-R, where R represents a primary alkyl group, react easily. If R is a secondary or tertiary alkyl group, the reaction is slower. Aromatic ketones and aliphatic-aromatic ketones do not react with hydrogen sulfites. Cyclohexanone and cyclopentanone undergo the reaction with bisulfite quite easily. It should not be forgotten that the addition of sulfite can also take place on an activated double bond. For example, crotonaldehyde reacts with two molecules of sodium hydrogen sulfite.

During the isolation work it is advantageous—especially for the separation of small amounts of high-boiling aldehydes and ketones—to convert them to water-soluble derivatives. Other substances present in the original mixture can be separated, for example, by extraction with ether, while the derivatives of carbonyl compounds remain in the aqueous phase. The solubility in water is attained because the reagent for the carbonyl group contains a functional group in its molecule which causes the solubility of the derivative, as, for example, the carboxyl, sulfo, or amino group. In addition to p-toluenesulfonate of the hydrazide of N-methylnicotinic acid (XIV) (68) and carboxyarylhydrazines (XV) (69), Girard reagents T (hydrochloride of the hydrazide of trimethylammoniumacetic acid – XVI) and P (hydrochloride of the hydrazide of pyridiniumacetic acid – XVII) (70 – 72), and hydrochloride of the hydrazide of N-dimethylaminoacetic acid (XVIII) (73) are utilized. For a review of the use of Girard reagents as well as for notes on their preparation see (74).

When Girard reagents are used the hydrazones formed are not isolated, but the carbonyl compounds are regenerated from them after the separation of other compounds by a change in pH (acidification) and extraction with ether (75). Girard reagents generally react well with ketones. If we wish the reaction to go well with aldehydes, it is necessary to use pure reagents (the presence of hydrazine hydrochloride must be avoided). However, the purification of the reagents is difficult and they are also very hygroscopic. The corresponding hydrazones are soluble in water at all pH values, and hence they are isolated with difficulty.

However, if reagents are used which have a tertiary amino group (XVIII) in the molecule instead of an ammonium group, all these difficulties vanish. The reagent (XVIII) (73) reacts both with aldehydes and ketones in alcohol and in water very rapidly, and the initial carbonyl compound is easily isolable by mere acidification. Neutralization of the solutions of hydro-chlorides of hydrazones prepared with the reagent (XVIII) brings about the separation of free hydrazones, which, as a rule, crystallize well and have sharp melting points, so that they are also suitable for identification.

Paper chromatography can be used for the identification of a mixture of aldehydes and ketons isolated by means of Grignard reagents and for the control of the purity of the derivatives obtained (76).

Unequal rates of hydrolysis of hydrazones of Girard reagents can be used for the separation of isomeric ketones (77).

p-Carboxyphenylhydrazones are derivatives which usually do not have sharp melting points (they melt at high temperatures with decomposition), but the presence of a carboxyl group in their molecule enables the determina-tion of the neutralization equivalent and hence also the molecular weight of the carbonyl compound. These derivatives can be prepared by mixing both components in an aqueous or aqueous-alcoholic medium. The derivative mostly starts to separate immediately and the reaction can be brought to completion by heating on a water bath. The use of this reagent has been studied for various carbonyl compounds (69, 78).

Regeneration of carbonyl compounds from their hydrazones and oximes is most commonly carried out by acid hydrolysis and the expulsion of the analyzed carbonyl compound by a more reactive carbonyl compound; in the case of oximes and in special circumstances oxidative cleavage can also be applied.

Derivatives of ketones and aromatic and unsaturated aldehydes are most easily hydrolyzed. In the case of aliphatic aldehydes the cleavage does not take place so easily. Of course, the substituents of the used hydrazine also play a role. For example, semicarbazones are hydrolyzed with difficulty; differences in the reactivity of semicarbazones were successfully applied for the separation of a mixture of α- and β-citral. For the separation of hydra-zones of α-, β-unsaturated aldehydes, oxalic or phthalic acid can be used; for easily hydrolyzable oximes, sulfurous acid may serve. More stable hydra-zones require mineral acids for their hydrolysis, often concentrated, and also elevated temperatures (79). It is advantageous to carry out the cleavage in an inert atmosphere to prevent unwanted side reactions (80).

The isolation of the liberated carbonyl compounds is carried out, depending on their character, by filtration, extraction, steam distillation, or other means. Hydrazones can also be split by the expulsion of the carbonyl compound by a more reactive aldehyde, as, for example, benzaldehyde, p-nitrobenzaldehyde, 2,4-dinitrobenzaldehyde, and formaldehyde. The use of nonvolatile levulinic acid (81) is advantageous. In cases where the alde-hyde formed on hydrolytic cleavage is unstable under the given conditions (for example, succinic dialdehyde), the cleavage of the oxime can be carried out oxidatively in an acid medium — with nitrous acid or with amyl nitrite, ferric chloride, or bromine in an aqueous medium. This method of cleavage can be used only for carbonyl compounds which are not susceptible to oxidation.

General Method of Regeneration of Carbonyl Compounds Based on the Use of Levulinic Acid

To 2,4-dinitrophenylhydrazone placed in a distillation flask a 50 — 200 fold excess of an aqueous solution of levulinic acid is added (nine parts of levulinic acid and one part of water) and the mixture is heated over a boiling water bath for 5 min. The carbonyl compound is then isolated either by extraction with light petroleum or by steam distillation.

Note: When carbonyl compounds with a conjugated double bond are regenerated the cleavage is carried out with a mixture of nine parts of levulinic acid and one part of 1N sulfuric acid.

References

1. Karaoglanov, Z.: Z. anal. Chem. **121**, 92 (1941).
2. Benedict, S. R.: J. Biol. Chem. **5**, 485 (1908).
3. Feder, E.: Arch. Pharm. **245**, 25 (1907).
4. Long, A. G., Quayle, J. R., and Stedman, R. J.: J. Chem. Soc. **1951**, 2197.
5. Sawicki, E., Miller, R., Stanley, T. W., and Hauser, T. R.: Anal. Chem. **30**, 1130 (1958).
6. Dutt, S.: J. Chem. Soc. **1924**, 802.
7. Pomeroy, J. H., and Pollard, C. B.: Quart. Journ. Fla. Acad. Sci. **10**, 13 (1948).
8. Kamel, M., and Wizinger, R.: Helv. Chim. Acta **43**, 594 (1960).
9. Levine, V. E., and Taterka, M.: Anal. Chim. Acta **15**, 237 (1956).
10. Rosenthaler, L., and Vegezzi, G.: Mitt. Gebiete Lebensm. u. Hyg. **44**, 475 (1953).
11. Sawicki, E., and Stanley, T. W.: Anal. Chem. **31**, 122 (1959).
12. Hünig, S., Utermann J., and Erlemann, G.: Chem. Ber. **88**, 708 (1955).
12a. Hünig S., and Utermann, J.: Chem. Ber. **88**, 1201, 1485 (1955).
13. Velluz, L., Amiard, G. and Pesez, M.: Bull. Soc. Chim. France **15**, 680 (1948).
14. Singleton, F. G., and Pollard C. B.: J. Am. Chem. Soc. **63**, 240 (1941).
15. Hörmann, H., Grassmann, W., and Fries, G.: Ann. **616**, 125 (1958).
16. Wislicenus, W., and Schäfer, R.: Ber. **41**, 4169 (1908).
17. Brandstätter, M.: Mikrochemie ver. Mikrochim. Acta **32**, 33 (1944).
    Szmant, H. H., and Planinsek, H. J.: J. Am. Chem. Soc. **72**, 4042 (1950).
    Roberts, J. D., and Green, C.: J. Am. Chem. Soc. **68**, 214 (1946).
    Nazarov, I. N., Kazicyna, L. A., and Zaretskaya, I. I.: Zhur. Obshch. Khim. **27**, 606 (1957).
18. Bryant, W. M. D.: J. Am. Chem. Soc. **54**, 3758 (1932); **55**, 3201 (1933); **58**, 2335 (1936).
    Mitchell, J., Jr.: Anal. Chem. **21**, 448 (1949).
19. Bell, F. O.: Biochem. J. **35**, 312 (1941).
    Clark, G. L., Kaye, W. I., and Parks, T. D.: Ing. Eng. Chem., Anal. Ed. **18**, 310 (1946).

Djerassi, C., and Ryan, E.: J. Am. Chem. Soc. **71**, 1000 (1949).

Gordon, B. E., Wopat, F., Jr., Burnham, H. D., and Jones, L. C., Jr.: Anal. Chem. **23**, 1754 (1951).

Malkin T., and Tranter T. C., J. Chem. Soc. 1951, 1178

20. Ross, J. H.: Anal. Chem. **25**, 1288 (1953).

21. Lenko, H. G., and Ford, J. A.: Anal. Chem. **35**, 1418 (1963).

22. Auvinen, E. M., and Favorskaya, I. A.: Vestn. Leningr. Univ. Ser. Fiz. Khim **2 (10)**, 122 (1963).

23. Schwartz, D. P., Johnson, G. R., and Parks, O. W.: Mikrochem. J. **6**, 37 (1962).

24. Bredereck, H.: Ber. **65**, 1833 (1932).

25. Bredereck, H., and Fritzsche, E.: Ber. **70**, 802 (1937).

26. Brunner, H., and Farmer, E. H.: J. Chem. Soc. **1937**, 1039.

27. Bryant, W. M. D.: J. Am. Chem. Soc. **60**, 2814 (1938).

28. Ingold, C. K., Pritchard, G. J., and Smith, H. G.: J. Chem. Soc. **1934**, 79.

29. Simon, E.: Biochem. Zeit. **247**, 171 (1932).

30. Braddock, L. I. et al.: Anal. Chem. **25**, 301 (1953).

31. Braddock, L. I., and Willard, M. L.: J. Am. Chem. Soc. **73**, 5866 (1951).

32. Brandstätter, M.: Mikchrochem. ver. Mikrochim. Acta **32**, 33 (1944).

33. Kuznetsov, N. V., Komarova, L. I., and Safronova, L. P.: Izv. Akad. Nauk SSSR, Otdel. Khim. Nauk **1963**, 750.

34. Hünig, S., and Utermann, J.: Chem. Ber. **88**, 423, 1485 (1955).

35. Gasparič, J., and Večeřa, M.: Chem. Listy **51**, 287 (1957); Collection Czech. Chem. Commun. **22**, 1426 (1957).

36. Šorm, F., Suchý, M., and Herout, V.: Chem. Listy **46**, 55 (1952).

37. Beckurst, H., Tröger, J., and Westerkamp, A.: Arch. Pharm. **247**, 657 (1909).

38. Badings, H. T., and Wassink, J. G.: Neth. Milk Dairy J. **17**, 132, (1963).

39. Bruemmer, J. M., and Mueller-Penning, T. J.: J. Chromatog. **27**, 290 (1967).

40. Ronkainen, P.: J. Chromatog. **27**, 380 (1967).

41. Struck, H.: Mikrochim. Acta **1956**, 1277.

42. Petrenko-Kritschenko, P., and Kantscheff, W.: Ber. **39**, 1452 (1906).

43. Veibel, S.: The Identification of Organic Compounds, G.E.C. Gad Publisher 1954.

44. Harries, C.: Ann. **330**, 185 (1903).

45. Trozzolo, A. M., and Lieber, E.: Anal. Chem. **22**, 764 (1950).

46. v. Richter, V., and Anschütz, R.: Chemie der Kohlestoffverbindungen, 12th Ed., Part I, p. 445. Akad. Verlagsges., Leipzig 1928.

47. D.R.P. (German Patent) 195, 655−57 (1906).

48. Nägeli, E., Ber. **16**, 494 (1883).

49. Posner, T.: Ber. **42**, 2523 (1909).

50. Riley, H. L., Morley, J. F., and Friend N. A. C.: J. Chem. Soc. **1932**, 1875.

51. Fecht, H.: Ber. **40**, 3893 (1907).

52. Perret, J. J.: Helv. Chim. Acta **34**, 1531 (1951).

53. Conant, J. B., and Bartlett, P. D.: J. Am. Chem. Soc. **54**, 2881 (1932).

54. Klein, G., and Linser, H.: Mikrochemie, Pregl's Festschr. **1925**, 204.

55. Sawicki, E., Noe, J., and Stanley, T. W.: Mikrochim. Acta, **1960**, 286.

56. Lewin, L.: Ber. **32**, 3388 (1899).

57. Simon, L.: Compt. Rend. **125**, 1105 (1897).

58. Fromageot, C., and Heitz, P.: Mikrochim. Acta **3, 52** (1938).

59. Fuson, R. C., and Bull, B. A.: Chem. Revs. **15**, 275 (1934).

60. Seelye, R. N., and Turney, T. A.: J. Chem. Educ. **36**, 572 (1959).
61. Gillis, B. T.: J. Org. Chem. **24**, 1027 (1959).
62. Hughes, E. E., and Lias, S. G.: Anal. Chem. **32**, 707 (1960).
63. Hinsberg, O.: Ann. **237**, 327 (1887).
64. Witter, R. F., Snyder, J., and Stotz, E. H.: J. Biol. Chem. **176**, 493 (1948).
65. Jaunin, R., and Godat, J. P.: Helv. Chim. Acta **44**, 95 (1961).
66. Kaiser, R.: Gas-Chromatographie. Akademische Verlagsges., Leipzig 1960.
67. Hunter, I. R., Dimick, K. P., and Corse, J. W.: Chem. and Ind. **1956**, 294.
68. Allen, C. F. H., and Gates, J. W.: J. Org. Chem. **6**, 596 (1941).
69. Veibel, S.: Acta Chem. Scand. **1**, 54 (1947); Monatsh. **81**, 330 (1950).
70. Girard, A., and Sandulesco, G.: Helv. Chim. Acta **19**, 1095 (1936).
71. Weissenberg, A., and Ginsburg, D.: Bull. Research Council **5A**, 268 (1958).
72. Petit, A., and Tallard, S.: Ind. Parfum. **3**, 75 (1948).
73. Viscontini, M., and Meier, J.: Helv. Chim. Acta **33**, 1773 (1950).
74. Wheeler, O. H.: Chem. Revs. **62**, 205 (1962).
    Gaddis, A. M., Ellis, R., and Currit, G. T.: J. Food Sci. **29**, 6 (1964).
    Ellis, R., and Gaddis, A. M.: Anal. Biochem. **13**, 565 (1965).
75. Utzinger, G. E.: Helv. Chim. Acta **35**, 1359 (1952).
76. Seligman, R. B., et al.: Chem. and Ind. **1954**, 1195.
77. Nigan, I. C., and Levi, L.: Anal. Chem. **35**, 1087 (1963).
78. Veibel, S., and Vrang, Th.: Dansk Tidskr. Farm. **17**, 112 (1943).
79. Bodforss, S.: Ber. **72**, 468 (1939).
80. Ramirez, F., and Kirby, A. F.: J. Am. Chem. Soc. **74**, 4331 (1952); **75**, 6026 (1953).
81. Keeney, M.: Anal. Chem. **29**, 1489 (1957).

CHAPTER XIII

CARBOXYLIC ACIDS
AND THEIR DERIVATIVES

1. Carboxylic Acids.
Detection of the Carboxy Group

The form in which acids are present in the analyzed material is important
to the analyst: whether they are free, in aqueous solution, in the form of
salts or solutions of these salts, or mixed with other neutral or acid substances.
The character of the sample determines the course of the analytical procedure.
It is also important to bear in mind that some carboxylic acids are volatile
(lower aliphatic monocarboxylic acids), some are volatilized by steam distillation (monocarboxylic acids), others are water soluble (lower monocarboxylic and certain dicarboxylic acids), and practically all are soluble in 5%
sodium hydrogen carbonate. All these properties offer a possibility of separating these substances from others. As carboxylic acids do not give characteristic color reactions, all general detections of this class of substance
are based on their acidity. This is determined in aqueous solutions by means
of indicators, and in the case of water-insoluble acids on the basis of their
solubility in 5% sodium hydrogen carbonate. In all instances, however, it
is advisable to complete the proof of acidity by titration with alkali; in the
case of pure acids by determining their NE (neutralization equivalent),
which is characteristic for each acid. In the case of more complicated mixtures titration affords information on the approximate amount of acids present in the sample and whether they represent a substantial fraction of it. If
acidity is found in the sample, it is also necessary to look for the presence of
sulfur, because the acidity could be caused by the presence of sulfonic acids.
However, as the detection of acidity need not be reliable in more complicated
cases, it is much safer to employ paper chromatography and to carry out
the acidity detection on the chromatogram. This method has the advantage
that it can also be carried out with salts.

In additon to using common indicators and indicator papers, the
detection of acidity can be performed by some other tests (1):

(a) In the presence of acids, disulfate of dihydroindanthroazine, colorless itself, is converted to indanthrene blue. This reaction is used in paper chromatography.

(b) The formation of diazonium salts and their coupling with aromatic amines take place only in acid media. A spot test based on this reaction was devised by Nomura (2).

Reagent: Solution A: 2.9 g of sodium sulfanilate (dihydrate) and 70. g of $NaNO_2$ are dissolved in 30 ml of water. Solution B: 1.8 g of α-naphthylamine are dissolved in 40 ml of ethanol. Both solutions are mixed shortly before use.

Procedure: Several drops of the reagent are placed on a spot-test plate and a drop of the sample solution, or a crystal of the sample or of an analyzed dry residue (after concentration of an extract) is added. Depending on the solubility of the sample and on its acidity, an orange to red color is produced either immediately or after several minutes.

The test is negative with phenols; salts of carboxylic and sulfonic acids also give no reaction. Amino acids react after addition of formaldehyde (which blocks the basic amino group). Positive reactions are also obtained with weak acids, which cannot be detected with indicators.

(c) In paper chromatography the fact that certain reducing substances such as glucose or phenol do not reduce ammoniacal silver nitrate in the presence of acids (3) is often used for the detection of acids.

In addition to the general reactions described, certain acids can be detected by specific reactions:

1. Formic acid reduces ammoniacal silver nitrate (see p. 210), and if reduced with magnesium in hydrochloric acid, it yields formaldehyde, which can easily be detected with chromotropic acid. In the presence of conc. sulfuric acid formic acid reacts with phenols with the formation of corresponding triphenylmethane dyes (see p. 196).

Reagents: powdered magnesium, dilute HCl, conc. H_2SO_4, chromotropic acid.

Procedure: A drop of the sample solution is mixed with one drop of HCl (1 : 1), and to this solution a small quantity of powdered magnesium is added in several portions until the evolution of gas ceases. A small quantity (on the tip of a knife) of chromotropic acid is then added to the mixture, followed by 3 ml of conc. sulfuric acid. If a violet color does not develop immediately — which would prove the presence of formic acid in the sample — the mixture should be heated at 60 °C in a water bath for 10 min.

2. Oxalic acid. Reduction of this acid with magnesium and chromotropic acid gives rise to glycolic acid, from which conc. H_2SO_4 liberates formaldehyde, which can be detected, for example, with chromotropic acid. When melted with diphenylamine or carbazole, oxalic acid gives triphenylmethane dyes.

Reagents: diphenylamine or carbazole, anhydrous zinc chloride, ethanol.

Procedure: One crystal of oxalic acid and a small amount of diphenyl-amine or carbazole (on the tip of a knife or spatula) are placed in a micro test tube and mixed with twice that amount of $ZnCl_2$. The tube is then heated over a microburner. The mixture first melts and then turns blue on further heating. After cooling, the melt is dissolved in 1 ml of ethanol, giving a blue solution.

3. Dicarboxylic acids: 1,2-Dicarboxylic acids, and also 1,2-sulfo-carboxylic acids, as well as their derivatives, esters, amides, and anhydrides, when melted with resorcinol in the presence of anhydrous zinc chloride yield the corresponding fluorescein derivatives (see p. 197).

Reagents: resorcinol, anhydrous $ZnCl_2$, 1% NaOH.

Procedure: In a micro test tube a small amount of the dicarboxylic acid is mixed with zinc chloride and a few crystals of resorcinol and the tube is heated at 130 °C for several minutes. After cooling, the melt is extracted with 1% NaOH, giving an orange solution with intense green or yellow fluorescence in ultraviolet light.

Identification of Carboxylic Acids

More than 70 reagents have been proposed for the identification of carboxylic acids. This illustrates both the great importance of this group of substances and the fact that at present no general method is available for their identification. The derivatives which have been proposed can be divided into 6 groups:

I. *Salts* (a) of heavy metals (for example, silver salts), (b) of amines and hydrazines (for example, phenylethylamine, phenylhydrazine), and (c) of S-alkylthiuronium (S-benzylthiuronium).

II. *Esters.* (a) Alkyl esters (for example, methyl esters), and (b) phena-cyl esters (for example, esters of *p*-bromophenacyl alcohol).

III. *Amides and hydrazides* (for example, *p*-toluidides, 2,4-dinitro-phenylhydrazides).

IV. *2-Alkylbenzimidazoles.*

V. *Ureides.*

VI. *Addition compounds with urea.*

The choice of suitable derivatives depends primarily on whether the acid is anhydrous (in such a case esters and amides are prepared) or in aqueous solution (identification in the form of salts, phenacyl esters). For the identification of lower acids, salts, esters, and 2-alkylbenzimidazoles are suitable, while higher acids are converted to hydrazides, ureides, and also addition compounds with urea. For microidentification S-1-naphthylmethylthiuronium salts are suitable, as are, in certain instances, *p*-toluidides. Paper, thin-layer, and gas chromatography (see p. 259) are also very suitable. For a review of the separation and identification of carboxylic acids by gas and paper chromatography see (4).

Salts of Carboxylic Acids

Phenylethylammonium, benzylammonium, and piperazinium salts were found useful for identification purposes; in addition, phenylhydrazine gives crystalline salts with carboxylic acids. Apart from a few exceptions, this method of identification is not very suitable, because the salts mentioned are generally too soluble and difficult to purify, and they often decompose if exposed to long contact with air. Some authors have proposed an identification based on the measurement of IR spectra of respective benzylammonium salts (5) or on the conversion of acids to tetraphenylstibonium salts (6). Stronger acids give salts of very limited solubility with certain metals, which are usually prepared by boiling a solution or suspension of the acid with a metal oxide (lead oxide, zinc oxide), filtering the solution, and allowing the salt to crystallize on cooling. Silver salts are prepared by precipitation of aqueous solutions of ammonium salts of carboxylic acids with 0.1 N $AgNO_3$ under vigorous stirring X-ray diffraction (7) is used as an identification constant, and the determination of Ag by ignition can serve as a suitable additional identification procedure. For identification in the form of salts S-alkylthiuronium salts of carboxylic acids are most suitable. S-Benzylthiuronium chloride has been proposed (8) as a reagent, sometimes substituted with bromine, chlorine, or a nitro group. The disadvantage of these salts—an appreciable solubility—does not exist in the case of S-1-naphthylmethylthiuronium chloride (9);

$$RCOONa + \left[C_{10}H_7CH_2SC \overset{NH_2}{\underset{NH_2}{\diagup}} \right]^{\oplus} Cl^- \rightarrow$$

$$NaCl + RCOO^{\ominus} \left[C_{10}H_7CH_2SC \overset{NH_2}{\underset{NH_2}{\diagup}} \right]^{\oplus}$$

The salts are best prepared by heating a small excess of S-1-naphthylmethylthiuronium chloride with the sodium salt of carboxylic acid in methanol. The salts precipitate almost immediately and in high yield. In many cases the products are practically pure derivatives or are of such purity that one crystallization suffices. The salt can be precipitated from aqueous solutions of carboxylic acid salts (neutral medium) with a saturated solution of the reagent. The precipitated salts crystallize from ethanol. The preparation of salts is rapid and easy, and the solubility allows the identification of even milligram quantities of acids. Disadvantages of this method are that the solubilities of salts of lower dicarboxylic acids are too high, and that the salts of higher acids have melting points which are too close (S-1-naphthylmethylthiuronium salts of lauric acid 155 °C; myristic acid 152.5 °C; palmitic acid 150.5 °C; stearic acid 150 °C). In addition, melting points determined in a capillary (9) and on a Kofler block differ (for a review see below), and the measurement of mixture melting points for the identification is not suitable (10). For the identification of acids Taborsky (11) recommends sharply melting salts of 5-methoxytryptamine, while Affsprung and Gainer (12) recommend tetraalkylstibonium salts.

S-1-Naphthylmethylthiuronium Salts

Method A

Reagents: 1-naphthylmethylthiuronium chloride, 10% sodium ethoxide (1 g of sodium metal is introduced with caution and cooling into 10 ml of abs. ethanol).

Procedure: The acid (0.1 g) is dissolved in 1 ml of ethanol, to which one drop of phenolphthalein is added; the solution is neutralized carefully with the solution of ethoxide. A solution of S-1-naphthylmethylthiuronium chloride in a minimum amount of hot ethanol is added to the dissolved sodium salt of the acid, and the mixture is heated to incipient boiling and filtered. After cooling, the separated salt is collected on a filtration crucible. Crystallization is carried out from ethanol.

A Review of Melting Points of S-1-Naphthylmethylthiuronium Salts

Acid	Method A Mp °C		Method B Mp °C on a Kofler block
	on a Kofler block	in a capillary	
Formic	163−166	175−176	163−164
Acetic	145−147	153−154	142−144
Propionic	122−124	118−119	129−130
Isovaleric	145−146	143−144	143−144
Benzoic	180−181	175−176	179−180
Phenylacetic	166−167	159−160	161−162
Palmitic	150−150.5		
Levulinic	140−141		

Method B

Reagent: saturated aqueous solution of S-1-naphthylmethylthiuronium chloride (approx. 1%).

Procedure: The sodium salt (0.1 g) and 1 ml of water are placed in a test tube and mixed with 15 ml of the reagent. After 1 hr standing the separated salt is filtered off using a filtration crucible and washed with 3 ml of water. The product is crystallized from ethanol.

Carboxylic Acid Esters

The identification of acids by their conversion to methyl or ethyl esters is carried out only when the esters formed have properties suitable for identification; they are of special value for the separation of acids by gas chromatography. In most cases the ester is converted to a further derivative. Esterification is carried out by heating the carboxylic acid (one part) under reflux with five parts of absolute methanol containing $3-5\%$ of dry HCl or H_2SO_4. After $3-5$ hr the excess methanol is distilled off and the remaining ester is freed from the acid by mixing it with a soda solution and extracting with ether. It is important to keep in mind that in certain cases the esterification can only take place slowly for sterical reasons and also that certain acids can add hydrogen chloride; in the presence of sulfuric acid, methanol can etherify even a phenolic hydroxy group to give phenol ethers. A simple and quick method of esterification consists in the reaction of carboxylic acids with diazomethane:

$$RCOOH + CH_2N_2 \rightarrow RCOOCH_3 + N_2$$

which can be achieved by simple mixing of the components in an ethereal, chloroformic, or acetonic medium. When diazomethane is used for esterification one should keep in mind that it also reacts with phenolic hydroxy groups, enolic hydroxyls, with aldehydes, α,β-diketones, thioketones, etc.

For identification purposes substituted benzyl and phenacyl esters of carboxylic acids are often used,

$$RCOONa + XC_6H_4Y \rightarrow XC_6H_4CH_2OCOR \text{ (or } XC_6H_4COCH_2OCOR) + NaCl \text{ (NaBr)}$$

where X = Cl, Br, NO_2, $C_6H_5N=N$, and Y = CH_2Cl, $COCH_2Br$. Although the utilization of the reactions mentioned is accompanied by a number of difficulties, especially when carried out by beginners (the reagents and some crude esters are lachrymatory or irritate the skin), it is a convenient procedure because it makes the reaction possible even in the presence of water, of which use is made for the preparation of derivatives of acids which can be isolated from aqueous solutions only with difficulty. The reaction is usually carried out by boiling the halogenide with the sodium salt of the acid in aqueous-alcoholic solution. The time of the reaction depends on the nature of the acid, and lasts $1-3$ hr (di- and tricarboxylic acids require a longer reaction time). Esters are isolated from the reaction mixtures by dilution with water and are crystallized from aqueous alcohol. It is important for the reaction medium to be neither alkaline nor too acid, to prevent side reactions. It is recommended that the pH of the reaction solution be $4-5$. Side reactions give rise predominantly to alcohol, both by direct reaction of the reagent and by the saponification of the formed ester. An ether formed by alcoholysis of phenacyl bromide can occur as an additional impurity:

$$ArCOCH_2Br + ROH \rightarrow HBr + ArCOCH_2OR$$

The danger of side reactions is especially great during the preparation of phenacyl esters of formic acid. This led to an mp of 140 °C being given for p-bromophenacyl formate when it was in fact that of p-bromophenacyl alcohol, the true melting point of the ester being 92 °C $(13-15)$. Keto acids (16) are also able to give by-products on reaction with phenacyl bromides. Phenacyl esters of carboxylic acids can also be prepared from p-bromo and p-phenyl-α-diazoacetophenone under the catalytic influence of copper (II) chloride (17). The reaction is carried out by gentle heating of the components in dioxane in the presence of cupric chloride. Berger (18) has also pointed out the possibility of alkylation of the phenolic group during the reaction of p-bromophenacyl bromide with aromatic hydroxy acids. Identification and colorimetric determination of carboxylic acids after their conversion to p-nitrophenacyl esters has been described by Bartos (19).

General Method of Preparation of Phenacyl Esters

The acid (1 mmole) is neutralized (to phenolphthalein) with N NaOH (from a burette or a graduated pipette) and the neutral solution (the red color after neutralization is eliminated by addition of a droplet of acid) is mixed with 8 ml of ethanol and 0.9 mmole of phenacyl halogenide. If the acid is water-insoluble, the neutralization is carried out in the presence of ethanol; the control of the amount of N NaOH used for neutralization serves to determine the precise quantity of phenacyl halogenide, which never should be present in excess. The reaction mixture is boiled under reflux. The reaction time is from 30 to 180 min, depending on the character of the acid. Phenacyl ester separates from the solution on cooling or after dilution with water. Esters often separate as oils, and long standing is necessary before they crystallize. Recrystallization is carried out from ethanol.

p-Bromophenacyl Ester of Benzoic Acid

Reagents: 5% NaOH, 95% ethanol, p-bromophenacyl bromide.

Procedure: Benzoic acid (100 mg) is dissolved in a minimum amount of ethanol and accurately neutralized (with respect to phenolphthalein) with 5% sodium hydroxide. A small crystal of benzoic acid is then added to the neutral solution followed by 5 ml of ethanol and 0.3 g of p-bromophenacyl bromide. The reaction is carried out in a 30-ml flask with a ground joint on a water bath and under reflux, for 1 hr. After cooling, it is allowed to stand in a refrigerator for 30 min. The separated ester is filtered off under suction. Yield, 0.14 g; mp, 114−115 °C. Crystallization from 4 ml of ethanol and 2 ml of water (added under heating to incipient turbidity) gives 0.13 g of product, mp 119 °C.

Amides and Hydrazides

The most convenient methods of identification of carboxylic acids via the preparation of their amides and hydrazides can be represented by the following scheme:

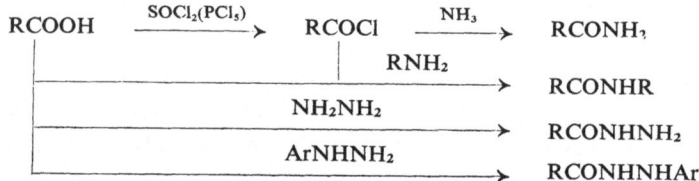

According to this scheme, the acid is first transformed to its chloride under the influence of $SOCl_2$ or PCl_5, and this is then reacted with ammonia or amine to give amide. It is also possible to transform the carboxylic group to an amidic one by direct reaction with an amine. Chlorides are prepared from free acids or from their salts. To carry out the reaction with phosphorus pentachloride, the anhydrous acid is mixed with the reagent under intense cooling. It can also be carried out in an inert solvent (benzene, chloroform, light petroleum). The reaction is completed by heating the mixture on a water bath. In this reaction $POCl_3$ is created simultaneously, which must be eliminated after the reaction has subsided by distillation under normal or reduced pressure (bp, 110 °C). It can also be eliminated by heating the mixture to 110−120 °C and blowing a stream of dry carbon dioxide through it. When PCl_5 is employed one should consider that in addition to the carboxylic group, strongly acid phenols (picric acid) and the carbonyl group react as well. The risk of side reactions can be avoided if thionyl chloride is used in place of phosphorus pentachloride. Carboxylic acids are converted to chlorides by refluxing them with a 20−50% excess of thionyl chloride for 30 min in a 60−70 °C warm bath. For the reaction of oxalic and acetic acids the reaction temperature should be decreased to $40-50$ °C and the reaction time increased to 45 min. Thionyl chloride is a suitable reagent for the preparation of chlorides of the majority of carboxylic acids, with the exception of aromatic acids with negative substituents, especially in the para position. Chlorides of these acids (for example, p-chloro-, p-bromo-, and p-hydroxybenzoic acids) should be prepared with PCl_5. Dicarboxylic acids with $4-5$ carbon atoms between the carboxyls can cyclize to ketones under the influence of thionyl chloride (for example, adipic acid gives cyclopentanone). Further, dichloro- and trichloroacetic acids cannot be transformed with thionyl chloride to corresponding chlorides, and in the reaction of amino acids with thionyl chloride the amino group is attacked. For anomalous reactions of thionyl chloride also see (20).

To carry out the conversion of chlorides into amides, the simple method of dropping the acid chloride into an aqueous solution of amine can be used successfully. However, heating the acid chloride with a benzenic solution of the base (in $50-100\%$ excess) under reflux for 10 min is used more commonly. The benzene solution is washed from the excess of the base with water and dilute hydrochloric acid (by shaking in a funnel), and the amide is isolated by distilling off the solvent. Anilides and toluidides of aliphatic acids up to C_{10} can also be prepared by direct reaction of the acid or of its sodium salt with the corresponding base. Anhydrous acid is heated carefully with a $2-2.5$ fold excess of the base first at 150 °C for 10 min and then at $180-190$ °C for 30 min. After cooling, the excess base is eliminated by extracting the reaction mixture with dilute hydrochloric acid. If it is not possible to isolate the free anhydrous acid, its sodium salt is prepared, which is then perfectly dried, ground thoroughly with the base (usually, a $2-2.5$ fold excess of p-toluidine is employed), and heated with 0.1 ml of conc. hydrochloric acid in the manner described above.

Phenylhydrazides of acids are prepared by boiling phenylhydrazine with excess acid, either without a solvent or in benzene solution. Acid hydrazide crystallizes after cooling or after diluting the reaction mixture with benzene. Derivatives of lower monocarboxylic acids are best crystallized from benzene, derivatives of higher mono and dicarboxylic acids from alcohol or alcohol-water mixtures. Dicarboxylic acids yield bis-phenylhydrazides.

Benzo-p-toluide

Reagents: thionyl chloride (pure; if necessary, distilled with quinoline), benzene, p-toluidine, 5% HCl, 5% NaOH, ethanol.

Procedure: Benzoic acid (200 mg) and 1.5 ml of thionyl chloride are placed in a 50-ml flask with ground-glass joint and a conical bottom, fitted with a reflux condenser the upper end of which is connected by a polyvinyl chloride tubing to a funnel immersed in a beaker filled with water. The flask is heated at 75 – 80 °C in a water bath for 30 min and 500 mg of *p*-toluidine in 50 ml of benzene are then added and the mixture refluxed for 15 min more. After cooling, the mixture is transferred into a separatory funnel and the flask is rinsed with benzene and 1 ml of ethanol. The solution is extracted successively with 5 ml of water, 5 ml of 5% HCl, 5 ml of 5% NaOH, and again with 5 ml of water. The benzene layer is poured back into a distillation flask, the separatory funnel is rinsed with 2 ml of ethanol, and the solvent is distilled off. The distillation residue is dissolved completely in 5 – 6 ml of boiling ethanol, charcoal is added, and the suspension is filtered, while hot, through a cotton wool plug in a funnel. The plug is washed with 1 – 2 ml of hot ethanol and the filtrate is diluted with 3 – 4 ml of water, which causes incipient precipitation. The turbidity is eliminated by heating the mixture, which is then allowed to stand and crystallize. The separated *p*-toluide is recrystallized from 70 – 80% aqueous ethanol: yield 175 mg; mp, 156 – 158 °C.

p-Toluides from Sodium Salts

Reagents: *p*-toluidine, conc. HCl, ether, 5% HCl, sodium sulfate, ethanol.

Procedure: A mixture of 50 mg of dry sodium salt of an acid, 125 mg of toluidine, and 0.1 ml conc. HCl is heated in a micro test tube in an oil bath at 140 °C for 10 min. The temperature is then increased to 180 °C and the mixture kept at this temperature for another 15 min. After being cooled the reaction mixture is diluted with 2 ml of ether, and stirred thoroughlyf and the contents are transferred into a separatory funnel using additional ether (approx. 10 ml). The ethereal solution is washed twice with 3 ml of 5% HCl and 3 ml of water. The ether solution is dried over sodium sulfate for 1 hr with occasional shaking and then filtered into a distillation flask. After distilling off the ether the residue is dissolved in hot ethanol (2 – 3 ml), filtered through a bit of cotton wool into a micro test tube. Water is added to the filtrate until it becomes slightly turbid, and the tube is set aside in a refrigerator for crystallization. The crystals are collected on a filtration tube. The yield is 30 mg if phenylacetic acid was used as starting material. Crystallization from 1 ml of 50% ethanol gives 22 mg of toluidide, mp, 134 °C. *p*-Toluides of propionic, isovaleric, and benzoic acids were prepared in the same way. After two crystallizations the yields were 5 – 10 mg and the melting points were in agreement with those in the literature.

Aceto-*p*-toluide

Reagents: *p*-toluidine, conc. HCl, 5% HCl, benzene, ethanol.

Procedure: 100 mg of dry, well-ground sodium acetate, 250 mg of p-toluidine, and 0.1 ml of conc. HCl are mixed in a micro test tube, and the mixture is heated in an oil bath at 150 °C for 10 min. The temperature is then increased within 15 min to 180 °C and the mixture is heated at this temperature for an additional 15 min. After cooling, the mixture is triturated with 3 ml of ether and transferred with the help of an additional 15 ml of ether into a small separatory funnel. The ethereal layer is washed twice with 3 ml of 5% HCl and once with 4 ml of water, dried over sodium sulfate, filtered into a distillation flask, and evaporated to dryness. The residue is dissolved in 3 ml of hot ethanol and filtered into a micro test tube through a cotton-wool plug. Water is added to the filtrate until slightly turbid, and crystallization is allowed to occur. The product is filtered on a filtration tube. Yield, 42 mg; mp, 145 – 147 °C. Crystallization from 0.3 ml of ethanol with the addition of water to incipient turbidity gave 16 mg of the product, mp 145 – 147 °C.

By the same procedure, the toluide was also prepared from 0.1 ml of acetic acid and 250 ml of *p*-toluidine (yield, 73 mg; mp, 145 – 147 °C); after crystallization from 0.5 ml of ethanol with water added just to incipient precipitation the yield is 50 mg and mp 145 – 147 °C.

2-Alkylbenzimidazoles

Reaction of carboxylic acids with *o*-phenylenediamine leads to 2-alkylbenzimidazoles which have properties suitable for identification (21, 22), i.e., they crystallize well and have sharp melting points which are not too close to each other (formic acid derivative, mp, 172 – 173 °C; stearic acid derivative, 93.5 – 94.7 °C). Another advantage consists in their forming another series of derivatives: they can be converted to picrates,

For paper chromatography of 2-alkylbenzimidazoles see (23).

General Method of Preparation of 2-Alkylbenzimidazoles

The acid (5 – 20 mmole) is mixed with an equivalent amount of *o*-phenylenediamine and 0.1 ml of 5% HCl and refluxed for 15 – 20 min. After cooling, the product is precipitated by alkalizing with aqueous ammonia. The derivatives are crystallized from alcohol.

Preparation of picrates: 2-Alkylbenzimidazole (50 – 200 mg) is dissolved in a minimum amount of hot ethanol, and the solution is added to a saturated solution of picric acid in ethanol.

Ureides, Semicarbazides, and Thiosemicarbazides

Ureides can be obtained by reacting urea with acid chlorides or with their esters (for the preparation of chlorides see p. 255; for the preparation of esters see p. 252). Ureides crystallize well and are suitable mainly for the identification of higher fatty acids.

$$RCOCl + NH_2CONH_2 \rightarrow RCONHCONH_2$$

A derivative of stearic acid can be obtained by 24 hr standing of a mixture of 4 g of ethyl stearate, 22 ml of 25% sodium ethoxide, 7 ml of pyridine, and 20 g of urea. The product is isolated by pouring the reaction mixture into water acidified with acetic acid and is then crystallized from ethanol. The yield is 90% (24); the excess of urea is necessary to prevent the formation of diacyl urea (25). There are other possibilities for identifying higher acids. For example, esterification (see p. 252), or conversion to hydrazides which can be converted further to another series of derivatives when reacted with phenylisocyanate or phenylisothiocyanate:

$$RCOOH \rightarrow RCOOR \xrightarrow{NH_2NH_2} RCONHNH_2 \xrightarrow{ArCNO} RCONHNHCONHAr$$

Detailed description of the preparation of these derivatives can be found in the original literature (26).

Addition Compounds with Urea (Clathrates)

Addition compounds with urea are especially suitable for the identification of higher-molecular acids (27). The addition compounds crystallize well and can be prepared easily in high yield; the material prepared by direct precipitation is usually sufficiently pure for analytical purposes. Their advantage consists in the fact that the weight ratio of urea: acid is 3 : 1, leading to an increase in the amount of the derivative with respect to the acid. Also, the regeneration of the acid is very easy. The clathrate decomposes on addition of water, in which urea dissolves. A disadvantage of clathrates from the point of view of identification consists in unsharp melting points (similar to that of urea) and the fact that they cannot be used for X-ray analysis, because X-ray diagrams of single homologs do not differ. The temperature at which clathrates decompose serves as an identification constant. This temperature is reproducible with $\pm 1\ °C$ deviation (27), which is fully sufficient or identification purposes.

Clathrate of Urea with Palmitic Acid

Reagents: urea, methanol.

Procedure: Palmitic acid (1 g) is dissolved in 20 ml of methanol containing 3 g of urea. The mixture is gently warmed over a water bath and set aside for crystallization. After 5 hr standing at room temperature the product is filtered off under suction. Yield, 2.0 g, decomposition from 115 °C.

Clathrate of Urea with Stearic Acid

The procedure is the same as for palmitic acid; the yield is 2.0 g, decomposition is from 125 °C.

Paper Chromatography

For the separation of acids by paper chromatography a number of procedures are available. However, most of them are not universal. For example, free acids can be chromatographed only if they are not volatile; if they are volatile, the acids have to be chromatographed in the form of their ammonium or ethylammonium salts. Generally, carboxylic acids cannot be chromatographed in a neutral medium, because their incomplete dissociation usually causes problems. This is why acids are always chromatographed preventing dissociation and hydrolysis, either in a medium containing stronger acids or in an alkaline medium in the form of theiι salts. These difficulties can be avoided if acids are converted to derivatives, as, for example, 2,4-dinitrobenzyl esters. Lower aliphatic acids are best chromatographed in the form of ammonium or ethylammonium salts in the following mixtures: n-butanol /ammonia, n-propanol-ammonia (2 : 1), or n-butanol/ 0.1 N aqueous ethylamine. This method is suitable for $C_1 - C_6$ acids, but the first two members, i.e., formic and acetic acids, usually do not separate. The acids are usually spotted on the chromatogram in the form of their ammonium or alkali salts (free acids are dissolved in a mixture of ethanol and ammonia) in an amount from 20 to 60 µg. In ammoniacal systems the detection is carried out by spraying the dry chromatogram with an indicator—for example, an 0.05% solution of bromophenol blue in ethanol. Acids appear as yellow spots on a green-blue background. For the ethylammonium system the chromatogram is thoroughly dried in a stream of air (or best by allowing it to hang in the air overnight), then sprayed with 0.1% solution of ninhydrin in ethanol acidified with a few drops of acetic acid, and heated to 100 C° until red-violet spots of ethylammonium salts appear. On more prolonged heating the background also assumes color. For the stabilization of the chromatogram see p. 283.

Formic acid can also be detected with ammoniacal silver nitrate solution. The R_F values of acids increase with the number of carbon atoms. Hence, the lowest value belongs to formic acid. Isomeric acids do not separate well (for the separation of lower fatty acids in the form of 2,4-dinitrobenzyl esters see below). Higher fatty acids may be chromatographed in their free form on papers impregnated with 5% solution of paraffin oil with 90%

acetic acid (or even glacial acetic acid) as the mobile phase. The acids are spotted on the chromatogram in the form of hexane solution in amounts of 50 µg and more. The detection is carried out as follows: After the chromatogram has been done it is dried and immersed into a very dilute aqueous copper acetate solution for 15 min, the paper is washed in running water for 30 min and is then immersed into a 0.1 % aqueous solution of potassium ferrocyanide for an additional 15 min. Finally, washing in running water is repeated (15 min). Higher fatty acids, from C_{10} up, appear as red-brown spots. The R_F values decrease with chain length.

Dicarboxylic acids can be chromatographed both in ammoniacal and in acid systems. For example, the homologous series of dicarboxylic acids, from oxalic to pimelic, can be separated in the solvent system n-propanol-ammonia (2 :1), maleic acid can be separated from fumaric acid in n-butanol-formic acid /water (4 : 1 : 5); detection can be carried out with indicators as in the case of fatty acids.

For aromatic acids the system n-propanol-ammonia (2 : 1) is also suitable, as are the detection reagents used for volatile acids.

In the case of aromatic dicarboxylic acids (phthalic, isophthalic, terephthalic) their p and m isomers practically cannot be resolved in the ammoniacal solvent systems. In an acid medium separation does take place, but it is difficult to detect. For the separation of all three acids a specially prepared paper (28) can be used, or they can be converted to nitroderivatives (29).

Chromatography of fatty acids in the form of 2,4-dinitrobenzyl esters (30 – 32): For the preparation of derivatives alkali salts of fatty acids are used as starting material. Into a micro test tube 5×10^{-4} mole of alkali salts of fatty acids is introduced, followed by 4×10^{-4} mole of 2,4-dinitrobenzyl bromide and 2 ml of an acetone-benzene-water (16 : 9 : 2) mixture. The tube is sealed and is heated for 1 hr in a boiling water bath. After cooling and opening the ampoule, the mixture can be applied directly on paper. Chromatography is carried out on papers impregnated with 70% dimethylformamide using a 35 : 1 mixture of cyclohexane-kerosene (bp, 190 – 220 °C) as solvent (for $C_4 - C_{14}$ acids) or on papers impregnated with 20% formamide in cyclohexane-benzene (25:1) (for acids up to C_7).

If the acids are in the form of esters, anhydrides, or chlorides, it is best to convert them to hydroxamic acids and to separate the latter on paper (see p. 277).

Gas Chromatography

The following stationary phases have been proposed for the separation of fatty acids or their esters: Apiezon L, M (33), benzyldiphenyl (33), dioctyl phthalate (33), paraffin oil (33), paraffin (33), pentachlorodiphenyl (34), silicone oils (34 – 38), and tritolyl phosphate (34). For the separation of higher boiling acids it is advantageous to convert them to methyl esters. Diazomethane gas introduced into a solution of acids in ether containing 10% of methanol is used for this. Methanol enhances the reaction appreciably; if the reaction is carried out only in pure ether, it is slow, not quantitative, and by-products are formed (39, 40). To esterify acids, the reaction with the corresponding alcohol in the presence of boron trifluoride (41) or in the presence of hydrochloric acid (42) may also be used. For the conversion of mono and dicarboxylic acids ($C_4 - C_{20}$) to propyl esters see (43). For gas chromatography of methyl esters of $C_{22} - C_{32}$ acids see (44).

Analysis of Salts of Carboxylic Acids

For identification purposes water-soluble salts of carboxylic acids are converted to poorly soluble S-l-naphthylmethylthiuronium salts (p. 251). Alkali salts of carboxylic acids can be used for identification either by reacting them with p-bromophenacyl bromide (p. 252), or by converting them to p-toluides (p. 255). If they are present in aqueous solutions, the water should be distilled off. Free acids can be set free from their salts with mineral acids and isolated either by steam distillation or extraction with ether; if they are solid and insoluble in water, they can be separated by filtration. For nonvolatile water-soluble acids, analyzed in the form of their salts, the use of ion exchangers is very advantageous if these salts are also soluble in water or in ethanol [see p. 58 and (45)]. This method is very simple and can be used for the isolation of the majority of acids. An aqueous or alcoholic solution of a salt (which may also be warm) is filtered through a short column of a cation exchanger, yielding an aqueous or alcoholic solution of the acid. Often the conversion can be carried out by merely mixing and stirring the salt solution with a cation exchanger. The conversion can be done even with a suspension of a salt of limited solubility.

Water-insoluble salts are dissolved in dilute acids and can be isolated by extraction, steam distillation, or even filtration if the liberated acid is solid and insoluble in water. Insoluble salts of alkaline-earth metals can be converted to soluble sodium salts by boiling with sodium carbonate solution. Insoluble silver, copper, lead, and similar salts are decomposed in an aqueous suspension with hydrogen sulfide and filtered from the precipitated metal sulfide. The acid can then be isolated from the filtrate.

2. Esters (Detection and Identification)

Esters of carboxylic acids react with hydroxylamine in alkaline media according to the equation

$$RCOOR' + NH_2OH \rightarrow RCONHOH + R'OH$$

giving rise to hydroxamic acids which, with ferric salts, give intensely colored iron complex salts (see p. 277).

Reagents: Equal parts of 12.5% methanolic NaOH and 12.5% methanolic hydroxylamine hydrochloride solutions (finely ground hydrochloride is dissolved in hot methanol) are mixed and the precipitated NaCl is filtered off; alcohol; 0.5 N HCl; 1% $FeCl_3$.

Procedure: To a few drops of an alcoholic solution of the tested substance are added several drops of the reagent, and the mixture is heated gently and, after a while, acidified with hydrochloric acid and mixed with a few drops of ferric chloride. If the mixture contains an ester, a red-violet color appears.

For more recent papers on this reaction see (46−48).

A simple detection procedure for esters is based on their hydrolysis. A drop of the tested ester is dissolved in 2−3 ml of ethanol and mixed with a solution of phenolphthalein and a drop of 0.1 N NaOH. If on heating the solution loses its color, the presence of an ester is indicated. The same reaction is also given by lactones.

In the identification of esters we are faced with problems of identifying their acid and alcoholic components. In principle, we can solve this problem in two ways − either by identifying the alcoholic component in one part of the sample and the acid in the other part, or by carrying out a hydrolysis of the ester and isolation of both reaction products, followed by their identification. The choice of method depends on the nature of the ester; preliminary physical tests can help in making the proper choice (melting point, boiling point, refractive index).

For the direct identification of the acid component exchange reactions are used in which the ester is transformed to an amide (using ammonia), or a substituted amide (using benzylamine) to hydrazide (with hydrazine − see p. 263) or morpholide (with N-β-aminoethylmorpholine) (49). The use of ammonia is limited by the solubility of amides. In addition, certain substituted esters may react in a different way (for example, haloesters react with benzylamine to give derivatives formed by the exchange of a halogen with the $C_6H_5CH_2NH-$ group). An advantage of the reaction of esters with hydrazine consists in the fact that the hydrazides can be transformed to a second series of derivatives by reaction with α-naphthylisocyanates (see p. 340). The direct identification of the acid component can also be effected by paper chromatography, after converting the esters to hydroxamic acids.

Paper Chromatography of Hydroxamic Acids

To 1 ml of the reagent (see the detection of esters) are added 1 − 2 drops of the tested substances, and the mixture is gently heated. The solution obtained can be spotted on the chromatogram directly (1 − 5 μl). Chromatography is carried out as described on p. 277.

General Method of Preparation of Hydrazides

The analyzed ester (1 ml) and 1 ml of 85 − 90% hydrazine are placed in a 15-ml flask and the mixture is refluxed for 10 − 15 min. Ethanol is then added to the mixture dropwise until complete dissolution. The heating is continued for 2 − 3 hr, the solvent is evaporated, and the crude hydrazide is recrystallized from aqueous ethanol.

Toluides or anilides can be prepared from esters of $C_1 - C_6$ fatty acids by making use of a Grignard reagent (50).

A direct method for the identification of the alcoholic component of esters consists in the reaction of the ester with 3,5-dinitrobenzoyl chloride in pyridine. In the case of simple esters (51) the reesterification can be carried out by heating the tested ester with 3,5-dinitrobenzoic acid under catalysis with sulfuric or p-toluenesulfonic acid.

General Method of Identification of the Alcoholic Components of an Ester by Reaction with 3,5-Dinitrobenzoyl Chloride

A solution of the ester (0.3 g) and 3,5-dinitrobenzoyl chloride (0.5 g) in pyridine (3 ml) in a 15-ml flask is refluxed for 3 hr and then allowed to cool. Dilute sulfuric acid (3 %; 10 ml) is then added to the mixture, and after cooling and thorough shaking this is extracted with 5 ml of ether (ethanol-free, dried over sodium). The ethereal layer containing 3,5-dinitrobenzoate is separated and washed successively with 5 ml of 3% sulfuric acid, 5 ml of 3% NaOH, and water. The ether is evaporated and the residue crystallized from aqueous alcohol (see p. 150).

A reliable method for the identification of the alcoholic component consists in the cleavage of the ester with hydriodic acid and conversion of the alkyl iodide formed either to S-alkylthiuronium 3,5-dinitrobenzoate or to an ester of 3,5-dinitrobenzoic acid. The procedure is described in detail on p. 201; simultaneously, the alcoholic component of the ester may be identified by paper chromatography. The advantage of this method consists in the small amount of substance (several milligrams) necessary for the analysis.

Hydrolysis of Esters

To carry out the cleavage of esters to their components, alkaline hydrolysis in an aqueous medium or in alcohol is commonly used; hydrolysis in ethylene glycol has certain advantages. Different procedures differ, depending on the properties of the acids and alcohols formed by the hydrolysis. The majority of esters with $C_1 - C_4$ alcohols can be hydrolyzed by boiling with a 25% NaOH solution for 30 min. Esters boiling above 200 °C require $2-3$ hr reaction time. The course of hydrolysis can be followed by observation of the layer of the ester, which disappears gradually; however, the alcohol set free must be water-soluble. The determination of the neutralization equivalent of the ester is a suitable guideline for the choice ohf ydrolytic procedure.

A very simple hydrolytic method consists in heating the ester with potassium hydroxide in ethylene glycol. The high reaction temperature, $180-200$ °C, causes a substantially higher rate of hydrolysis ($5-10$ min) than in the usual procedure. In addition the hydrolysis takes place in a homogeneous medium.

When esters with other functional groups are to be identified one should not forget that some other reactions could also take place; for example, when β-ketoesters are hydrolyzed they are further cleaved (decarboxylation), esters of halocarboxylic acids give hydroxy acids, etc.

Reactivity of Esters to Alkaline Hydrolysis

Most esters of carboxylic acids are hydrolyzed in an alkaline medium so that the bond between the acyl group and the oxygen atom of the RO— group is broken. The reaction can be represented by the following equation:

$$OH-\underset{R}{\underset{|}{\overset{O}{\overset{\|}{C}}}}-OR' \;\rightleftharpoons\; \left[HO\ldots\underset{R}{\underset{|}{\overset{O}{\overset{\|}{C}}}}\ldots OR'\right]^{\ominus} \;\rightleftarrows\; HO-\underset{R}{\underset{|}{\overset{O}{\overset{\|}{C}}}}+R'O^{\ominus} \;\rightarrow\; RCOO^- + R'OH$$

With respect to this mechanism, the rate of hydrolysis is influenced by electrophilic properties of the carbonyl carbon atom (attached substituents attracting electrons increase the reactivity of this carbon), and by sterical hindrance, which prevents a smooth access of the hydroxyl ion and decreases the hydrolysis rate. If the ester molecule contains groups stabilizing the carbonyl group by conjugation, the rate of hydrolysis diminishes (benzoates; esters; of α,β-unsaturated acids).

By way of illustration, the relative rates of alkaline hydrolysis of some esters in water [cf. (52)], referred to methyl acetate, are given in parentheses: CH_3COOCH_3 (1.00); $CH_3COOC_2H_5$ (0.601); $CH_3COOCH(CH_3)_2$ (0.146); $CH_3COOC(CH_3)_3$ (0.0084); $CH_3COOC_3H_7$ (0.549); $CH_3COOCH=CH_2$

(57.7); $CH_3COOC_6H_5$ (7.63); $CH_3COOCH_2CH_2OH$ (1.52); $HCOOCH_3$ (223); $ClCH_2COOC_2H_5$ (761); $HOCH_2COOC_2H_5$ (11.9).

It should be pointed out that the reactivity of tertiary alcohols to alkaline hydrolysis is very low. In such cases acid hydrolysis is more advantageous.

Further Procedure after the Hydrolysis of Esters

Generally, for the identification of acids and alcohols obtained on hydrolysis the same reactions are used as described on pp. 148 and 247. The simplest case of identification is an ester in which the alcohol can be steam-distilled. After alkaline hydrolysis the alcohol is steam-distilled and identified in the distillate as 3,5-dinitrobenzoate (see p. 150). The acid in the alkaline residue after steam distillation is identified by one of the following procedures: (1) the acid is precipitated by acidification (insoluble acids, for example, aromatic), (2) it is precipitated from neutralized solution with S-1-naphthylmethylthiuronium chloride (see p. 251), or (3) the acid is distilled off from the acid medium by steam distillation, the distillate is neutralized and evaporated to dryness, and the sodium salt is converted either to the p-bromophenacyl ester (p. 253) or p-bromoanilide.

If the alcohol formed on hydrolysis is not volatile with steam, it can be extracted with ether and converted to a 3,5-dinitrobenzoate (see p. 150) either directly, or after the evaporation of ether. If both the alcohol and the acid are soluble in water and are not volatile with steam, the hydrolyzate is divided into two portions; in the first the acid is identified (by precipitation with S-1-naphthylmethylthiuronium chloride) and in the second the alcohol is identified (by reaction with 3,5-dinitrobenzoyl chloride (see p. 150). For the latter case the direct identification of the acid and the alcoholic components previous to hydrolysis is suitable.

The hydrolyzate can also by utilized for the identification of the acid and the alcohol by paper chromatography. For the identification of the acid the hydrolyzate is applied directly (see p. 259) and the alcoholic component is chromatographed after its conversion to 3,5-dinitrobenzoate.

The following procedures illustrate with practical examples the identification of the most important types of esters.

Butyl Acetate

Reagents: 3,5-dinitrobenzoyl chloride, 5% H_2SO_4, 50% NaOH, ether, ethanol, anhydrous pyridine, S-1-napthylmethylthiuronium chloride, 0.1 N NaOH.

Procedure: A mixture of 0.1 ml of butyl acetate and 0.5 ml of 50% NaOH is placed in an ampoule of 4 ml volume which is then sealed and heated in a boiling water bath for 2 hr. The ampoule should be shaken occasionally. For safety it should be wrapped in a piece of cloth. After cooling, the ampoule is opened and the contents transferred into a distillation flask by rinsing the ampoule with 15 ml of water. A part (10 ml) of the liquid is then distilled off, 5 ml of water are added, and an additional 5 ml are distilled off. In the combined distillates the alcohol can be identified by converting it to an ester of 3,5-dinitrobenzoic acid. Yield, 54 mg; mp, 55—57 °C. After crystallization from 5 ml of ethanol with the addition of two drops of water 10 mg of the product were obtained, mp 66 °C. The residue in the flask is neutralized with 5% sulfuric acid and an excess of sulfuric acid (5 ml) is added to it. Two-thirds of the contents of the flask are then distilled off and the distillate is titrated with 0.1 N NaOH. The distillation should be carried out carefully to prevent traces of sulfuric acid from passing into the distillate. The consumption of 0.1 N NaOH is 5.5 ml. Then 40 mg of S-1-naphthylmethylthiuronium chloride are added to the distillate and the mixture heated until a clear solution results. After cooling in a refrigerator, the separated salt is isolated by filtration using a filtration tube (67 mg; mp 142 to 144 °C) and is then crystallized from 3 ml of ethanol. Yield, 26 mg; mp, 145 °C.

Diethyl Adipate

Reagents: solid KOH, diethylene glycol, 20% sulfuric acid.

Procedure: One pellet of KOH and 0.15 ml of diethylene glycol, dissolved in 2.2 ml of water, are placed in a 5-ml flask (with a ground joint), and, after cooling, 0.5 ml of diethyl adipate are added to the mixture, which is then heated under reflux for 10 min (with occasional shaking; the ester layer disappears eventually). The mixture is then cooled and diluted with 3 ml of water, the flask is connected with a condenser, and 3 ml of liquid are distilled off. An additional 2 ml of water are added to the mixture in the flask and 2 ml of distillate are collected again. The residue in the flask is neutralized (with respect to Congo paper) with 20% sulfuric acid and cooled in a refrigerator. The separated crystals are collected on a funnel (under suction) and washed twice with 0.5 ml of water. Yield of the crude product is 0.35 g, mp 146—151 °C. After crystallization from 3 ml of water the yield of the pure product is 0.10 g, mp 151—152 °C. Combined distillates are diluted with water to make the volume 10 ml and they are then worked up as described on p. 150 for the identification of aqueous solutions of alcohols; yield, 236 mg; mp, 91.5—92 °C.

Methyl Salicylate
(Identification of the Acid Component)

Reagents: 5 N NaOH, ethylene glycol, 5% sulfuric acid, ethanol, S-1-naphthylmethylthiuronium chloride.

Procedure: Methyl salicylate (0.05 ml) in a test tube is mixed with 0.2 ml of 5 N NaOH and 1 ml of ethylene glycol and the mixture boiled for 5 min. After dilution with 3 ml of water the mixture is neutralized with sulfuric acid (decoloration of phenolphthalein). The neutral solution is mixed with 5 ml of a saturated solution of S-1-naphthylmethylthiuronium chloride and, with warming, ethanol until dissolution is complete. The mixture is then filtered through a small folded filter and the filtrate is cooled in a refrigerator. The crystallized salt is collected (with suction) on a filtration tube; yield, 50 mg; mp, 167 – 168 °C.

Methyl Salicylate

(Identification of the Alcoholic and the Acid Components)

Reagents: ethanolamine, HCl 1 : 1, 50% aqueous ethanol, 3,5-dinitrobenzoyl chloride.

Procedure: A mixture of 1 g of methyl salicylate and 3 ml of ethanolamine in a 10-ml flask is refluxed for 1 hr. After cooling, the mixture is carefully acidified with hydrochloric acid (test with Congo paper) and 5 ml of liquid are distilled off. The residue in the flask is cooled in a refrigerator, and the separated amide is collected by filtration on a filtration tube and then washed twice with 1 ml of water. Yield, 0.56 g; mp, 110 °C. Crystallization from 10 ml of 50% ethanol gives 0.26 g, mp 115 – 116 °C. The distillate is worked up as in the procedure given on p. 152 for the identification of alcohols in aqueous solutions by reaction with 3,5-dinitrobenzoyl chloride.

Benzyl Benzoate

Reagents: 50% KOH, methanol, 3,5-dinitrobenzoyl chloride, pyridine, benzene, cyclohexane, 5% sulfuric acid, 5% NaOH, potassium carbonate, ether, S-1-naphthylmethylthiuronium chloride, anhydrous sodium sulfate.

Procedure: 0.1 ml of benzyl benzoate, 0.5 ml of 50% KOH, and 0.5 ml of methanol are mixed in a 10-ml flask and refluxed for 1 hr (oil bath). After cooling, the mixture is diluted with 10 ml of water and then extracted in a separatory funnel three times with 5 ml of ether. The combined ethereal extracts are dried with sodium sulfate and filtered into a filtration flask, and ether is distilled off. The flask with the residue is then heated for 10 min in a drying oven at 105 – 110 °C, and after a short cooling a solution of 0.5 g of 3,5-dinitrobenzoyl chloride in 2 ml of benzene and 0.1 ml of pyridine is added to it. Potassium carbonate (11 g) is then added to the mixture while cooling externally with ice, to make the solution saturated, and the mixture is shaken for 3 min. Ether (30 ml) is then added and allowed to stand with occasional shaking. The ethereal layer is separated and washed successively with two 5-ml portions of 5% sulfuric acid, 5 ml of water, two 5-ml portions of 5% NaOH, and twice with water. The ethereal layer is dried with sodium sulfate, filtered, and distilled off. The residue is crystallized from cyclohexane, yielding 50 mg of product, mp 112–113 °C. After extraction of benzyl alcohol with ether the alkaline aqueous solution is neutralized (to phenolphthalein) with 5% sulfuric acid, mixed with 5 ml of a saturated aqueous solution of S-1-naphthylmethylthiuronium chloride, and then with hot ethanol until homogeneous. The mixture is filtered through a small folded filter and the filtrate cooled. Yield, 106 mg of crystals; mp, 173–174°C

Ethyl Acetate

Reagents: 50% KOH, S-1-naphthymethylthiuronium chloride, HCl 1:1.

Procedure: A mixture of ethyl acetate (0.2 ml) and 50% KOH (1 ml) is placed into an ampoule, which is sealed and heated in a boiling water bath for 30 min (the ampoule should be wrapped in a piece of cloth). After partial cooling, the ampoule is thoroughly shaken and again put into the water bath for an additional 30 min. The cooled ampoule is opened and the contents transferred, rinsing the ampoule with 5 ml of water, into a distillation flask. After collecting 5 ml of distillate the flask is replenished with 5 ml of water and distilled again to collect an additional 5 ml of the distillate. This is then worked up as in the procedure for the identification of alcohols in aqueous solutions described on p. 150. Yield, 91 mg; mp 91–92 °C. After crystallization from 0.5 ml of 80% ethanol, 83 mg of the product are obtained, mp 91–92 °C. The residue in the flask is neutralized with hydrochloric acid and the pH is adjusted to 3–4; 0.1 g of S-1-naphthylmethylthiuronium chloride is added to the solution, which is heated until dissolution is complete. The addition of water may also help. After filtration through a folded filter the filtrate is cooled for crystallization. The separated product is collected on a filtration tube, washed twice with 0.5 ml of water, and dried. Yield, 0.25 g; mp, 146–148 °C; after crystallization from 5 ml of ethanol, the yield is 0.14 g, mp 146–147 °C.

3,5-Dinitrobenzoate of 3,4-Xylenol

(Isolation of the Phenolic Component from Its Ester)

Reagents: 10% NaOH, 10% H_2SO_4, anhydrous sodium sulfate, ethanol, ether.

Procedure: A mixture of 1 g of 3,5-dinitrobenzoate of 3,4-xylenol and 50 ml of 10% NaOH is placed into a 100-ml flask fitted with a reflux condenser and boiled for 1 hr. The cooled solution is saturated with carbon dioxide and extracted consecutively with 10 ml and twice with 5 ml of ether. The pooled extract is dried over sodium sulfate, filtered into a distillation flask, and concentrated to a volume of approx. 5 ml. The residue is transferred gradually into a Hickmann flask (each time the remaining ether is evaporated) and distilled carefully. The distillate (0.2 g of 3,4-xylenol) is sucked into a capillary tube.

3. Acid Chlorides and Anhydrides (Detection and Identification)

The high reactivity of chlorides is utilized for the preparation of carboxylic acid derivatives by first converting the acids to chlorides and then allowing them to react with amines or alcohols to give suitable derivatives. For the identification of acid chlorides, reactions described on p. 254 are employed. A simple method of preparation of amides from acid chlorides is illustrated by the preparation of stearoylamide. Other useful amines are aniline or p-toluidine; the reaction of amines with chlorides is usually carried out by heating the components in an inert solvent (benzene).

Stearoylamide (Identification of stearoyl chloride)

Reagents: ammonia, ethanol.

Procedure: Stearoyl chloride (100 mg) and 1.5 ml of ammonia are well mixed in a 10-ml test tube by stirring with a glass rod, and the mixture is allowed to stand in a refrigerator for 30 min. The precipitated amide is filtered off, washed twice with 0.5 ml of water, and dried in a stream of dry air. The yield of the product was 0.1 g, mp 93 – 96 °C. To crystallize the crude amide, it is dissolved in 1 ml of hot ethanol, water is added to incipient turbidity, the mixture is heated again until it is homogeneous, and it is then allowed to cool slowly. Yield, 85 mg; mp, 99 – 99.5 °C.

Should the chloride molecule contain certain groups which would cause the corresponding amide to have too high a melting point (nitro groups, halogens), an ester is prepared from the chloride by reaction with methanol or ethanol (see p. 252). Chlorides of aromatic acids yield solid acids on hydrolysis, which makes it possible simply to heat the tested substance with an aqueous solution of soda until dissolved and to precipitate the acid by acidification of the solution.

Hydrolysis of 3,5-Dinitrobenzoyl Chloride

Reagents: 5% NaOH, 10% HCl, ethanol.

Procedure: 10 mg of 3,5-dinitrobenzoyl chloride and 1 ml of 5% NaOH are mixed in a micro test tube and heated until dissolution is complete. The mixture is acidified with hydrochloric acid (test with Congo paper) and allowed to stand for 20 min in a refrigerator. The separated acid is filtered off using a filtration tube and washed with 0.5 ml of 10% HCl, twice with 0.5 ml of water, and 0.2 ml of ethanol. Yield, 6 mg; mp 210–214 °C; after crystallization from 0.5 ml of ethanol the yield is 3 mg; mp 210–211 °C.

Reaction with Hydroxylamine

Acid anhydrides and chlorides react with hydroxylamine in the same way as esters, with the formation of hydroxamic acids. The color reaction and paper chromatographic analysis can be carried out as described on p. 262 for esters. The reaction conditions can be adjusted so that anhydrides can be detected or identified in the presence of esters. However, acid chlorides react under both sets of conditions (53, 54).

Additional Color Reactions

Anhydrides of dicarboxylic acids give, when melted with resorcinol and zinc chloride, corresponding fluoresceins (p. 197). Acyclic anhydrides of carboxylic acids react with α-(p-nitrobenzoyl) aminophenylacetic acid with the formation of a blue color (55).

Acyclic acid anhydrides react with alcohols (phenols) to give esters and carboxylic acids; acid esters (monoesters) are formed from cyclic anhydrides;

$$(RCO)_2O + R'H \rightarrow RCOOR' + RCOOH$$

$$R\begin{array}{c}CO\\ \diagdown \\ CO\end{array}O + R'OH \rightarrow R\begin{array}{c}COOH\\ \diagdown \\ COOR'\end{array}$$

In the first case, reaction with amines leads to amides; amido acids formed when amines react with cyclic anhydrides can be recyclized to cyclic imides (p. 344). When mixed acyclic or asymmetrical cyclic anhydrides react, a mixture of products can be formed, although the stronger acid usually gives an ester (or amide) (56). Reaction of alcohols with anhydrides very often takes place even in the cold; certain anhydrides, however, are resistent to heat and can be crystallized from ethanol (57); they can be cleaved by heating with sodium ethoxide in ethanol or benzene. If the acids formed by hydrolysis are solid and suitable for identification, this procedure can be considered as the simplest for identification purposes. When anhydrides of liquid acids are to be identified, the reaction with aromatic amines is generally employed; mixed anhydrides are best identified by chromatography of acids formed on alkaline hydrolysis.

Hydrolysis of Phthalic Anhydride

Reagents: 10% NaOH, HCl 1 : 1.

Procedure: In a micro test tube 100 mg of phthalic anhydride are dissolved in 2 ml of 10% NaOH by gentle heating, and the mixture is acidified, after cooling, with hydrochloric acid (1 : 1) (test with Congo paper) and allowed to stand in a refrigerator for 30 min. The separated acid is collected on a filtration tube, washed twice with 0.5 ml of water, dried, and weighed. Yield, 120 mg; mp, 213 °C. After crystallization from 1 ml of water the yield was 80 mg, mp 213—214 °C.

4. Acid Amides
(Detection and Identification)

Acid amides give a color reaction with fluorescein chloride which has already been described in the discussion on amines (p. 324). The biuret reaction is given by compounds which contain a group in which two carbonamide groups are bound to one carbon or nitrogen atom, as, for example, in malonamide (I), biuret (II), or oxamide (III); in an alkaline medium these substances give a red-violet complex compound with cupric hydroxide.

$$CH_2 \Big\langle \begin{smallmatrix} CONH_2 \\ CONH_2 \end{smallmatrix} \qquad NH \Big\langle \begin{smallmatrix} CONH_2 \\ CONH_2 \end{smallmatrix} \qquad \begin{smallmatrix} CONH_2 \\ | \\ CONH_2 \end{smallmatrix}$$

$$\text{(I)} \qquad\qquad\qquad \text{(II)} \qquad\qquad \text{(III)}$$

If two hydrogens of the NH_2 groups in these compounds are substituted, the reaction does not take place. A positive reaction is given by all compounds having at least two $-CONH_2$, $-CSNH_2$,

$$\text{or} \quad -C \Big\langle \begin{smallmatrix} NH \\ NH_2 \end{smallmatrix}$$

groups in position next to the copper atom entering the molecule. Thus, a positive biuret reaction is given by proteins and the products of their degradation to tetrapeptides. In the case of tri- and dipeptides the reaction is not always clear-cut, and it is not given by amino acids.

Reagents: 5% NaOH; 1% $CuSO_4 \cdot 5\,H_2O$ solution.

Procedure: Excess sodium hydroxide is added to an aqueous solution or suspension of the tested amide in a test tube, followed by dropwise addition of copper (II) sulfate. After each drop the mixture should be shaken. The cupric hydroxide formed temporarily dissolves gradually and a solution is formed which is first pink, and then changes to violet and blue-violet.

The reaction with hydroxylamine, leading to hydroxamic acids, takes place less well with amides than with other derivatives of carboxylic acids. It must be carried out at elevated temperatures, for example, in propylene glycol (58, 59). The detection can be carried out in two ways:

a) *Reagents:* methanolic 0.5 N hydroxylamine hydrochloride solution; aqueous 10% $FeCl_3$ solution.

Procedure: Approximately 50 mg of the tested amide are boiled for 3 min in 1 ml of hydroxylamine hydrochloride solution. After cooling, a few drops of $FeCl_3$ solution are added and the formation of a red color is observed.

b) *Reagents:* 1 M hydroxylamine hydrochloride in propylene glycol; 5% $FeCl_3$ solution in alcohol.

Procedure: Approximately 30 mg of the tested amine are boiled in a test tube with 2 ml of hydroxylamine hydrochloride solution. After cooling, 1 ml of the $FeCl_3$ solution is added: a red color is formed.

Paper chromatography of fatty acid amides from C_{10} up was carried out by Kaufmann (60) on papers impregnated with undecane in 70% acetic acid as the mobile phase. Detection is carried out with 3% mercuric acetate followed by diphenylcarbazone.

Among direct methods of identification of acid amides, the reaction with xanthydrol (61, 62) should be mentioned, leading to N-xanthylamides, as should the reaction with phthalyl chloride (N-acylphthalimides are formed) and with diphenylmethanol [which leads to N-(diphenylmethyl) amides] (63);

$$RCONH_2 + \text{(anthracenol)} \rightarrow \text{(N-acyl anthracene)} + H_2O$$

$$RCONH_2 + \text{(benzene-COCl, COCl)} \rightarrow \text{(CO, CO)NCOR} + 2\,HCl$$

Finally, we should like to point out the possibility of preparing oxalates (64) and Hg salts (65). The use of these methods, among which the preparation of xanthyl derivatives and oxalates will be illustrated by practical examples, is limited. Xanthyl and phthaloyl derivatives can be prepared from derivatives containing a $-CONH_2$ group; oxalates of amides have been prepared only from low-molecular-weight amides.

A general method of identification of amides (as well as of esters) consists in using their hydrolysis and the identification, separately, of the acid and the basic components. For hydrolysis both alkalies (aqueous or alcoholic KOH or NaOH solutions) and acids (HCl, HBr, $40-70\%$ H_2SO_4, orthophosphoric acid) are used at temperatures depending on the stability of the amide toward the hydrolysis. An increase in the temperature of hydrolysis can be achieved either by carrying out the hydrolysis in a higher-boiling solvent (KOH in glycol or glycerol makes it possible to carry out the hydrolysis at 200 – 300 °C), or by carrying it out in a sealed tube (hydrolysis with hydrochloric acid). Depending on the mode of cleavage, the final identification of the hydrolysis products can be carried out in several ways. The simplest method of hydrolysis of amides consists in heating the amide (3 – 5 mmole) in a distillation flask with 15 – 20 ml of 20% NaOH and collecting the distillate with simultaneous replacement of the evaporated water in the flask. The hydrolysis is interrupted as soon as the distillate is no longer alkaline. The distillate is neutralized with N HCl (which also serves for the determination of the quantity in moles of the base) and worked up by converting the base to some of the derivatives listed on p. 326. It is also possible to evaporate water first and to work up the hydrochloride of the base. The procedure for the isolation and identification of the alkaline solution containing the acid is analogous to that described for the identification of esters (p. 265).

The simplest method of acid hydrolysis of amides consists in heating the amide (3 – 5 mmole) in a distillation flask with 15 – 20 ml of 40% sulfuric acid. The distillate is analyzed for the presence of volatile acids (a test of the absence of sulfuric acid has to be carried out); after the neutralization of the distillate one of the methods described on p. 265 is applied. On the hydrolysis of an amide of a nonvolatile acid the latter either separates out from the solution or may be extracted with ether. An amide of a nonvolatile, water-soluble acid and an equally nonvolatile, water-soluble amine (for example, ethanolamine) presents a more complicated case. Here either a derivative of the original amide is prepared (such amides usually contain groups enhancing their solubility, for example, hydroxyl groups, which can be acetylated or benzoylated), or ion exchangers are used (66) for the isolation of hydrolytic products. The basic procedure for the hydrolysis of amides and the identification of the hydrolytic products will be described in practical examples. Paper and thin-layer chromatography are powerful tools for the identification of the acid and alcoholic components of the esters and the acid and basic components of the amides. The hydrolysis products can usually be used for the preparation of derivatives suitable for chromatography without previous isolation (see pp. 336, 260). In some cases the acid and the alkaline hydrolysis of amides on a microscale can be carried out by heating in a sealed ampoule, and the hydrolysis product can be identified directly by paper or thin-layer chromatography (67 – 69).

On lithium aluminum hydride reduction in ether, amides yield the corresponding amines,

$$R_1CONR_2R_3 \xrightarrow{\text{LiAlH}_4} R_1CH_2NR_2R_3$$

This procedure is useful because the reaction takes place under very mild conditions; however, the absence of other functional groups reducible with $LiAlH_4$ is a necessary condition.

Hydrolysis of Acetamide

Reagents: 2 N NaOH, 1% formaldehyde in water, 2 N HCl, S-1-naphthylmethylthiuronium chloride, ethanol.

Procedure: In a flask fitted with a reflux condenser 0.2 g of acetamide and 10 ml of 2 N NaOH are heated and distilled slowly. The distillate is collected in a flask containing 10 ml of 1% formaldehyde. The amount of liquid distilled off is made up (restored) from a dropping funnel until 20 ml of the distillate is collected. Eventually, the residue in the distillation flask is concentrated to 5 ml, neutralized with hydrochloric acid to pH 3−4, and mixed with 0.1 g of S-1-naphthylmethylthiuronium chloride. The mixture is heated until dissolved, and filtered through a folded filter, and the filtrate is allowed to cool in a refrigerator. The separated salt is filtered and washed twice with 0.5 ml of water (yield, 0.11 g; mp, 142−143 °C). After crystallization from 4 ml of methanol the yield of pure product was 50 mg, mp 143 to 145 °C. The distillate collected in the formaldehyde solution is concentrated to 5 ml, mixed with 5 ml of a saturated picric acid solution in ethanol, and allowed to stand in a refrigerator. The picrate of hexamethylenetetramine is isolated by filtration and washed twice with 0.5 ml of water. Yield, 64 mg; mp, 180−185 °C. After crystallization from 20 ml of ethanol the product (20 mg) melted at 183 °C.

Hydrolysis of Formanilide

Reagents: 10% NaOH, p-toluenesulfonyl chloride, acetone, HCl 1 : 1, ethanol, S-1-naphthylmethylthiuronium chloride.

Procedure: Formanilide (1 g) is refluxed in a 25-ml flask with 10 ml of 10% NaOH for 1 hr. The reflux condenser is then substituted by a normal one and the mixture is distilled. After 5 ml of condensate have been collected 10 ml of water are added to the distillation flask and the distillation is continued until the volume is only ~ 5 ml. The distillate is collected in a 200-ml ground-joint Erlenmeyer flask and 10 ml of 10% NaOH and 5 ml of an acetone solution of 2 g p-toluenesulfonyl chloride are added. The mixture is shaken vigorously for 10 min and the solution is filtered through a folded filter. The filtrate is acidified with HCl until clearly acid and the solution is allowed to stand in a refrigerator for 30 min. The separation of crystals is enhanced by scratching and the precipitated sulfonylamide is filtered off and washed with 5 ml of 50% aqueous ethanol. Yield, 0.9 g; mp, 99−100 °C; crystallization from 13 ml of 50% aqueous ethanol gave 0.3 g of product, mp, 101 °C. The residue in the flask is acidified with hydrochloric acid to pH 3—4, then diluted with 20 ml water, and filtered, and 1 g of S-1-naphthylmethylthiuronium chloride added to the filtrate. The mixtureis heated and diluted with water until the dissolution is complete, and is then allowed to crystallize. Yield, 0.7 g; mp, 166—167 °C. After crystallization from 55 ml of ethanol the yield of the pure product is 0.3 g, mp, 167 to 169 °C.

Hydrolysis of 3,5-Dinitrobenzamide

Reagents: solid KOH, diethylene glycol, 1% HCl, ether.

Procedure: In a 5-ml ground-joint flask one pellet of KOH is dissolved in 0.3 ml of diethylene glycol and 0.2 ml of water. After cooling, 100 mg of 3,5-dinitrobenzamide are added and the flask is provided with a dropping funnel containing water. The mixture is boiled for 10 min and the condensate is collected in a flask containing 10 ml of 1% HCl. From the dropping funnel 3 ml of water are added to the distillation flask and the distillation is continued. This procedure is repeated until the distillate is no longer alkaline. Evaporation of the distillate gives ammonium chloride.

Hydrolysis of N-Methylsuccinimide

Reagents: 10% NaOH; suspension of a cation exchanger in water; 5% HCl.

Procedure: In a 10-ml flask provided with a dropping funnel 200 mg of N-methylsuccinimide are mixed with 2.5 ml of 10% NaOH. A condenser is connected to the flask and the mixture is slowly distilled until the distillate is no longer alkaline. The distillate is collected in a flask containing 10 ml of 5% HCl. The liquid distilled off from the flask is made up from the dropping funnel. The mixture in the distillation flask is then transferred into a 50-ml beaker, diluted with 15 ml of water, and mixed with 5 ml of the cation-exchanger suspension. After standing for 30 min (with occasional shaking) the liquid is decanted into another beaker, where it is mixed with 2 ml of catex suspension and heated on a water bath for 10 min at 50—60 °C. It is then filtered through a folded filter and the filtrate is evaporated on a porcelain dish to dryness. The residue is dried at 105 °C; yield, 180 mg; mp, 184—186 °C. The distillate containing methylamine hydrochloride is worked up to 3,5-dinitrobenzamide (see p. 336).

Formamide Oxalate

Reagents: anhydrous oxalic acid, anhydrous ethyl acetate.

Procedure: Anhydrous oxalic acid (0.9 g) and formamide (0.45 g) are mixed in a test tube and heated in a water bath at 60—65 °C for 10 min. Ethyl acetate (45 ml) is added and the melt is stirred to pass it into solution while heating it in the water bath. The warm solution is filtered through a folded filter, and the crystals separated after cooling are filtered off under suction and washed twice with 0.5 ml ethyl acetate. Yield, 0.65 g; mp, 109–110 °C. After crystallization from 15 ml of ethyl acetate the product (0.49 g) melted at 110 °C.

Xanthyl Derivative of Formamide

Reagents: xanthydrol (5% solution in glacial acetic acid); 70% dioxane. If xanthydrol is not completely soluble in acetic acid, it should be crystallized from ethyl acetate.

Procedure: Formamide (70 mg) is mixed in a 5-ml micro test tube with 2 ml of a 5% xanthydrol solution in glacial acetic acid and the tube is immersed in a water bath at 80 °C for 20 min. It is then allowed to cool in a refrigerator for 20 min and the product separated is filtered under suction on a filtration tube. Yield, 0.13 g; mp, 184–187 °C. After crystallization from 4 ml of 70% dioxane the product (0.10 g) melted àt 183 to 185 °C. If the product does not separate after prolonged standing in the refrigerator, the mixture should be diluted with a few drops of water.

Reduction of Acetamide with $LiAlH_4$

Reagents: 10% NaOH, ethereal $LiAlH_4$ solution (25 mg/1 ml).

Procedure: A dry, ground-glass-joint, 100-ml flask provided with a reflux condenser and a protecting tube filled with calcium chloride and solid sodium hydroxide is filled with 0.1 g of acetamide, 5 ml of ether, and 10 ml of the $LiAlH_4$ solution (measured quickly with a pipette and added through the reflux condenser). The mixture is refluxed for 15 min and then allowed to stand for an additional 20 min. Water is then added dropwise through the condenser until all excess hydride is decomposed, followed by 10 ml of 10% NaOH. A descending condenser is then substituted for the reflux condenser and the mixture is distilled to eliminate all the ether and 10 ml of the water. Five milliliters of water may be added to the distillation flask. The distillate containing the volatile amine is worked up as described on p. 335 to 3,5-dinitrobenzamide. Yield, 40 mg; mp, 121–124 °C.

5. Hydroxamic Acids

Hydroxamic acids RCONHOH are best identified by means of paper or thin-layer chromatography. On hydrolysis in an acid medium hydroxamic acids yield corresponding carboxylic acids; the isolation and identification of carboxylic acids obtained on hydrolysis of hydroxamic acids are the same as discussed earlier in the case of the hydrolysis of esters.

Reactions with Metal Salts

Hydroxamic acids give with cupric salts colored insoluble salts (70); with ferric salts in an aqueous or alcoholic medium they give intense red or violet-red salts. Their composition changes according to the ratio of Fe^{3+} salt and hydroxamic acid used, and according to the acidity and the solvent used (71). The procedure for the color reaction is described in the discussion of esters on p. 226.

Paper Chromatography

Lower hydroxamic acids $C_1 - C_6$ are best chromatographed in n-amyl alcohol—acetic acid—water (6 : 2 : 1) and the higher members of the homologous series on papers impregnated with lauryl alcohol and using the following mobile phases: ethanol — acetic acid — water (2 : 1 : 3) or (2 : 1 : 6.) The amount of substance chromatographed may be 5 – 40 μg. Detection is carried out by spraying the chromatogram with 1% ferric chloride in ethanol. Hydroxamic acids appear as violet spots on a yellow background. A chromatogram of hydroxamic acids is shown in Fig. 33.

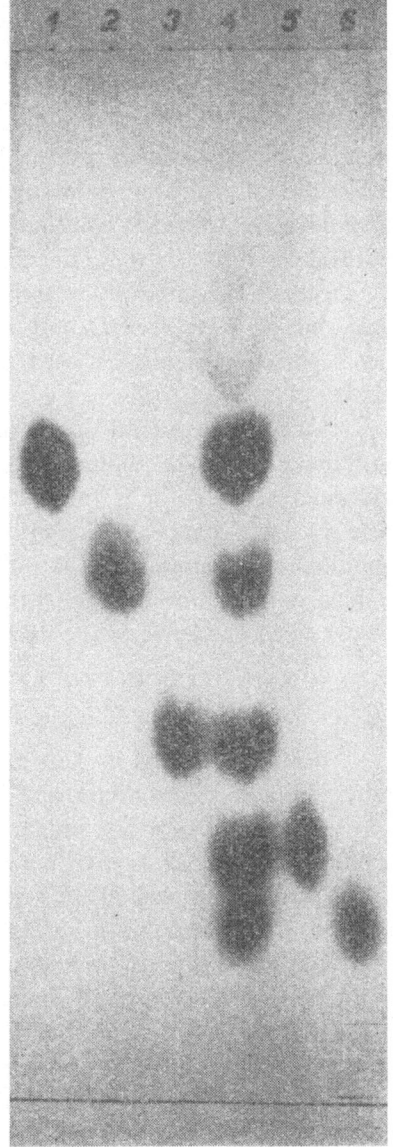

Fig. 33. Chromatogram of fatty acids converted to hydroxamic acids. Solvent system amyl alcohol-acetic acid- water (6 : 2 : 1). (1) Formic acid, (2) acetic acid, (3) propionic acid, (4) mixture, (5) butyric acid, (6) valeric acid.

6. Amino Acids

Detection and Identification Possibilities

Amino acids differ characteristically from one another depending mainly on the character of the amino group. Aliphatic amino acids are essentially different from those amino acids whose amino group is bound to an aromatic nucleus. Aromatic amino acids usually do not present a problem of identification, and methods described for the identification of amino compounds can well be used in this case too (p. 326). Characteristic representatives of aliphatic amino acids are α-amino acids obtained on hydrolysis of proteins. α-Amino acids are nonvolatile, water-soluble compounds, melting usually above 200 °C and over a pretty wide interval. Therefore, the melting point of an amino acid cannot be considered as a suitable identification constant, and it must always be compared with an authentic sample, or the influence of the rate of heating on its value must be checked. This property is ascribed to the dipolar character of the molecule, $NH_3^+ - CHR - COO^-$. It is important to stress that paper or thin-layer chromatography is used for the reliable identification of amino acids.

Aliphatic α-amino acids, and also aromatic ones with the amino group bound on an aliphatic side chain, give a color reaction with ferric chloride: a red color is formed, disappearing on addition of acid. They also react with copper (II) chloride (for example, if an aqueous or alcoholic solution of an amino acid is boiled with copper (II) oxide or carbonate, a blue color is formed) or with alkali hypochlorites (aliphatic α-amino acids), which give rise to the next lowest aldehyde, for example (72),

$$NH_2CH_2COOH + NaClO \rightarrow HCHO + NH_3 + CO_2 + NaCl$$

which can be detected with Schiff's reagent (p. 217).

a) *Reagents:* 0.5% aqueous solution of sodium 1,2-naphthoquinonesulfonate; dilute acetic acid 1 : 4; 5% sodium dithionite.

Procedure: Freshly prepared reagent solution (1 ml) is added to a dilute aqueous solution of the α-amino acid (5 ml) and allowed to stand for 15 min in the dark. Dilute acetic acid (1 ml) is then added, followed by 1 ml of dithionite solution; a red color is formed (73, 74).

b) *Reagents:* sodium hypochlorite solution; Schiff's reagent (p. 217).

Procedure: To a small amount of α-amino acid a few drops of sodium hypochlorite solution are added and the mixture is gently heated. After cooling, an excess of Schiff's reagent is added; a red color is formed.

Among other reactions of α-amino acids the reaction with *p*-nitrobenzoyl chloride can be mentioned, which is carried out in a hot alkaline medium (75) or in the presence of pyridine in the cold (blue color) (76); azoxolone derivatives are formed (77).

Reagents: 5% *p*-nitrobenzoyl chloride in benzene; pyridine.

Procedure: To a dilute aqueous α-amino acid solution (1 ml) pyridine (5 ml) is added and, after stirring, 1 ml of the reagent: various colors are formed—for example, orange with glycin, tryptophan, and α-alanine; violet with cystine.

Much more important than these reactions is the color test of amino acids with ninhydrin (78−81). The reactions with isatin or alloxan (82) belong to the same group. The course of the reaction of amino acids with ninhydrin is not yet completely understood. At present it is supposed that in the first step Strecker's degradation of amino acids takes place accompanied by the formation of carbon dioxide, ammonia, and an aldehyde possessing one carbon atom less than the original amino acid. Ninhydrin (I) is supposed to be reduced simultaneously to diketohydrindol (II), which reacts with the liberated ammonia and a second molecule of ninhydrin to form the intensely blue anion of diketohydrindenediketohydrindamine (III).

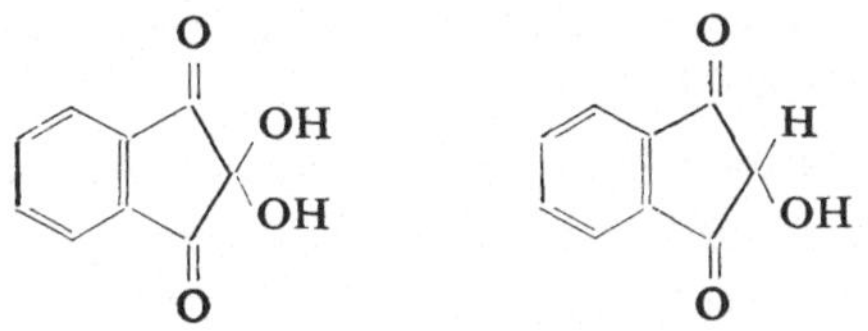

(I)   (II)

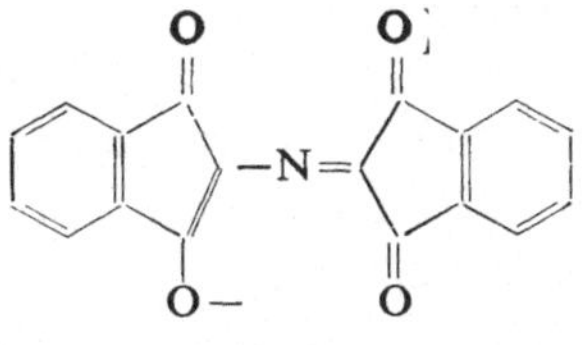

(III)

This reaction is also given with various reducing agents and with ammonia in the absence of amino acids. However, the course of the reaction with amino acids is probably more complicated, because aliphatic amines react as well and different amino acids give somewhat differing colors. More attention was given to reactions of ninhydrin or isatin with cyclic imino acids, which first give a yellow color (83). The reaction is also given with a series of aliphatic amino compounds and proteins, as well as the products of their cleavage. For a comparison of practical tests with ninhydrin, isatin, and alloxan see (84).

Reagent: Ninhydrin (0.1 g) dissolved in 300 ml of water.

Procedure: Two drops of the reagent are added to 1 ml of a neutral amino acid solution and the mixture is heated over a water bath. Red to violet color is formed with most amino acids; proline and oxyproline give yellow colors.

Derivatives for the identification of amino acids are prepared by substituting the amino group. In addition to the acetylation (85, 86), benzoylation (87), and 3,5-dinitrobenzoylation (88), the reaction with 2,4-dinitrofluorobenzene (89) is also suitable for the preparation of derivatives. This reagent reacts with amino acids at room temperature.

General Method of Preparation of Acetyl Derivatives

To a solution of 1 g of amino acid in a minimum amount of water are added 4 ml of acetic anhydride, and the mixture is vigorously shaken in a separatory funnel for 5 min. If the liquid gets warm spontaneously, the shaking is prolonged for an additional 20 min. In the opposite case, the solution is refluxed for 30 min over a boiling water bath. The solution is then cooled with ice (or with cooling mixture) and the separated acetyl derivative is crystallized from water or aqueous alcohol.

General Method of Preparation of Benzoyl and 3,5-Dinitrobenzoyl Derivatives of Amino Acids

In a 10-ml test tube containing 1 mmole of the acid dissolved in 3 ml of 1 N NaOH, 1 mmole of benzoyl chloride is added (or 1 mmole of 3,5-dinitrobenzoyl chloride dissolved in 2 ml of benzene). The test tube is stoppered with a rubber stopper and shaken thoroughly for 20 min in a shaker. The reaction mixture is acidified with 10% HCl to pH $4-5$ (test with Congo red). The derivative separated is filtered off with suction, washed with 25% aqueous methanol, and crystallized from aqueous ethanol. To eliminate benzoic acid, it is recommended that the product be washed (by shaking it) with a few milliliters of light petroleum.

α-Naphthylureido derivatives of amino acids are prepared by reacting amino acids with α-naphthylisocyanate; a brief heating in an acid medium causes cyclization with the formation of hydantoins which crystallize well and have sharp melting points (90):

$$NH_2-CH_2-COOH + ArNCO \rightarrow ArNH-CO-NH \rightarrow Ar-N-CO-NH + H_2O$$
$$HO-CO-CH_2 \qquad CO\text{——}CH_2$$

General Method of Preparation of α-Naphthylureido Derivatives

Three millimoles of amino acid in 3 ml of N NaOH are mixed in a 15-ml test tube with 0.6 ml of α-naphthylisocyanate and 5 ml of water. After closing the tube with a rubber stopper the contents are thoroughly shaken for

20 min in a shaker. After filtering off α-naphthylurea the pH value of the filtrate is adjusted to 4−5 (test with Congo red) and the separated ureido derivative is filtered off after 1 hr standing. It is crystallized from 90% ethanol.

The procedure for the preparation of phenylhydantoins is the same as described above; 0.4 ml of phenylisocyanate is used (beware, poison!). The ureide separated is filtered off with suction and then boiled in a test tube with 5 ml of 10% HCl for 3−5 min. After cooling, the crystallized hydantoin is filtered off and recrystallized (91).

Ureido Derivative of DL-Alanine

Reagents: α-naphthylisocyanate, N NaOH, indicator papers, 50% aqueous ethanol, 2 N HCl.

Procedure: 10 mg of alanine are dissolved in a test tube in 10 ml of water and 1.1 ml of N NaOH, and 0.3 ml of α-naphthylisocyanate is added to the solution. The test tube is stoppered (rubber stopper) and shaken for 3 min and then allowed to stand with occasional shaking for 45 min. The α-naphthylurea formed is filtered off using a filtration tube and the filtrate is acidified with hydrochloric acid (to pH 4−5; control with Congo red paper). After 1 hr standing in a refrigerator the derivative is filtered off and washed twice with 1 ml of water. Yield, 87 mg; mp, 183−186 °C. Crystallization from 10 ml of 50% aqueous ethanol gave 35 mg of product, mp 196 to 197 °C.

Ureido Derivative of ε-Aminocaproic Acid

Reagents: α-naphthylisocyanate, N NaOH, pH papers, 50% NaOH, 2 N HCl.

Procedure: In a 10-ml test tube 50 mg of ε-aminocaproic acid are dissolved in 0.6 ml of NaOH and the solution is mixed with 10 ml of water and 0.15 ml of α-naphthylisocyanate. The test tube is closed with a rubber stopper, shaken vigorously for 2 min, and then allowed to stand with occasional shaking for 45 min. The separated α-naphthylurea is filtered off under suction using a filtration tube and washed with 1 ml of water, and the filtrate is acidified to pH 4−5. After 1 hr standing in a refrigerator the separated derivative is filtered off and washed twice with 1 ml of water. Yield, 20 mg; mp, 156−157 °C. Crystallization from 1.5 ml of 50% aqueous ethanol gave 9 mg of product, mp 165−166 °C.

Salts of Amino Acids

For identification−especially under the microscope−salts of amino acids with dilituric acid or picrolonic acid (92) or 2-nitroindane-1,3-dione (93 to 95) are used.

DL-Alanine Diliturate

Reagent: dilituric acid (3-nitrobarbituric acid).

Procedure: Dilituric acid (200 mg) is dissolved in the necessary amount of hot water, and 100 ml of alanine are added to the solution. The mixture is boiled briefly and then allowed to cool slowly. The separated crystals are filtered off and washed twice with 1 ml of water. Yield, 0.2 mg. The melting point is not sharp. After crystallization from 7 ml of water 0.15 g of the product is obtained, which does not melt sharply.

Chromatography of Amino Acids

Amino acids were the first group of substances analyzed using paper chromatography. A large proportion of papers concerning this method are devoted to the problems of separation, identification, and determination of amino acids, mainly from the biochemical point of view. In biochemical analyses of amino acids the first and most important step consists in the preparation of a sample, i.e., in the separation of small amounts of amino acids from an excess of proteins, lipids, sugars, and salts. All these compounds can, in excess, cause interference. In such cases, where a large number of amino acids (in a mixture) are to be separated and identified, the use of two-dimensional chromatography has been found very suitable. A number of solvent systems and their combinations have been proposed for this purpose. Only when the analyst knows the qualitative composition of his samples can he use one-dimensional chromatography. For the differentiation of certain pairs (or trios) of amino acids special solvents are used. A more detailed review of these problems is, however, beyond the scope of this book, and for more thorough treatment of these problems reference should be made to the monograph by Hais and Macek (96).

Paper chromatography has been most successful in the study of protein structures, which are submitted, after purification to hydrolytic cleavage and subsequent identification by paper chromatography of components set free on hydrolysis.

In industrial practice, the identification of amino acids may be important, for example, for the detection of substances of protein character in some industrial products (glue, for example). According to our experience, this is best carried out by paper chromatographically detecting the presence of amino acids in their acid (HCl) hydrolyzates. The same operations should be carried out with standard, authentic samples. The following solvent systems should serve as well-established practical examples: phenol saturated with water; *m*-cresol-water (1 : 1) containing traces of cupron (α-benzoinoxime) and with 0.1% ammonia in the atmosphere; isobutyric acid-water (4 : 1); amyl alcohol-pyridine-water (35 : 35 : 30); sec-butanol- 3%

ammonia (4 : 1); water-saturated benzyl alcohol; n-butanol-ethanol-water (4 : 1 : 1); n-butanol-acetic acid-water (9 : 1 : 1) or (4 : 1 : 1); benzyl alcohol-acetic acid-water (50 : 10 : 13); ethanol-water (95 : 5); n-butanol-borate buffer pH 8.3 (paper impregnated with this buffer). For thin-layer chromatography on silica gel G the following solvents are suitable: ethanol-water (7 : 3); ethanol-ammonia (7 : 3); n-propanol-water (7 : 3), and n-propanol-water (1 : 1) (97). For two-dimensional chromatography the combination n-butanol-acetic acid-water (4 : 1 : 1) and phenol-water (75 : 25 parts by weight, 20 mg of NaCN per 100 g of the mixture) may be used.

Amino acids are best detected with ninhydrin, isatin, or sodium 1,2-naphthoquinone-4-sulfonate. For certain amino acids specific detections are suitable which are made possible by the presence of other functional groups in the molecule than the amino group.

Detection with ninhydrin: The chromatogram is sprayed with a 0.2% ninhydrin solution in ethanol to which a few drops of acetic acid have been added. After spraying, the chromatogram is heated briefly at 60 °C, care being taken to see that the background does not turn dark. Another very common and frequently recommended method consists in dipping the chromatogram into a 0.25% solution of ninhydrin in anhydrous acetone and heating it as in the preceding example. The spots on the chromatogram can be stabilized by spraying it with a solution prepared by mixing 1 ml of a saturated copper (II) nitrate solution with 0.2 ml of 10% HNO_3 and methanol up to 100 ml. The chromatogram is then exposed to ammonia vapors. Pinkish-red spots of amino acids appear which are indefinitely stable.

Detection with isatin: The chromatogram is sprayed with a 0.4% solution of isatin in ethanol containing 4% of acetic acid, or it can be dipped in a 0.2% solution of isatin in anhydrous acetone containing 4% of acetic acid. The chromatograms are then heated at 100−110 °C for 10 min.

Detection with sodium 1,2-naphthoquinone -4-sulfonate: The chromatogram is sprayed with a 0.2% solution of sodium 1,2-naphthoquinone-4-sulfonate in 5% sodium carbonate solution. Various amino acids give various colors even in the cold.

A very effective separation technique for amino acids is ion-exchange chromatography, on the principles of which automatic analyzers have been constructed (98−102). Gas chromatography can be used for the separation of amino acids after their conversion to suitable derivatives (103−110).

7. Lactams

Lactams are intramolecular amides of amino acids (especially γ-, δ, -ε-amino acids) formed similarly to lactones from hydroxy acids. After hydrolysis the acid formed is identified.

Hydrolysis of Caprolactam

Reagents: HCl (1 : 3); 10% Na_2CO_3; a suspension of Ag_2O prepared in the following manner: to 1.5 g of $AgNO_3$ dissolved in a 250-ml Erlenmeyer flask in 20 ml of water 30 ml of 10% NaOH are added with vigorous stirring, and the mixture is diluted with water to the final volume of 200 ml.

Procedure: Caprolactam (1 g) is refluxed in a 30-ml flask with 10 ml of dilute HCl (1 : 3) until a clear solution results (approx. 1 hr). The solution is evaporated on a shallow dish to dryness and the residue is dried at 105°C to eliminate residual HCl. The residue is then dissolved in 30 ml of water and added to the suspension of Ag_2O in an 80-ml Erlenmeyer flask. A strong stream of hydrogen sulfide gas is then introduced into the solution for 5 min, the precipitate is filtered through folded filter paper, and the filtrate is evaporated to 5 ml final volume. The precipitated acid is filtered off under suction on a filtration crucible and crystallized from water. Yield, 0.30 g; mp, 125 – 130 °C.

8. Hydroxy Acids

The difficulties inherent in the identification of these compounds consist in the facts that during the working up of the reaction mixtures the hydroxy group of one molecule is esterified by the carboxyl group of another molecule, and that intramolecular esterification with the formation of lactones can take place as well. Lactones are formed in particular from γ-, δ-, and ε-hydroxy acids.

Generally, the most suitable method of identification of hydroxy acids consists in the esterification of the carboxyl group. It is also possible to oxidize the hydroxy group – after its presence has been detected – by means of hydrogen peroxide in the presence of ferrous sulfate to the corresponding keto acid (111); in the case of α-hydroxy acids, however, the oxidation may lead to a further degradation (112). α-Hydroxy acids react at higher temperatures with chloral, with the formation of chloralides,

$$\begin{array}{c} R\ OH \\ \backslash\ | \\ C-COOH + CCl_3CHO \\ / \\ R' \end{array} \rightarrow \begin{array}{c} R \quad\ O\!-\!-\!CH\!-\!CCl_3 \\ \backslash\ / \quad\ | \\ C \quad\quad\ | \\ /\ \backslash \quad\ | \\ R' \quad CO\!-\!O \end{array} + H_2O$$

In addition, hydroxy acids also form derivatives based on the reactivity of the carboxyl group. For example, the most common hydroxy acids, glycolic, lactic, and α-hydroxy-isobutyric, give anilides, toluides, and *p*-bromophenacyl esters (113).

For gas chromatography of aromatic hydroxy acids on silicone rubber plus polyethylene glycol as stationary phase see (114).

Detection of α-Hydroxy Acids

When boiled with MnO_2, PbO_2, or H_2O_2, α-hydroxy acids undergo oxidative cleavage, with the formation of carbon dioxide, while the acids are transformed to aldehydes or ketones with one carbon atom less in the molecule than they had originally. The reaction is carried out in the device depicted in Fig. 31. Carbon dioxide and the aldehyde or ketone are detected in the condenser test tube.

Reagents: powdered PbO_2; syrupy phosphoric acid; barita water; paraffin oil; saturated solution of 2,4-dinitrophenylhyrazine in 2 N HCl.

Procedure: Into the larger test tube of the device in Fig. 31 a solution of the sample is introduced (50 − 100 mg of α-hydroxy acid), followed by 1 g of PbO_2 and 2 ml of H_3PO_4. The connecting glass tube reaches below the surface of the barita water in the condenser test tube. A thin layer of paraffin oil prevents the barita water from absorbing carbon dioxide from the air. The reaction mixture is heated slowly to boiling point and a few drops are distilled. A white precipitate in the barita solution represents a positive reaction of CO_2. The supernatant aqueous solution is then mixed with the 2,4-dinitrophenylhydrazine solution, and the turbidity or precipitate formed is extracted with a few drops of benzene, and the extract is applied on a chromatogram (p. 222).

Under the influence of warm concentrated sulfuric acid α-hydroxy acids are cleaved to corresponding lower aldehydes or ketones and formic acid ($CO + H_2O$). The aldehydes, or ketones, react with simultaneously added reagents, usually with some phenol (p. 213), as for example, 2,7-dihydroxynaphthalene or chromotropic acid in the case of formaldehyde, or guaiacol, carbazole, or *p*-hydroxydiphenyl in the case of acetaldehyde (115 − 117).

Detection of Glycolic Acid

Reagent: 0.01% 2,7-dihydroxynaphthalene in conc. sulfuric acid.
Procedure: A few drops of an aqueous solution of glycolic acid are heated with 2 ml of the reagent on a boiling water bath for 20 min; a red-violet color is formed.

Detection of Lactic Acid

Reagents: conc. H_2SO_4, 5% ethanolic guaiacol solution.
Procedure: 2 ml of conc. H_2SO_4 are added to 0.2 ml of a solution of lactic acid and the mixture is heated on a water bath for 2 min. After cooling,

two drops of the guaiacol solution are added to the mixture; a red color is formed.

Among other color reactions of α-hydroxy acids, the formation of a yellow color on reaction of glycolic, lactic, malic, tartaric, citric, and mandelic acids with an almost colorless ferric chloride solution (118) should be mentioned, as well as the reaction with sodium nitroprusside after their oxidation with potassium permanganate (119).

p-Bromophenacyl Ester of Lactic Acid

Reagents: 10% NaOH, 95% ethanol, p-bromophenacyl bromide.

Procedure: Lactic acid (0.5 g) is neutralized (with respect to phenolphthalein) in a 50-ml flask with 10% sodium hydroxide, and a drop of lactic acid is then added to the neutralized solution followed by 10 ml of 95% ethanol, and 1 g of p-bromophenacyl bromide. The flask is connected with a reflux condenser and the mixture boiled on a water bath for 1 hr. After cooling, the mixture is allowed to stand in a refrigerator for 30 min and the precipitate is filtered off. Yield, 0.85 g; mp, 100 °C. Crystallization from 15 ml of ethanol and 4 ml of water gave 0.22 g of a product melting at 104 − 105 °C.

Hydroxy Polycarboxylic Acids

Malic Acid. When heated with conc. sulfuric acid and a small amount or resorcinol or orcinol, malic acid gives derivatives of umbelliferone, which fluoresce strongly in alkaline media. A very strong fluorescence is also observed on the reaction of malic acid with 2-naphthol (120).

a) *Reagents:* resorcinol, conc. H_2SO_4, ammonia.

Procedure: Resorcinol (10 mg) and 2 ml of conc. sulfuric acid are added to 2 − 3 drops of a malic acid solution and the mixture is heated on a water bath for 5 min. After cooling, the reaction mixture is poured into 10 ml of water and alkalized with excess ammonia; a strongly fluorescing (blue) solution is formed.

b) *Reagent:* 0.01% 2-naphthol in conc. H_2SO_4.

Procedure: A few drops of malic acid solution are evaporated to dryness, 1 ml of the reagent is added, and the mixture is poured into water; the solution is observed under ultraviolet light (strong blue fluorescence).

Tartaric Acid. Fenton's color test is characteristic of tartaric acid (121). This is based on the reaction of the acid with a ferrous salt and hydrogen peroxide, and eventual alkalization. Characteristic colors are also formed on warming tartaric acid with resorcinol, pyrogallol, 2-naphthol, gallic acid, etc. (122). On warming tartaric acid with sulfuric acid containing 2,2′-dinaphthol a green fluorescence appears. Under the conditions of the reaction glycolic and glyoxylic acids give red-brown to brown, glyceric and mesoxalic acids give gray, and gluconic, glucuronic, dihydroxy tartaric, tartronic, and malic acids give green colors. Other acids do not interfere.

Reagent: 0.05% 2,2'-dinaphthol solution in conc. H_2SO_4.

Procedure: A drop of the tested solution or a small amount of the solid tartaric acid and 1 ml of the reagent are heated on a water bath (85 — 90 °C) for 20 — 30 min. During the heating the violet fluorescence of the reagent gradually disappears and a brillant green appears, which intensifies after cooling.

Citric Acid. After oxidation with permanganate, citric acid gives color reactions with sodium nitroprusside (123) or resorcinol (124). However, the detection based on its conversion to citrazinic acid, whose ammonium salt shows a strong blue fluorescence, is much more important (125).

Reagent: thionyl chloride, conc. ammonia, conc. H_2SO_4.

Procedure: A solution of citric acid is evaporated to dryness and the residue (in a microcrucible) is mixed with a few drops of thionyl chloride and evaporated to dryness again. Eight drops of ammonia are added to the residue and the mixture is heated gently over a microflame until only two drops of liquid are left in the crucible. After cooling, six drops of H_2SO_4 are added and the mixture is heated until fumes of sulfuric acid appear. The contents of the crucible are transferred and rinsed into a test tube, excess ammonia is added, and the blue fluorescence is observed under ultraviolet light.

Polyhydroxy Aldehydo Acids

The best-known reaction of uronic acid is that with naphthoresorcinol in conc. hydrochloric acid. The dyes formed are probably xanthene derivatives soluble in benzene (in contrast to derivatives of pentoses and hexoses) (126). Characteristic colors are also formed with carbazole in sulfuric acid, based on the formation of 5-formylfuroic acid, which subsequently reacts with carbazole.

Reagents: 38% HCl, naphthoresorcinol, benzene.

Procedure: A few milligrams of uronic acid are dissolved in 4 ml of HCl and 4 ml of water, and the mixture is boiled for a few moments; 4 ml of the clear solution are mixed with ~ 100 mg of naphthoresorcinol and the mixture is boiled again for 30 min, then cooled and extracted with an equal volume of benzene; the benzene layer assumes a violet color.

Paper Chromatography

In contrast to volatile aliphatic acids, hydroxy acids can also be chromatographed in an acid medium; ammoniacal systems are used only for checking the results. Among acid systems, the most often used is the mixture n-buta-nol − formic acid − water (4 : 1 : 5 or 10 : 2 : 5). The most common ammo-niacal system is n-propanol-ammonia (2 : 1). The amount of acids or of their salts applied on the paper should be between 20 and 100 µg. For detection all general methods for acids (p. 259) can be used. For certain acids suitable color reactions can also be used. For example, citric acid, aconitic acid, and tartaric acid can be detected on the chromatogram by spraying it with a 4% p-dimethylaminobenzaldehyde solution in acetic anhydride and subsequent heating to 150 − 160 °C. Red to orange-red colors are formed.

9. Keto Acids

Identification of keto acids can be carried out by the preparation of their insoluble salts with S-1-naphthylmethylthiuronium chloride or by taking advantage of the reactivity of the carbonyl group or also by derivatization of the carbonyl group.

Color Reaction of Keto Acids, and Paper Chromatography

Pyruvic acid dissolved in dilute acetic acid gives a blue color with sodium nitroprusside and ammonia (127). Pyruvic acid can be also reduced, like formic acid (p. 248), with magnesium in sulfuric acid to the corresponding aldehyde, and this can be detected by the addition of p-hydroxydiphenyl (128). When heating with 1-naphthol in conc. sulfuric acid pyruvic acid gives a red to blue color. Acetoacetic acid and its esters give a blue to red color with ferric chloride solution in methanol (reaction of the enol form). With sodium nitroprusside solution in alkaline medium a red color is given. For the color reaction with p-benzoquinone see p. 300. Heating acetoacetate with resorcinol and conc. hydrochloric acid, followed by alkalization produces a strong blue fluorescence (129).

Keto acids can be chromatographed both as free acids and in the form of their derivatives. For free acids the solvent system n-butanol − acetic acid − water (4 : 1 : 5) may be used, and they can be detected with 2,4-dinitrophenylhydrazine or p-phenylenediamine (p. 302). It is more suitable to chromatograph keto acids in the form of their 2,4-dinitro-phenylhydrazones in the system n-butanol − ethanol − water (4 : 1 : 5) or n-butanol − ethanol − 0.5 M ammonia (7 : 1 : 2) and to detect them as des-cribed on p. 222 [or by thin-layer chromatography (130, 131)].

For the preparation of S-1-naphthylmethylthiuronium salts see p. 251.

10. Halo Acids

The detection of halo acids is carried out by combining the reactions for the detection of the carboxyl group with that of halogens. In addition, special reactions for single acids are also available. Monochloro- and mono-bromoacetic acids can be detected by using the reaction with ammonium thiocyanate (132). Rhodanine is formed, which gives a color reaction with sodium 1,2-naphthoquinone-4-sulfonate (see also p. 320). Paper chromatography of halo acids can be carried out in the system n-propanol – ammonia (7 : 3). The amount of the acids applied on the paper should be 50 – 100 µg (133). Iodinated acids can be detected by suspending the wet chromatogram in a cylinder containing chlorine gas for a few seconds; under the influence of chlorine, iodine is liberated, which reacts with the residues of ammonia in the paper, giving rise to yellow spots on a white background. Brominated acids are detected by spraying the chromatogram with 0.1% fluorescein solution in 75% ethanol and suspending it in a cylinder containing chlorine gas. The liberated bromine reacts with fluorescein, and red spots appear on an almost colorless background.

From the list of melting points of derivatives of halo acids given in Table 10 it is evident that for the identification of halo acids, amides, anilides and p-toluides are suitable; p-bromophenacyl esters were prepared only in a few cases. If satisfactory results cannot be obtained by the application of procedures given for the preparation of carboxylic acid derivatives (see p. 249), original literature should be consulted. Identification of aromatic, halo acids is easy because they are well-crystallized solids, and, in addition the preparation of derivatives, for example, p-bromophenacyl esters, causes no difficulty.

11. Lactones

Lactones are formed predominantly from γ-, δ-, and ε-hydroxy acids. Most lactones of the aliphatic series are converted easily under the influence of water to corresponding hydroxy acids. For the hydrolysis of aromatic lactones heating with alkalies is necessary. The reactions of lactones with ammonia and phenylhydrazine, which can be used in certain cases for identification purposes, are described in greater detail in a special monograph (134). Aliphatic γ-lactones react in benzene solution with thionyl chloride, with the formation of hydroxy acid chlorides, from which suitable derivatives can be prepared (135).

Derivatives of Halo Acids   *Table 10*

Acid	Boiling or melting point of the acid, °C	Melting point			
		Amides	Anilides	p-Toluides	p-Bromo-phenacyl ester
Difluoroacetic	134—135	52	—	—	—
α-Chloropropionic	186	80	92	124	99
Dichloroacetic	194	98	118	153	—
α-Bromopropionic	205[a]	123	99	125	—
α-Bromobutyric	217[a]	112	98	92	—
Fluoroacetic	32[b]	108	—	—	—
α-Bromoisovaleric	44[b]	133	116	124	—
Dibromoacetic	48[b]	156	—	—	—
α-Bromoisobutyric	49[b]	148	83	93	—
Bromoacetic	50[b]	91	131	—	—
Trichloroacetic	58[b]	141	97	113	—
β-Bromopropionic	62.5[b]	111	—	—	—
Chloroacetic	63[b]	121	134	162	104

[a] Bp with decomposition.
[b] Mp.

References

1. Schlögl, K.: Naturwissenschaften **46**, 447 (1959).
2. Nomura, Y.: Bull. Chem. Soc. Japan **32**, 536 (1959).
3. Löffler, J. E., and Reichl, E. R.: Mikrochim. Acta **1953**, 79.
4. Churáček, J.: Sbor. Věd. Pr., Vys. Šk. Chem. Technol. Pardubice **14**, 85 (1966).
5. Cheeseman, G. W. H., and Poller, R. C.: Analyst **86**, 256 (1961).
6. Affsprung, H. E., and May, H. E.: Anal. Chem. **32**, 1164 (1960).
7. Matthews, F. W., Warren, G. G., and Michell, J. H.: Anal. Chem. **22**, 514 (1950).
8. Donleavy, J. J.: J. Am. Chem. Soc. **58**, 1004 (1936).
9. Bonner, W. A.: J. Am. Chem. Soc. **70**, 3508 (1948).
10. Berger, J. and Uldall, I.: Acta Chem. Scand. **17**, 1939 (1963).
11. Taborsky, R. G.: Anal. Chem. **36**, 1663 (1964).
12. Affsprung, H. E., and Gainer, A. B.: Anal. Chim. Acta **27**, 578 (1962).
13. Shriner, R. L., and Fuson, R. C.: The Systematic Identification of Organic Compounds. J. Wiley, New York 1948.
14. Neish, A. C., and Lemieux, R. U.: Can. J. Chem. **30**, 454 (1952).
15. Berger, J.: Analyt. Chemistry 1962 (Proc. Feigl Anniv. Symp., Birmingham), p. 48.
16. Langenbeck, W., and Baehren, F.: Ber. **69**, 514 (1936).
17. Erickson, J. L. E., Dechary, J. M., and Kesling, M. R.: J. Am. Chem. Soc. **73**, 5301 (1951).

18. Berger, J.: Acta Chem. Scand. **17**, 1943 (1963).
19. Bartos, J.: Talanta **8**, 556 (1961).
20. Stollé, R., and Wolf, F.: Ber. **46**, 2248 (1913).
21. Pool, W. O., Harwood, H. J., and Ralston, A. W.: J. Am. Chem. Soc. **59**, 178 (1937).
22. Brown, E. L., and Campbell, N.: J. Chem. Soc. **1937**, 1699.
23. Stránský, Z., Stužka, V., and Růžička, E.: Mikrochim. Acta **1966**, 77.
24. Stendahl, M. N.: Compt. rend. **196**, 1811 (1933).
25. Stoughton, R. W.: J. Org. Chem. **2**, 514 (1938).
26. Buu-Hoi, N. P., Xuong, N. D., and Lescot, E.: Bull. Soc. chim. France **1957**, 441.
27. Knight, H. B., Witnauer, L. P., et al.: Anal. Chem. **24**, 1331 (1952).
28. Micheel, F., and Schminke, W.: Chem. Ber. **91**, 984 (1958).
29. Franc, J.: Chem. Listy **51**, 2041 (1957).
30. Jureček, M., Churáček, J., and Červinka, V.: Mikrochim. Acta **1960**, 102.
31. Churáček, J.: Mikrochim. Acta **1961**, 65.
32. Churáček, J., and Vaněk, J.: Sb. věd. prací, VŠCHT Pardubice **1**, 61 (1963).
33. James, A. T., and Martin, A. J. P.: Biochem. J. **63**, 144 (1956).
34. Peyrot, P.: Chim. et Industrie **78**, 3 (1957).
35. Cropper, F. R., and Heywood, A.: Nature **174**, 1063 (1954).
36. Berthuis, R. L., and Keppler, J. G.: Nature **179**, 731 (1957).
37. Desty, D. H.: Gas Chromatography, p. 316. Butterworths, London, 1958.
38. Dijkstra, G., Keppler, J. G., and Schools, J. A.: Rec. Trav. chim. Pays Bas **74**, 805 (1955).
39. Schlenk, H., and Gellerman, J. L.: Anal. Chem. **32**, 1412 (1960).
40. Chang, T.−L., and Sweeley, C. C.: J. Lipid Res. **3**, 170 (1962).
41. Jones, E. P., and Davison, V. L.: J. Am. Oil Chem. Soc. **42**, 121 (1965).
42. Chisholm, M. J., and Hopkins, C. Y.: Can. J. Chem. **38**, 805 (1960).
43. Appleby, A. J., and Mayne, J. E. O.: J. Gas Chromatog. **5**, 266 (1967).
44. Hewett, O. R., Kipping, P. J., and Jeffery, P. G.: Nature **192**, 65 (1961).
45. Večeřa, M., and Friedrich, K.: Chem. Listy **51**, 283 (1957).
46. Goddu, R. F., LeBlanc, N. F., and Wright, C. M.: Anal. Chem. **27**, 1251 (1955).
47. Pilz, W.: Z. anal. Chem. **162**, 81 (1958).
48. Goldenberg, V., and Spoerri, P. E.: Anal. Chem. **30**, 1327 (1958); **31**, 1735 (1959).
49. Bost, R. W., and Mullen, L. V.: J. Am. Chem. Soc. **73**, 1967 (1951).
50. Hardy, D. V. N.: J. Chem. Soc. **1956**, 398.
51. Renefrow, W. B., and Chaney, A.: J. Am. Chem. Soc. **68**, 150 (1946).
52. Hammett, L. P.: Physical Organic Chemistry, McGraw-Hill, New York, 1940.
53. Goddu, R. F., LeBlanc, N. F., and Wright, C. M.: Anal. Chem. **27**, 1251 (1955).
54. Morgan, K. J.: Anal. Chim. Acta **19**, 27 (1958).
55. Davidson, D.: Anal. Chem. **26**, 576 (1954).
56. Graebe, C., and Leonhardt, M.: Ann. **290**, 217 (1896).
57. Staudinger, H.: Ber. **38**, 1735 (1905).
58. Soloway, D., and Lipschitz, A.: Anal. Chem. **24**, 898 (1952).
59. Davidson, D.: J. Chem. Educ. **17**, 81 (1940).
60. Kaufmann, H. P., and Skiba, K. J.: Fette, Seifen, Anstrichmittel **60**, 261 (1958).
61. Adriani, W.: Rec. Trav. chim. Pays Bas **35**, 180 (1915).
62. Phillips, R. F., and Pitt, B. M.: J. Am. Chem. Soc. **65**, 1355 (1943).

63. Cheeseman, G. W. H., and Poller, R. C.: Analyst **87**, 366 (1962).
64. MacKenzie, C. A., and Rawles, W. T.: Ing. Eng. Chem., Anal. Ed. **12**, 737 (1940).
65. Williams, J. W., et al.: J. Am. Chem. Soc. **64**, 1738 (1942).
66. Večeřa, M., and Friedrich, K.: Chem. Listy **51**, 283 (1957).
67. Gasparič, J., and Klouček, B.: Collection Czech. Chem. Commun. **31**, 106 (1966).
68. Borecký, J.: Mikrochim. Acta **1966**, 279.
69. Gasparič, J., and Marhan, J.: Collection Czech. Chem. Commun. **27**, 46 (1962).
70. Feigl, F., Anger, V., and Frehden, O.: Mikrochemie **15**, 9 (1934).
71. Bayer, E., and Reuther, K. H.: Ber. **89**, 2541 (1956).
72. Frehden, O., and Goldschmidt, L.: Mikrochim. Acta **2**, 184 (1937).
73. Folin, O.: J. Biol. Chem. **51**, 377, 393 (1922).
74. Furman, N. H., Morrison, G. H., and Wagner, A. F.: Anal. Chem. **22**, 1561 (1950).
75. Waser, E., and Brauchli, E.: Helv. Chim. Acta **7**, 740 (1924).
76. Edlbacher, S., and Litvan, F.: Zeit. physiol. Chem. **265**, 241 (1940).
77. Karrer, P., and Keller, R.: Helv. Chim. Acta **26**, 50 (1943).
78. Ruhemann, S.: J. Chem. Soc. 97, 1438 (1910); **99**, 792, 1306, 1486 (1911).
79. Abderhalden, E., et al.: Zeit. physiol. Chem. **77**, 249 (1912); **85**, 143 (1913).
80. McCaldin, D. J.: Chem. Revs. **60**, 39 (1960).
81. Friedman, M., and Sigel, C. W.: Biochemistry **5**, 478 (1966).
82. Abderhalden, R.: Zeit. physiol. Chem. **252**, 81 (1938).
83. Johnson, A. W., and McCaldin, D. J.: J. Chem. Soc. **1957**, 3470; **1958**, 817.
84. Opienska-Blauth, J.: Chem. analit. (Warsaw) **2**, 123 (1957).
85. Fischer, E.: Ber. **39**, 2320 (1906).
86. Herbst, R. H., and Shemin, D.: Organic Syntheses, Vol. 19, Wiley, New York, 1939.
87. Fischer, E.: Ber. **32**, 2451 (1899); **35**, 3779, 3784 (1902).
88. Saunders, B. C.: J. Chem. Soc. **1938**, 1397.
89. Sanger, F.: Biochem. J. **39**, 507 (1945).
90. Neuberg, C., and Rosenberg, E.: Biol. Z. **5**, 456 (1907).
91. Patten, A. J.: Zeit. physiol. Chem. **39**, 350 (1903).
92. Levene, P. A., and van Slyke, D. D.: J. Biol. Chem. **12**, 127 (1912).
93. Larsen, J., Witt, N. F., and Poe, C. F.: Mikrochemie **34**, 1 (1949).
94. Wanag, G.: Ber. **69**, 1066 (1936).
95. Wanag, G., and Lode, A.: Ber. **70**, 547 (1937).
96. Hais, I. M., and Macek, K.: Paper Chromatography, p. 437. Publ. House Czechoslov. Acad. Sci., Prague, 1963.
97. Stahl, E.: Dünnschicht-Chromatographie, p. 705, Springer-Verlag, Berlin—Heidelberg—New York, 1967.
98. Spackman, D. H., Stein, W. H., and Moore, S.: Federation Proc. **15**, 358 (1956); Anal. Chem. **30**, 1190 (1958).
99. Moore, S., Spackman, D. H., and Stein, W. H.: Federation Proc. **17**, 1107 (1958); Anal. Chem. **30**, 1185 (1958).
100. Orten, A. U., Doppke, H. J., and Spurrier, H. H.: Anal. Chem. **37**, 623 (1965).
101. Hacket, N., Mathias, A. P., and Pennington, F.: Anal. Biochem. **12**, 367 (1965).

102. Beckman Instruments Inc., Spinco Div., Application Data Sheet 245, Palo Alto, Calif.
103. Hagen, P., and Black, W.: Federation Proc. 23, 371 (1964); J. Chromatog. 16, 574 (1964).
104. Vitt, S. V., Saporovskaya, M. B., and Belikov, V. M.: Izvestiya Akad. Nauk SSSR, Otd. khim. Nauk 1964, 947; Zhur. Anal. Khim. 21, 227 (1966).
105. Cruickshank, P. A., and Sheehan, J. C.: Anal. Chem. 36, 1191 (1964).
106. Weygand, F., et al.: Zeit. Naturforsch. 18b, 93 (1963).
107. Smith, E. D., and Sheppard, H., Jr.: Nature 208, 878 (1965).
108. Gehrke, Ch. W., and Shahrokhi, F.: Anal. Biochem. 15, 97 (1966).
109. Pollock, G. E., and Oyama, V. I.: J. Gas Chromatography 4, 126 (1966).
110. Kesner, L., Muntwyler, E., Griffin, G. E., and Abrams, J.: Anal. Chem. 35, 83 (1963).
111. Fenton, H. J. H., and Jones, H. O.: J. Chem. Soc. 77, 77 (1900).
112. Dakin, H. D.: J. Biol. Chem. 4, 91 (1908).
113. Cheronis, N. D., and Entrikin, J. B.: Semimicro Qualitative Organic Analysis, Interscience Publishers, New York, 1958.
114. Mendez, J., and Stevenson, F. J.: J. Gas Chromatography 4, 483 (1966).
115. Denigés, G.: Bull. Trav. Soc. Pharm. Bordeaux 49, 193 (1909); Zeit. anal. Chem. 50, 189 (1911); Bull. Soc. chim. France 5, 647 (1909).
116. Eegriwe, E.: Zeit. anal. Chem. 89, 121 (1932).
117. Dische, Z.: Biochem. Zeit. 189 77, (1927).
118. Berg, M. A.: Bull. Soc. Chim. France 11, 883 (1894).
119. Caron, H., and Raquet, D.: J. Pharm. Chim. 2, 335 (1942).
120. Leininger, E., and Katz, S.: Anal. Chem. 21, 1375 (1949).
121. Fenton, H. J. H.: Chem. News 33, 190 (1876); Zeit. anal. Chem. 21, 123 (1882).
122. Eegriwe, E.: Zeit. anal. Chem. 89, 121 (1932).
123. Caron, H., and Raquet, D.: J. Pharm. Chim. 2, 232 (1942).
124. Arreguine, V.: Bull. Soc. Chim. biol. 11, 242 (1929).
125. Feigl, F., Anger, V., Frehden O.: Mikrochemie 17, 29 (1935).
126. Momose, T., Ueda, Y., and Iwasaki, M.: Pharm. Bull. (Japan), 3, 321 (1955).
127. Simon, L. J., and Piaux, L.: Bull. Soc. Chim. biol. 6, 477 (1924).
128. Eegriwe, E.: Zeit. anal. Chem. 95, 323 (1933).
129. Arreguine, V., and Garcia, E. D.: Ann. Chim. anal. 2, 36 (1920).
130. Eistofarx, E., and Egli, R. H.: Chromatogr. Methods Immed. Separ., Proc. Meet., Athens, 1965, 2, p. 95.
131. Häkkinen, H. M., and Kulonen, E.: J. Chromatog. 18, 174 (1965).
132. Feigl, F., and Gentil, V.: Anal. Chem. 29, 1715 (1957).
133. Hashmi, M. H., and Cullis, C. F.: Anal. Chim. Acta 14, 336 (1956).
134. Houben-Weyl: Methoden der organischen Chemie. Herausgegeben von E. Müller. Vol. VI, Lactone. Thieme, Stuttgart, 1953.
135. Barbier, P., and Locquin, R.: Bull. Soc. Chim. France (4) 13, 223, 229 (1913).

CHAPTER XIV

SOME OXYGEN-CONTAINING COMPOUNDS

1. Enols

Enols are usually identified in the form of carbonyl compounds. As the enol and the keto forms usually exist in rapidly attainable equilibrium, the whole amount of the compound is eventually converted to a derivative of the keto form:

$$-\underset{\underset{\text{OH}}{|}}{\text{C}}=\text{CH}- \quad \rightleftharpoons \quad -\underset{\underset{\text{O}}{\|}}{\text{C}}-\text{CH}_2- \xrightarrow{\text{reagent for carbonyl compounds}} \text{derivative}$$

If the equilibrium is shifted completely to the left side of the equation, reactions of hydroxyl groups are used. Phenols represent such an extreme case of enols stabilized by resonance. When derivatives are prepared using reagents for carbonyl groups one should keep in mind that the enolization is often caused by the presence of another carbonyl group (β-dicarbonyl compounds), and that in such reactions consecutive reactions or ring closures take place (for example, the formation of pyrazolones in reactions with phenylhydrazine). The problem of the identification of enols is therefore limited to the detection of the presence of an enol group in the molecule. This is carried out by chemical as well as by spectral methods. Reference should be made to the literature dealing with these problems (1).

For the characteristic detection of compounds containing a "labile hydrogen" in the methylene group see (2).

Detection of Enols with Ferric Chloride

In order to detect cis-enols of β-dicarbonyl and α-dicarbonyl compounds, the formation of yellow-red, red to blue-violet, and blue colors with ferric chloride solution in absolute methanol is commonly used. Chelates of the following type are formed:

$$R-C\overset{\overset{\displaystyle R}{\displaystyle |}}{\underset{\underset{\displaystyle R}{\displaystyle |}}{\overset{\displaystyle C=O}{\underset{\displaystyle C-O}{}}}}FeCl_2$$

Of course, phenols also react with $FeCl_3$ (p. 188). However, in methanolic solution only those phenols react which carry in the ortho position of the phenolic hydroxyl a group capable of the formation of a chelate bond, as, for example, in catechol or salicylic acid. Trans-enols do not produce a color in methanolic solutions, but do in aqueous solution. For more detailed information of the relationship between the reactivity and the color and the structure of enols see (3).

Reagents: 1% solution of anhydrous ferric chloride in absolute methanol; absolute methanol.

Procedure: A sample of cis-enol is dissolved in a few milliliters of methanol and mixed with a few drops of the reagent; a characteristic color is formed which in certain slowly enolizable compounds develops rather slowly.

2. Enediols

An important and common property of enediols is their reducing power; therefore, they are also called reductones. On oxidation (even by atmospheric oxygen) in acid, neutral, and alkaline media they are converted to α-dicarbonyl compounds:

$$\underset{\underset{\displaystyle OH\;OH}{}}{R-C=C-R} \xrightarrow{-2H^{\oplus}} \underset{\underset{\displaystyle O\;\;O}{}}{R-C-C-R} \xrightarrow{-2\ominus} \underset{\underset{\displaystyle O^{\ominus}\;\;O^{\ominus}}{}}{R-C=C-R}$$

3-Carbonyl-1,2-enediols having a carbonyl group (for example, ascorbic acid) in the molecule in addition to two enolic hydroxyls also belong to this group of compounds. The simplest representative of this group is triose reductone (a), formed during the alkaline degradation of sugars (4). A very simple method of identification of these compounds consists in using their reaction with aromatic primary amines, with which they combine in equimolar proportions and in the cold to yield Schiff's bases (b), while in the presence of excess amine and at elevated temperatures azomethine dyes are formed (c):

$$\begin{array}{ccccc}
CHO & & CH=NAr & & CHNHAr \\
| & & | & & \| \\
COH & \xrightarrow{ArNH_2} & COH & \rightleftharpoons & COH \\
\| & & \| & & | \\
CHOH & & CHOH & & CHO \\
(a) & & (b) & &
\end{array} \xrightarrow{ArNH_2}$$

$$
\begin{array}{ccc}
\text{CHNHAr} & & \overset{\oplus}{\text{CH}}=\text{NHAr} \\
\parallel & & \mid \\
\text{COH} & \rightleftharpoons & \text{COH} \\
\mid & & \parallel \\
\overset{\oplus}{\text{CH}}=\text{NHAr} & & \text{CHNHAr} \\
& (c) &
\end{array}
$$

For more detailed information reference should be made to the original literature (5—7). All enediols regularly give stronger colors with ferric chloride in methanolic solution than phenols and simple enols. The colors formed are blue or blue-green and are not very stable. In the case of certain strongly reducing enediols they are scarcely observable; however, they can be stabilized to a certain extent by the addition of pyridine or by buffering with sodium acetate (8—10). Enediols also give a characteristic color reaction with titanium tetrachloride, and in methanol with the addition of pyridine (11).

3. Ketenes

Ketenes are a group of substances with the characteristic grouping $R_2C=C=O$. Ketenes are very unstable compounds (especially aldoketenes $R-CH=C=O$), which do not give carbonyl-compound reactions, but which are very reactive, easily giving peroxides with oxygen and polymerizing. For the identification of ketenes their conversion to anilides is commonly used. The procedure is very simple: gases containing ketenes are introduced into a solution of aniline in xylene (12):

$$ ^{\cdot}R_2C=C=O + ArNH_2 \rightarrow ArNHCOCHR_2 $$

The procedure is suitable for the identification of carbon suboxide; when reacting with aniline, dianilide of oxalic acid is formed. Simple ketene gives acetanilide.

Acetanilide from Ketene and Aniline

Reagents: aniline and xylene.

Procedure: To 1% ketene solution in dry xylene (100 ml) 3 g of aniline are added and the mixture is allowed to stand in a refrigerator for 3 hr. The separated acetanilide is filtered off and dried. Yield, 0.83 g; mp, 114 to 115 °C.

4. Acetals, Ketals, and Orthoesters

Identification of acetals and ketals is carried out by acid hydrolysis and the subsequent preparation of derivatives of the carbonyl compounds and alcohols formed. Hydrolysis takes place on heating acetals or ketals with dilute acids (N HCl, N H_2SO_4),

$$RCH(OR_1)_2 \xrightarrow[H\oplus]{H_2O} RCHO + 2\,R_1OH$$

for $3-5$ min in the case of low-molecular compounds, while higher-molecular compounds should be heated for longer periods. Substitution with negative substituents of the aldehyde part inhibits the hydrolysis. For example, dimethyl acetal of monochloroacetaldehyde is hydrolyzable only with difficulty. Water-insoluble acetals should be hydrolyzed in a mixture of water and dioxane. For identification purposes 2,4-dinitrophenylhydrazones or semicarbazones are prepared after hydrolysis from the carbonyl components, while the alcoholic components are converted to 3,5-dinitrobenzoates. See (13) for the identification of acetals of the tetrahydrofuran series by gas chromatography with the utilization of Kovats indices.

General Method for the Identification of Carbonyl Components of Acetals and Ketals

The simplest method consists in mixing the acetal or the ketal with a solution of 2,4-dinitrophenylhydrazine in 2 N HCl (see p. 218). It is also possible to hydrolyze $0.1-0.2$ ml of the sample first by refluxing it with 5 ml of N H_2SO_4 for $10-15$ min.

General Method of Identification of the Alcoholic Components of Acetals and Ketals

a) The sample ($0.1-0.2$ ml) is hydrolyzed under reflux with 5 ml of N H_2SO_4 and 5 ml of dioxane for $10-15$ min. The mixture is concentrated by distilling off $4-5$ ml of the liquid, and the alcohol present in the distillate is converted to 3,5-dinitrobenzoate as described on p. 150.

b) 3,5-Dinitrobenzoyl chloride (1.0 g) and $2-3$ ml of acetal or ketal in a 25-ml flask are boiled with reflux for $5-60$ min. The reaction time depends on the boiling point of the sample: samples boiling up to 60 °C are boiled for 60 min, for bp $60-100$ °C 30 min is needed, for bp above 100 °C $5-10$ min is enough. If the sample darkens on boiling, the reaction should be stopped. After cooling, 10 ml of a 5% sodium hydrogen carbonate solution are added; the solid phase is separated and triturated with 10 ml of 5% $NaHCO_3$, the suspension is warmed to $\sim 45-50$ °C, and the crude ester is washed with water, dried in air, and crystallized from 95% ethanol.

Diethyl Acetal. Identification of the Aldehydic and the Alcoholic Components

Reagents: 2,4-dinitrophenylhydrazine (a saturated solution in 2 N HCl); ethanol.

Procedure: 10 ml of the reagent are added into a 20-ml test tube containing 0.2 ml of diethyl acetal, and the mixture is allowed to stand for 1 hr. The separated 2,4-dinitrophenylhydrazone is filtered off and dried. Yield, 45 mg; mp, 157−158 °C. After crystallization from 8 ml of ethanol 30 mg of pure product were obtained, mp 159−160 °C. After the isolation of the hydrazone the filtrate is transferred into a distillation flask, diluted with 10 ml of water, and distilled. Ten milliliters of the condensate are collected and the distillate is worked up as described on p. 150, preparing a 3,5-dinitrobenzoate. Yield, 98 mg; mp, 91 °C.

Ketals and orthoesters are hydrolyzed with dilute acids even more easily than acetals. Usually, a short, 1−3 min heating with 1−3% HCl or H_2SO_4 is sufficient. Orthoesters are quite stable to alkalies (in contrast to esters of carboxylic acids). The procedure for the identification of the alcoholic component is identical with that described in the preceding paragraph. The acid component of the orthoester is identified, depending on the character of the acid, by one of the procedures described on p. 249. The problems of isolation and identification of the acid component are identical with those discussed on p. 265 (identification of carboxylic acid esters).

$$RC(OR_1)_3 \xrightarrow[H_2O]{H^{\oplus}} 2\,R_1OH + RCOOR_1$$

$$RCOOR_1 \xrightarrow[H_2O]{H^{\oplus}} RCOOH + R_1OH$$

The conditions of acid hydrolysis and the properties of the acid and the alcohol determine whether the hydrolysis will proceed to the first or the second step. In special cases reference should be made to the original literature (14, 15).

Hydrolysis of Ethyl Orthoformate

Reagents: N H_2SO_4, 30% NaOH, S-1-naphthylmethylthiuronium chloride, ethanol.

Procedure: In a 10-ml flask provided with a reflux condenser 0.1 ml of ethyl orthoformate is heated with 5 ml of N H_2SO_4 until dissolved. After cooling, the mixture is alkalized with sodium hydroxide to a clearly alkaline reaction, and 10 ml of the liquid are distilled off. If necessary, the amount of liquid distilled off is replaced by water. The distillate is worked up as on p. 150 (to 3,5-dinitrobenzoate). Yield, 10 mg; mp, 90−91 °C. The residue in the flask (after distillation) is converted to naphthylmethylthiuronium salt (p. 251).

5. Quinones

The properties of quinones are to a certain extent similar to those of carbonyl compounds. They form oximes, thiosemicarbazones, and phenylhydrazones. The rate of the reaction of quinones with hydroxylamine varies, and a number of quinones do not react with it at all. Substituents already present in the quinone molecule play a predominant role (16). In acid media hydroxylamine can have an oxidative influence, parallel to oximation, for example in the case of quinol, which is converted to quinone dioxime (17); free hydroxylamine can, however, have a reductive influence: it reduces quinone to hydroquinone (18).

Quinone oximes are prepared, for example, by heating quinone with a double excess of hydroxylamine hydrochloride in a mixture of ethanol and chloroform (6 : 1). Under these circumstances phenanthrene quinone yields a monoxime, while anthraquinone does not react even after prolonged heating. Quinones also react with phenylhydrazine in various ways: α- and β-naphthoquinone as well as phenanthrenequinone yield condensation products with phenylhydrazine which can be formulated as azo compounds (19).

Anthraquinone does not react with phenylhydrazine. On the other hand, benzoquinone, toluquinone, and xyloquinone oxidize phenylhydrazine and are themselves reduced to hydroquinones. Quinones of the benzene and naphthalene series react with 2,4-dinitrophenylhydrazine in acetic acid with the formation of corresponding 2,4-dinitrophenyhldrazones or the tautomeric 2,4-dinitrophenylazo compounds. These can be applied (for example, after dilution with water and extraction with benzene or chloroform) directly onto the chromatograms and identified by paper or thin-layer chromatography (20). For the color reaction of 2-methyl-1,4-naphthoquinone with phenyl-p-carboxy- or p-sulfophenylhydrazine with alkali hydroxide see the original literature (21, 22).

Benzenesulfinic acid reacts easily with o- and p-quinones with the formation of sulfones (23); the hydroxy compounds formed yield crystalline benzoyl derivatives. For special cases the condensation reaction with aniline may be important, giving rise to anils (24) and quinhydrones, i.e., molecular compounds of quinones and phenols (25). For the detection of quinones a series of color reactions has been proposed.

Reaction with Dithionite and Alkali
(Vat Dyeing Test)

Anthraquinone, its derivatives, and a series of polycyclic quinones (Vat dyes) are reduced with an alkaline solution of sodium dithionite heated to 80 °C to dihydroxy derivatives. Solutions of alkaline salts of quinones are usually of a different color (for example, red for anthraquinone). After dropping the sample on filter paper the original color returns slowly in consequence of air oxidation.

Reagents: 20% sodium dithionite solution; 5% NaOH.

Procedure: The sample is mixed with 1 ml of sodium dithionite solution and several drops of alkali, and the mixture is heated at 80 °C. In the case of anthraquinone a red solution is obtained; if a drop of this solution is spotted on the filter paper, the spot again becomes yellow. With other polycyclic quinones the heating produces other colors, but after air oxidation the original color is always restored.

Reaction with Cyanoacetate and Acetoacetate

Compounds with active methylene groups (cyanoacetate, acetoacetate, malonate, acetylacetone) give a blue color with a series of quinones in ammoniacal ethanol. A positive reaction was observed with *p*-benzoquinone, chloranil, α-naphthoquinone, and thymoquinone. The reaction is not observed with β-naphthoquinone, anthraquinone, an dphenanthrenequinone (26 – 28). Although the reaction of quinones with compounds containing an active methylene group has been studied thoroughly, the mechanism of the color reaction is not yet known. Blue substances have so far only been isolated from the reaction of malonate with 1,4-naphthoquinone, and the following structure was proposed for them (28):

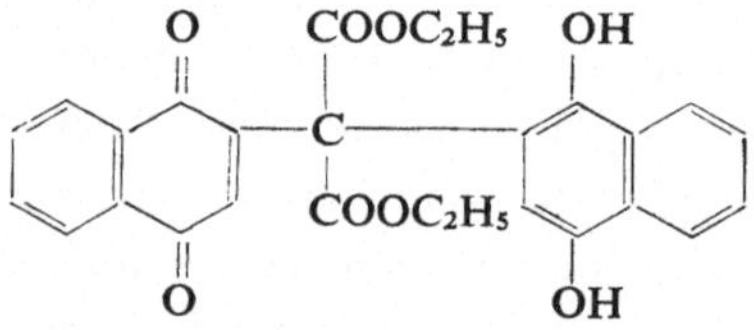

Recently, quinaldinium salts or 3-ethylrhodanin were proposed as reagents for quinones with a terminal quinoid benzene nucleus. A greater sensitivity is achieved, but the course of the reaction must be followed spectroscopically at 500 – 800 nm (30).

Reagents: ethanol; ethyl cyanoacetate or acetoacetate; ammonia.

Procedure: A small amount of quinone is dissolved in 1 – 2 ml of ethanol, and two drops of cyanoacetate and 2 – 3 ml of an ethanol-ammonia 1 : 1 mixture are added. A blue color occurs.

Among other color reactions of quinones the color of their solutions in conc. sulfuric acid should be mentioned. 1,2-, 1,4-, and other isomeric quinones give distinctly different colors; they change from yellow to green and blue (31). A simple detection scheme for quinones consists in bringing them into contact with an alkali, which leads to the formation of a brown to black color. Oxidative properties of quinones are also very important: p-benzoquinone and 1,4-naphthoquinone liberate iodine from an acidified potassium iodide solution according to the equation

$$C_6H_4O_2 + 2\ HCl + 2\ KI \rightarrow C_6H_4(OH)_2 + 2\ KCl + I_2$$

Reduction to quinol can also be carried out with powdered zinc. In 75% sulfuric acid, oxidation of diphenylbenzidine with quinones is accompanied by the formation of a blue-violet color. Anthraquinone, diacetyl, and benzil do not react, but with all other organic oxidants the reaction is smooth (32). In a similar manner the reaction with p-dimethylaminodiphenylmethane can also be used for the detection of chloranil (33). Quinones, similar to 1,2-diketones, catalyze the slow reaction of formaldehyde with o-dinitrobenzene, in which the violet salt of the acid form of o-isonitrosobenzene (34) is formed. It seems, however, that certain phenols react as well (35).

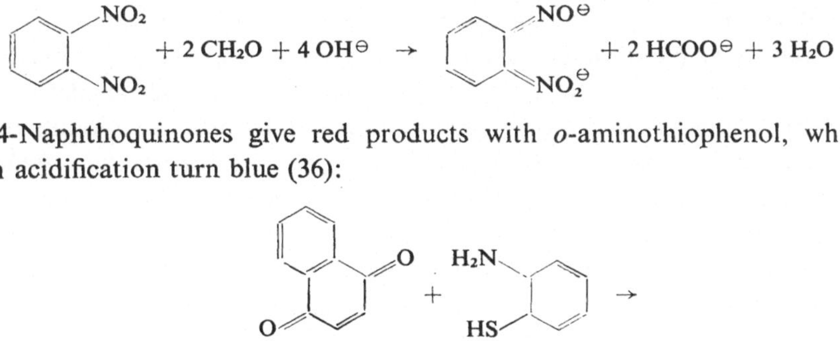

$$\text{(o-dinitrobenzene)} + 2\ CH_2O + 4\ OH^{\ominus} \rightarrow \text{(o-isonitrosobenzene salt)} + 2\ HCOO^{\ominus} + 3\ H_2O$$

1,4-Naphthoquinones give red products with o-aminothiophenol, which on acidification turn blue (36):

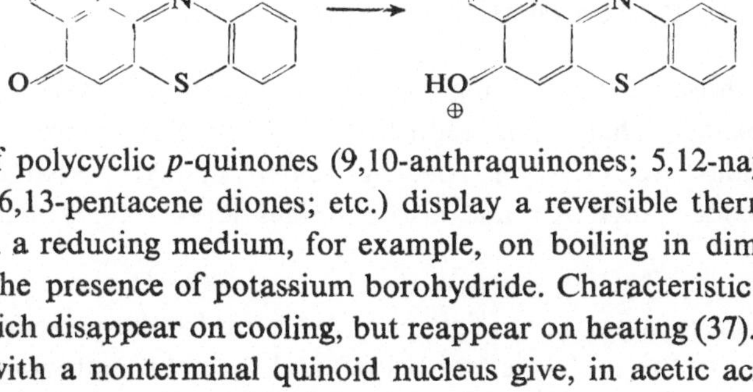

A series of polycyclic p-quinones (9,10-anthraquinones; 5,12-naphthacene-quinones; 6,13-pentacene diones; etc.) display a reversible thermochromic reaction in a reducing medium, for example, on boiling in dimethylformamide in the presence of potassium borohydride. Characteristic colors are formed which disappear on cooling, but reappear on heating (37). Polycyclic quinones with a nonterminal quinoid nucleus give, in acetic acid solution with aromatic amines (especially 3,4-dimethoxyaniline) and on heating, blue

to green color with the formation of the corresponding anils. 1,4-Naphtho-
quinone gives a purple color under identical conditions (38). Similar to
1,2-diketones, o-quinones give fluorescing phenazines with o-diamines. For
the reaction of quinones with ethylenediamine see (39). A fluorescence
test with o-phenylenediamine has been described (40) for the detection of
2-hydroxy-1,4-naphthoquinone, which does not give a positive reaction
with o-aminothiophenol or methylquinaldinium salt.

Paper Chromatography

Quinones of the naphthalene series and certain simpler polycyclic quinones
can be separated in two solvent systems: $25-50\%$ dimethylformamide/he-
xane and 1-bromonaphthalene/90% acetic acid (41). The former system is
suitable for naphthoquinones and rapid preliminary analyses of polycyclic
quinones, the latter is useful because of its suitability for the chromatographic separation of larger amounts of sample. This depends not only on
the nature of the separated substance, but also on the sensitivity of the
detection and on the system itself. For example, in the first system anthraquinone spots form tails at a concentration of 10 µg per spot, while in the
second its spots are round even at a concentration of 50 µg per spot. The
advantage of both systems consists in the fact that they separate quinones
well from their parent hydrocarbons, which is very important for practical
purposes. The detection of quinones is carried out under ultraviolet light.
A series of quinones fluoresce, especially after irradiation. Naphthoquinones
appear as dark spots which turn brown after prolonged irradiation. o-Quinones can be detected after spraying with 0.5% o-phenylenediamine solution
in 10% trichloroacetic acid and brief heating at 100 °C. A strong yellow
fluorescence appears under ultraviolet light. For the chromatography of
aminoanthraquinones see p. 349, for hydroxyanthraquinones see (42), and
for 2,4-dinitrophenylhydrazones of quinones see p. 299.

6. Peroxides and Peroxy Acids

Organic compounds containing two mutually bound oxygen atoms are
called peroxides. Their main representatives are alkyl peroxides ROOH,
acyl peroxides AcOOH (i.e., peroxy acids), dialkyl peroxides ROOR,
diacyl peroxides AcOOAc, and also ozonides $R-O-R$. When peroxides
$$\begin{array}{c} \diagdown \qquad \diagup \\ O-O \end{array}$$
are analyzed it is important to first carry out a test for the presence of a per-
oxidic group in the molecule (see below) and then chromatograph (43) and
measure IR spectra (44). Isolation and identification of decomposition

products of peroxides (catalytically, reductively, oxidatively, hydrolytically) which, in a number of cases, do not give single products, afford other possibilities for the structure determination of this class of substances (43).

Explosive decomposition on heating is a typical property of peroxides. This requires careful manipulation. All peroxy compounds liberate iodine from acidified iodide solution, but the rate of this reaction is different in different types of peroxides. At room temperature iodine is not liberated by di-tert-butyl peroxide, trimeric ketone peroxides, and ditrityl peroxides. (45).

Color Reactions with Titanium (IV) Sulfate

Almost all peroxides react with titanium (IV) sulfate in 62% sulfuric acid with the formation of a yellow color. Ditrityl peroxide is an exception. Aromatic diacyl peroxides react slowly and only on heating.

Reagents: TiO_2 (1 g) and 20 g of $KHSO_4$ are melted in a platinum crucible until the melt is transparent. The yellow melt is cooled by rapid immersion of the crucible into cold water, and is ground to small pieces. It is then dissolved with shaking in a mixture of 250 ml of water and 250 ml of conc. H_2SO_4; the solution is filtered through a sintered-glass filter. Peroxide-free ether.

Procedure: The reagent is added to a grain of the compound or to its ethereal solution (a control experiment is indispensable!) in a test tube and gently warmed if necessary. The formation of a yellow color means a positive reaction. Only those peroxy compounds giving a yellow color, which liberate hydrogen peroxide in a weakly acid medium, react with titanium (IV) sulfate in 1.6% H_2SO_4 (46). Hydroxy- and dihydroxyalkyl peroxides react immediately in the cold, while peroxy acids and certain alkylhydroxy peroxides react slowly. Dialkyl peroxides react only after many hours or even days. Water-insoluble peroxides can be dissolved in methanol.

Procedure: TiO_2 (1 g) is melted with 20 g of $KHSO_4$ as in the preceding example, and the melt is dissolved in 5 ml conc. H_2SO_4 and diluted with water to make the volume 500 ml. The solution should not be heated, to prevent the separation of polytitanic acid. The detection is carried out as in the preceding example.

Reaction with Iron Pentacarbonyl

All hydroperoxides, peroxy acids, peroxy acid esters, and diacyl peroxides give a positive reaction with iron pentacarbonyl with the evolution of heat and CO_2 and Fe_2O_3 separation:

$$2\ Fe(CO)_5 + 13\ ROOH \rightarrow Fe_2O_3 + 13\ ROH + 10\ CO_2$$

Dialkyl peroxides, ozonides, and dimeric and trimeric ketone peroxides do not react.

Reagents: iron pentacarbonyl; peroxide-free light petroleum or benzene.

Procedure: 1 ml of iron pentacarbonyl is dissolved in 24 ml of light petroleum. A sample of the tested substance is mixed with 1 − 2 ml of the reagent.

Color Reaction with Phenolphthalin

Phenolphthalin is very sensitive to hydrogen peroxide and peroxy acids in alkaline media. It is oxidized to violet-red phenolphthalein. Dialkyl peroxides and hydroperoxides do not react.

Reagent: Phenolphthalein (1 g) is refluxed with 10 g of NaOH, 5 g of zinc powder, and 20 ml of water for ∼2 hr until the solution is decolorized. After cooling, the solution is filtered through a sintered-glass filter and the filtrate is made up with water to 50 ml. It is kept in a dark place in the presence of small amounts of zinc. To carry out the test, 10 ml of the reagent are diluted first with 30 ml of water.

Procedure: A small amount of peroxy acid or of its solution is mixed in a test tube with a few drops of the reagent; a violet-red color is formed.

Reactions of Peroxy Compounds   *Table 11*

Peroxy compound	Phenol-phthalin	TiO_2 in 1.6% H_2SO_4	Iron penta-carbonyl	Lead tetra-acetate
Hydroxyalkyl hydroperoxides	−[a]	+	+	+
Dihydroxyalkyl peroxides	−[a]	+	+	+
Peroxy acids	+	+	+	+[b]
Alkyl hydroperoxides	−[c]	−[d]	+	+
Hydroxydialkyl peroxides	−	−	+	+
Diacyl peroxides	+	−	+	−
Peroxo esters	+	−	−[e]	−
Dialkyl peroxides	−	−	−	−
Dimeric and trimeric keto peroxides	−	−[f]	−	−
Ozonides	+	−[g]	−	−

[a] For a short time the reaction is partly positive. [b] Peroxo formic acid gives a positive reaction. [c] Tertiary hydroperoxides give a positive reaction. [d] Tert-butyl hydroperoxide and decalin peroxide give a positive reaction, cyclohexane and tetralin peroxide react very slowly, primary alkyl hydroperoxides react negatively. [f] Trimeric acetone peroxide give a slow positive reaction. [g] Partly positive, partly negative.

To carry out a similar reaction, leucomethylene blue or the more stable benzoyl-leucomethylene blue can also be used. The first reagent is oxidized by peroxides to a blue dye (47) and the second to the cation of methylene blue.

Di-tert-butyl peroxide does not react (48). Hydroperoxide can be distinguished from other types of peroxides by a positive reaction with lead tetraacetate, which is accompanied by the evolution of oxygen. It is carried out with a solution of the tested substance in glacial acetic acid by addition of a grain of lead tetraacetate. A list of reactions of peroxy compounds is given in Table 11.

Paper and Thin-Layer Chromatography

As peroxy compounds differ from one another in solubility, it is obvious that several solvent systems are necessary for their separation. The following solvent systems have been found suitable (49—51): dimethylformamide/ /hexane to benzene; ethylene glycol/hexane; formamide/hexane; n-butanol-ethanol-water (45 : 5 : 50) or (10 : 10 : 1). For detection, sensitive down to 1 μg, the following sprays have been recommended: (a) 0.1% p-amino-dimethylaniline hydrochloride; and (b) a mixture of 3 ml of glacial acetic acid, 2 ml of a saturated potassium iodide solution, and 5 ml of starch solution. For the chromatography of organic peroxides on thin layers see (52, 53).

7. Sugars

Although the sugar molecule always contains reactive groups which are suitable both for detection and identification purposes (preparation of derivatives), the identification of this class of substances represents a difficult problem.

First, the properties of saccharides do not differ too much from one to another, second, the identification of less-pure preparations (usually containing substances of similar character as impurities) is very tedious, and, finally, the melting or decomposition points of sugars and also of their derivatives are usually much dependent on the method by which they are determined. The rate of heating can be a cause of variability of melting points, the difference sometimes being 10—20 °C. It is therefore recommended that the last step of identification be carried out by comparing the melting point (decomposition point) of the identified substance with that of an authentic sample.

The literature gives a large number of color reactions for sugars, which can be classified in a few groups: (1) reductive reactions, (2) reactions based on the oxidation of α-diol groups, (3) reactions based on the formation of furan derivatives which react with other reagents, and (4) other reactions.

The progress in the use of color reactions of sugars was due mainly to biochemistry, which needed sensitive methods for the detection and colorimetric determination of sugars in biological material. For a review of such reactions see (54).

Reductive Reactions (Detection)

Aldoses and ketoses, similar to aldehydes, reduce heated Tollens' reagent (see p. 210), Nessler's reagent (p. 211), Fehling's reagent [p. 210; see also (55)], and also Nylander's reagent (an alkaline solution of a bismuth salt) and a solution of ammonium molybdate. The reaction with nitro compounds should also be mentioned here.

Detection with Nylander's Reagent

Reagent: Basic bismuth nitrate (2.0 g) and 4.0 g of Seignett salt are dissolved in 100 ml of 8% NaOH.

Procedure: One milliliter of the reagent is added to a few milliliters of a dilute sugar solution and the mixture is boiled briefly. In the presence of a reducing sugar the solution first turns brown, and later on a black precipitate separates.

Reaction with Nitro Compounds (Detection)

A number of polynitro derivatives can be reduced under the influence of monosaccharides. For example, picric acid is reduced to picraminic acid (56, 57).

Among other nitro compounds used for the reaction, 2,4-dinitro-phenol (58), 3,5-dinitrosalicylic acid (reduced to 3-amino-5-nitrosalicylic acid) (59, 60), 4,6-dinitroguaiacol (61), sodium-1-nitroanthraquinone-5-sulfonate (62), 3,5-dinitrophthalic acid (63), 3,4-dinitrobenzoic acid (64, 65), and m-dinitrobenzene (66) should be mentioned.

o-Dinitro compounds also give red and violet-to-blue colors in an alcoholic solution under the influence of reducing sugars. Fructose gives an especially strong color (67). Ascorbic acid also gives a positive reaction.

Detection with o-Dinitrobenzene

Reagents: 1% *o*-dinitrobenzene solution, 25% sodium carbonate solution.

Procedure: A drop of an aqueous solution of sugar is mixed with a drop of *o*-dinitrobenzene solution and 0.5 ml of sodium carbonate solution and the mixture is diluted with 1 ml of ethanol. After heating the mixture over a small flame the color appears.

Detection with Picric Acid

Reagent: Picric acid (0.2 g) and 0.4 g of anhydrous sodium carbonate are dissolved in 100 ml of water.

Procedure: An aqueous solution of sugar (approx. 1 ml) in a test tube is additioned with an equal amount of the reagent. A yellow color is formed which changes to red-brown.

Reactions Based on the Oxidation of α-Diol Groups (Detection)

The oxidation of α-diol groups of saccharides with periodic acid (see p. 177) or lead tetraacetate belong to this class of test. The aldehydes formed react with other reagents, for example, with Schiff's reagent and phenols. For paper chromatography use is made of the fact that the unreacted periodate can oxidize to benzidine blue, while iodate (p. 313) cannot. Of course, other α-diols which are not sugars also give positive reactions with these reagents.

Reactions Based on the Formation of Furan Derivatives (Detection)

Reducing sugars are dehydrated in the presence of hot mineral acids with simultaneous cyclization to furan derivatives; pentoses give fural, methylpentoses give 5-methylfural, and hexoses yield 5-hydroxymethylfural:

$$\begin{matrix} CH(OH)-CH(OH) & & CH-CH \\ | \quad\quad\quad | & \rightarrow & \| \quad\quad \| \\ (HO)CH_2-CH(OH) \quad CH(OH)-CHO & & (HO)CH_2-C \quad\quad C-CHO + 3\,H_2O \\ & & \diagdown_O\diagup \end{matrix}$$

Corresponding sugar acids behave in the same manner, giving rise to 5-formylfuroic acid (68). For the formation of the furan ring it is necessary for the sugar molecule to contain an alcoholic group (OH) in α-position to the aldehyde group. Therefore, the dehydration of 2-desoxy-D-ribose leads to levulinic acid only. Fural or its derivatives can then be detected by reaction with some phenol, aromatic amine, etc. The rate of dehydration is dependent on the type of sugar, on the kind and the concentration of the acid, on the length of heating, and on the temperature. The rate of condensation of the fural derivative is dependent roughly on the same factors. As pentoses and ketoses react much more rapidly than aldohexoses, the conditions can often be chosen in such a manner that the reaction is selective for the first group of sugars. In certain cases different types of sugars give different colors; sometimes spectrophotometry is used for their differentiation (69).

Dehydration is carried out with sulfuric acid, hydrochloric acid, or phosphoric acid. Among phenols, phenol, thymol, guaiacol, resorcinol, orcinol, catechol, phloroglucinol, α-naphthol, and naphthoresorcinol have been proposed. Among amines, aniline, p-anisidine, benzidine, 3,3'-dimethoxybenzidine, diphenylamine, indole and its derivatives, carbazole, N-ethylcarbazole, α,α-dinaphthylamine, and 1,2,7,8-dibenzocarbazole have been found valuable. In its classical form (Molisch reaction) the test for saccharides is carried out with 1-naphthol and conc. sulfuric acid (70). It is supposed that under the conditions of the reaction corresponding triphenylmethane dyes are formed, although the course of the reaction might be more complicated (71).

The reaction of saccharides with anthrone in conc. sulfuric acid (72 to 74) also probably belongs to this group of reactions. Saccharides give a green color, uronic acid a pink to red one. For the mechanism of this reaction see (75). Ascorbic acid and derivatives of fural also give a positive reaction.

Test with Anthrone

Reagent: 0.1% anthrone solution in dilute H_2SO_4 (76 ml conc. H_2SO_4 a re poured into water to make the volume 100 ml).

Procedure: The aqueous sugar solution (2 ml) is heated with 10 ml of the reagent over a boiling water bath: green color.

Molisch Reaction

Reagents: 3% alcoholic 1-naphthol solution; conc. sulfuric acid.

Procedure: To a dilute aqueous monosaccharide solution (2 ml) 1-naphthol solution (five drops) is added, and then several milliliters of conc. sulfuric acid are added carefully down the sides of the test tube. A violet or red color is formed on the surface of the sulfuric acid. After mixing, the liquid assumes a dark violet color and after dilution a blue-violet precipitate separates which is soluble in ether with a yellow color. A control experiment is carried out simultaneously without 1-naphthol, as a series of substances gives similar colors with sulfuric acid alone.

Detection of Pentoses

Reagent: Orcinol (0.25 g) is dissolved in 125 ml of 25% HCl and mixed with five drops of $FeCl_3$ solution.

Procedure: One milliliter of 0.1% pentose solution is mixed with a double volume of the reagent and boiled for 1−2 min. A blue dye is formed which can be extracted with amyl alcohol. Aldohexoses and ketoses give a brown color.

Detection of Ketoses with Resorcinol

Reagents: Ethanolic solution of HCl, saturated at 0 °C; resorcinol.

Procedure: A small amount of saccharide in a test tube is mixed with 3−4 ml of hydrochloric acid solution and ∼50 mg of resorcinol. In the presence of ketose a cherry-red color appears within 3 min.

Detection of Ketoses with Diphenylamine

Reagents: 2% diphenylamine solution in ethanol; conc. HCl.

Procedure: 2 ml of diphenylamine solution and 2 ml of conc. HCl are added to a few milligrams of ketose and the mixture is warmed over a water bath: violet-blue color.

Osazones (Identification)

Of a number of proposed derivatives, the products of the reaction of sugars with phenylhydrazine or with substituted phenylhydrazines are the most suitable. The reaction is carried out in 50% sulfuric acid at room temperature. Phenylhydrazones are formed.

$$
\begin{array}{ccc}
\underset{\substack{\displaystyle | \\ \vdots}}{\overset{\displaystyle C\diagup^{H}_{\diagdown O}}{H-C-OH}} & \xrightarrow{ArNHNH_2} & \underset{\substack{\displaystyle | \\ \vdots}}{\overset{\displaystyle H-C=N-NHAr}{H-C-OH}} \xrightarrow{ArNHNH_2}
\end{array}
$$

$$
\underset{\substack{\displaystyle | \\ \vdots}}{\overset{\displaystyle H-C=N-NHAr}{C=N-NHAr}} \xrightarrow[H^{\oplus}]{Cu^{2\oplus}} \underset{\substack{\displaystyle | \\ \vdots}}{\overset{\displaystyle H-C=N}{C=N}}\!\!\!\diagdown N-Ar
$$

With excess phenylhydrazine in aqueous media buffered with acetate buffer osazones are formed, which are more suitable for identification in spite of the fact that their melting or decomposition points depend on the rate of heating. On mild oxidation in acid media osazones are converted to triazoles, which are characterized by sharp melting points and thus serve for further characterization of the osazones.

In addition to phenylhydrazine, *p*-nitrophenylhydrazine and methylphenylhydrazine are primarily used for the formation of derivatives because they enable differentiation between aldoses and ketoses; they react only with keto sugars. For example, D-fructose reacts easily with this reagent, while D-glucose and D-mannose do not.

When osazones are used as derivatives for identification it should be borne in mind that saccharides differing in configurations only at the last two carbon atoms afford identical osazones (for example, D-glucose, D-mannose, D-fructose).

The rate of formation of osazones and their solubility, which differ substantially from case to case, can also serve as characteristic features useful in the identification of sugars. For example, osazones of glucose and fructose precipitate shortly after the heating of the solution, while saccharose requires 20 – 30 min of heating. Osazones of lactose and maltose precipitate only after cooling. In addition, phenylosazones of D-glucose, D-fructose, and D-mannose (which are identical) are insoluble in cold acetone, in contrast to phenylosazones of xylose, arabinose, rhamnose, and fructose.

General Method of Preparation of Phenylosazones

The tested substance (3 mmole) is dissolved in 20 ml of water, mixed with 1 g of phenylhydrazine hydrochloride and 1.5 g of sodium acetate, and heated for 3 min. After cooling, the separated phenylosazone is filtered off (if no precipitate is formed, it is advisable to stopper the flask and to allow the reaction mixture to stand for 24 hr), washed with 20 ml of water, and recrystallized from dilute or concentrated ethanol.

Phenyl-D-glucosazone

Reagents: phenylhydrazine (freshly distilled), glacial acetic acid, sodium acetate, 5% acetic acid, methanol.

Procedure: Phenylhydrazine ($\sim$200 mg) in a 10-ml test tube is mixed with 0.12 ml of glacial acetic acid and 50 mg of sodium acetate, and 100 mg of glucose dissolved in 4 ml of water are added to this mixture. The test tube is immersed into a boiling water bath for 30 min, 1 ml of water is added, and the mixture is allowed to cool in a refrigerator. The separated osazone is filtered on a filtration tube, washed with 1 ml of 5% acetic acid and 1 ml of water, and dried in a stream of dry air. Yield, 0.1 g; mp, 208 – 209 °C. For crystallization the product is dissolved in 15 ml of hot methanol, and water is then added dropwise to the solution to incipient turbidity. The mixture is then heated and allowed to crystallize (85 mg; mp, 208 °C).

Phenyl-D-glucosotriazole

Reagents: 6 N H_2SO_4, $CuSO_4 \cdot 5 H_2O$, isopropanol, 30% ethanol.

Procedure: To glucose phenylosazone (50 mg) in a 25-ml flask fitted with a reflux condenser are added 4 ml of water, 3 ml of isopropanol, one drop of 6 N H_2SO_4, and 0.15 g of copper sulfate, and the mixture is boiled (boiling stone!) for 1 hr. It is then concentrated to 2 ml and cooled in a refrigerator for 1 hr. The precipitated product is filtered off under suction on a filtration tube and the crude product is dissolved in 6 ml of boiling water. Charcoal is added to the hot solution and the mixture is filtered. The filtrate is heated and filtered again, and this is repeated several times. Eventually, the filtrate is allowed to stand in a refrigerator overnight and the crystallized derivative is filtered off and washed with 0.5 of water. Yield, 3 mg; mp, 197 – 198 °C.

Acylation and Aroylation (Identification)

The reactions of acylating agents (acetic anhydride) and aroylating agents (benzoyl chloride) with sugars take place easily, but have only limited use, because, depending on the reaction conditions, either the α or β forms, or a mixture of them, can be formed:

$$-\overset{\displaystyle -O\rceil}{\underset{\displaystyle OAc}{C-C-H}} \qquad -\overset{\displaystyle -O\rceil}{\underset{\displaystyle H}{C-C-OAc}}$$

α-form   β-form

The general method of preparation of acetates is demonstrated by the preparation of D-glucose pentaacetate.

D-Glucose Pentaacetate

Reagents: acetic anhydride, sodium acetate (freshly melted), methanol.

Procedure: Acetic anhydride (2.5 ml) is pipetted into a 10-ml flask, and 0.25 g of freshly melted sodium acetate is added. The mixture is heated, and 0.5 g of glucose is added in several portions. The flask is then connected with a reflux condenser and the mixture is heated for 1 hr on a boiling water bath. After cooling, 2.5 ml of water are added, the mixture is cooled in a refrigerator, and the separated oil is brought to crystallization by scratching with a glass rod. The acetate is filtered off and washed twice with 1 ml of water. Yield, 0.54 g; mp, 131 – 132 °C; after crystallization from 20 ml of methanol: 0.1 g, mp 131.5 °C.

The preparation of esters (76) of sugars with *p*-phenylazobenzoyl chloride is of interest mainly from the point of view of their isolation and separation from mixtures, the colors of the esters making it possible to separate them and detect them by column chromatography.

Other Methods of Identification

We first note the special monographs (77, 78) in which the problem of the identification of sugars is discussed to a greater detail.

For the identification of certain sugars (for example, galactose) their oxidation to carboxylic acids is useful. This is carried out by evaporation on a water bath of an aqueous sugar solution acidified with nitric acid (for 0.1 g of sugar 2 ml of a 20% HNO_3 are used). Galactose and lactose give muconic acid on oxidation.

The hydroxy groups of sugars can be further alkylated, tosylated (reaction with p-toluenesulfonyl chloride (79), or tritylated. In the latter case only primary hydroxy groups react (80).

Very important methods for the identification of sugars are paper and thin-layer chromatography (see below), which also make the control of the purity of individual preparations possible.

Among physical methods, measurement of the optical rotation is commonly used for the identification of sugars. This procedure requires very pure substances (control by paper or thin-layer chromatography); the purity of the identified preparations can be checked by the determination of their optical rotation.

Sugars can also be identified by studying their crystallographic properties under the microscope (81, 82).

Paper and Thin-Layer Chromatography

Sugars are a class of compounds for which the method of paper chromatography is especially thoroughly elaborated. Especially well known are their identification in natural materials and the course of their various reactions and transformations. This method is also very important for the determination of the structure of polysaccharides. In view of the good solubility of sugars and their derivatives in water these compounds can be chromatographed on untreated papers by means of a great number of solvent systems. In addition, a large number of chemical as well as enzymatic methods of detection are also available.

Simple Sugars

For the chromatography of simple sugars the following solvent systems can be chosen from the wide choice quoted in the literature:

1. Phenol saturated with water or with the addition of 15% of water only (with ammonia or HCOOH in the atmosphere of the chromatographic chamber).

2. Organic (upper) phase of the n-butanol — acetic acid — water (4 : 1 : 5) mixture.

3. n-Butanol — formic acid — water (12 : 1 : 1), diluted after 1 hr standing with 6 parts of water, and separated after 24 hr.

4. Ethyl acetate — acetic acid — water (3 : 1 : 3).

5. n-Butanol — butyric acid — water (1 : 1 : 1).

6. *n*-Butanol — pyridine — water (6 : 4 : 3).

7. Ethyl acetate — pyridine — water (2 : 1 : 2).

8. *n*-Butanol — acetone — water (2 : 7 : 1), for crude development.

The amount of sugar suitable for application is $10 - 100$ µg dissolved in water or aqueous ethanol.

For thin-layer chromatography kieselguhr or silica gel may be used, preferably impregnated with buffer. As mobile phases, the mixtures ethyl acetate — 65% isopropanol (65 : 35 or similar) or benzene — acetic acid — methanol (1 : 1 : 3) are suitable (83 — 85).

Detection can be carried out in the following ways:

1. Ammoniacal silver nitrate solution (see p. 210), with heating.

2. Triphenyltetrazolium chloride: 2% aqueous triphenyltetrazolium chloride is mixed before use with an equal amount of N NaOH. After spraying, the chromatogram is allowed to hang in darkness at room temperature or slightly higher. Substances containing a reducing group appear as red spots on a white background.

3. 3,5-Dinitrosalicylic acid: The chromatogram is sprayed with a 0.5% solution of 3,5-dinitrosalicylic acid in 4% NaOH. After brief drying in air the chromatogram is dried in a drying oven at 100 °C. Reducing sugars give brown spots on a yellow background.

4. Periodic acid and benzidine: The chromatogram is sprayed with a saturated solution of potassium metaperiodate and allowed to stand for 6 min. It is then sprayed with a solution of 0.1 M benzidine in 50% methanol with acetone and 0.2 N HCl (10 : 2 : 1). Sugars with free α-diol groupings appear as white spots on a blue background.

5. Naphthoresorcinol (or resorcinol, orcinol, phloroglucinol, or 1-naphthol): 0.2% naphthoresorcinol solution in acetone is mixed shortly before use with $3 \text{ N } H_3PO_4$ in the proportion 5 : 1. The chromatogram is sprayed with this reagent and then dried in an oven at about 90 °C. Ketoses are detected as colored spots.

6. Anilin hydrogen phthalate: 0.93 g of aniline and 1.66 g of phthalic acid are dissolved in 100 ml of water-saturated n-butanol. The chromatogram is sprayed with this mixture and then dried for several minutes at 105 °C. Aldopentoses and 2-ketohexonic acids give red spots, aldohexoses and 5-ketohexonic acids brown spots. The sensitivity of the detection is greater when the spots are observed under UV light.

Sugar Alcohols

As the behavior of sugar alcohols on the chromatogram is similar to that of the corresponding monosaccharides, the systems used for their separation are also similar:

9. *n*-Butanol – acetic acid – water (5 : 2 : 1).

10. *n*-Propanol – ethyl acetate – water (7 : 1 : 2).

11. tert-Amyl alcohol – *n*-propanol – water (4 : 1 : 1.5).

Detection can be carried out using methods 1 and 2 described above (for monosaccharides).

Polysaccharides

One of the greatest successes of paper chromatography is in the study of polysaccharide structure. The structure of these compounds is best determined, after their isolation from natural material and after partial or total purification by hydrolysis, by periodate oxidation and total methylation followed by total hydrolysis. In all instances the reaction products are then identified by paper chromatography.

A more extensive description of the problems of the paper chromatography of sugars is beyond the scope of this treatise, and therefore the reader should refer to specialized literature (86).

References

1. Henecka, H.: Chemie der β-Dicarbonylverbindungen. Springer, Berlin 1950.
2. Vengrinovich, L. M., Vladzimirskaya, E. V., and Turkevich, N. M.: Zhur. Anal. Khim, **22**, 1429 (1967).
3. Henecka, H.: Ber. **81**, 179 (1948).
4. v. Euler, H., and Martius, C.: Ann. **505**, 73 (1933).
5. v. Euler, E., and Hasselquist, H.: Reduktone, Sammlung chemischer und techn. Vorträge. Vol. 50. Enke, Stuttgart, 1950.
6. Reichstein, T., and Demole, V.: Barell Festschrift der Firma Hoffmann – La Roche. Basel, 1936.
7. Eistert, B., and Hasselquist, H.: Ark. Kemi **4**, 233 (1952).
8. Fenton, H. J. H.: J. Chem. Soc. **65**, 899 (1894).
9. Nef, J. H.: Ann. **357**, 290 (1907).
10. Arndt, F., Loewe, L., and Ayca, E.: Ber. **85**, 1150 (1952).
11. Weygand, F., and Csendes, E.: Ber. **85**, 45 (1952).
12. Houben – Weyl: Methoden der organischen Chemie, Bd. II, Analytische Methoden, p. 485. Thieme, Stuttgart, 1953.
13. Janák, J., Jonas, J., and Kratochvíl, M.: Collection Czech. Chem. Commun. **30**, 265 (1965).
14. Post, H. W.: The Chemistry of Aliphatic Orthoesters. Reinhold Publ. Corp., New York, 1943.
15. Grummitt, O., and Stearns, J. A., Jr.: J. Am. Chem. Soc. **77**, 3136 (1955).
16. Kehrmann, Fr., et al.: Ber. **23**, 3557 (1890); **21**, 3315 (1888); **27**, 217 (1894).
17. Nietzki, R., and Kehrmann, Fr.: Ber. **20**, 613 (1887).
18. Hantzsch, A., and Schniter, K.: Ber. **20**, 2279 (1887).
19. Zincke, T., and Bindewald, H.: Ber. **17**, 3026 (1884).
20. Gemzová, I., and Gasparič, J.: Collection Czech. Chem. Commun. **32**, 2740 (1967).
21. Novelli, A.: Science **93**, 358 (1941).

22. Novelli, A., and Conticello, J. S.: J. Amer. Chem. Soc. **66**, 842 (1944).
23. Hinsberg, O., et aj.: Ber. **27**, 3259 (1894); **28**, 1315 (1895); **29**, 2019 (1896).
24. Zincke, T.: Ber. **14**, 1493 (1881).
25. Pffeifer, P.: Organische Molekülverbindungen, p. 199. Enke, Stuttgart, 1922.
26. Craven, R.: J. Chem. Soc. **1931**, 1605.
27. Kesting, W.: Angew. Chem. **41**, 358 (1928); J. prakt. Chem. **138**, 215 (1933).
28. Kofler, M.: Helv. Chim. Acta **28**, 702 (1945).
29. Pratt, E. F., and Boehme, W. E.: J. Am. Chem. Soc. **73**, 444 (1951).
30. Sawicki, E., Stanley, T. W., and Hauser, T. R.: Anal. Chim. Acta **21**, 392.(1959).
31. Sawicki, E., and Elbert, W.: Anal. Chim. Acta **22**, 448 (1960).
32. Anger, V.: Mikrochim. Acta **1959**, 386.
33. Feigl, F., Gentil, V., and Maris, J. E. R.: Anal. Chim. Acta **13**, 210 (1955).
34. Feigl, F., and Neto, C. C.: Anal. Chem. **28**, 397 (1956).
35. Marzat, J., and Mesnard, P.: Bull. Soc. Pharm. Bordeaux **95**, 170 (1956).
36. Sawicki, E., and Elbert, W.: Anal. Chim. Acta **23**, 205 (1960).
37. Sawicki, E., Stanley, T. W., and Hauser, T. R.: Anal. Chem. **30**, 2005 (1958).
38. Sawicki, E., and Elbert, W.: Anal. Chim. Acta **22**, 448 (1960).
39. Carius, H., and Mapstone, G. E.: Chem. and Ind. **1956**, 266.
40. Sawicki, E., and Elbert, W.: Anal. Chim. Acta **23**, 205 (1960).
41. Gasparič, J.: Mikrochim. Acta **1958**, 681.
42. Stárka, L., and Vystrčil, A.: Chem. Listy **51**, 378 (1957).
43. Eggergluss, W.: Organische Peroxyde. Supplement to Angew. Chem. 1951.
44. Shreve, O. D., Heether, M. R., Knight, H. B., and Swern, D.: Anal. Chem. **23**, 282 (1951).
45. Criegee, R., Schnorrenberg, W., and Becke, J.: Ann. **565**, 7 (1949).
46. Lepper, W.: Chemiker Ztg. **66**, 314 (1942).
47. Ueberreiter, K., and Sorge, G.: Angew. Chem. **68**, 352 (1956).
48. Eiss, M. I., and Giesecke, P.: Anal. Chem. **31**, 1558 (1959).
49. Taylor, G. W.: Can. J. Chem. **36**, 1213 (1958).
50. Milas, N. A., and Belič, I.: J. Am. Chem. Soc. **81**, 3358 (1959).
51. Cartlidge, J., and Tipper, C. F. H.: Anal. Chim. Acta **22**, 106 (1960).
52. Hayano, S., Otta, T., and Fukushima, Y.: Bunseki Kagaku **15**, 365 (1966).
53. Sorokina, A. N., Batog, A. E., Romantsevich, M. K.: Zhur. Obshch. Khim. **37**, 766 (1967).
54. Khrustaleva, V. N., and Kozlov, V. V.: Uspekhi khimii **27**, 752 (1958).
55. Paleg, L. G.: Anal. Chem. **31**, 1902 (1959).
56. Braun, C. D.: J. prakt. Chem. **96**, 411 (1865); Zeit. anal. Chem. **4**, 185 (1865).
57. Dehn, W. M., and Hartman, F. A.: J. Am. Chem. Soc. **36**, 403 (1914).
58. Poe, C. F., and Edson, F. G.: Ind. Eng. Chem., Anal. Ed. **4**, 300 (1932).
59. Sumner, J. B., and Graham, W. A.: J. Biol. Chem. **47**, 5 (1921).
60. Miller, G. L.: Anal. Chem. **31**, 426 (1959).
61. Sumner, J. B.: J. Biol. Chem. **46**, xxi (1921).
62. Milroy, J. A.: Biochem. J. **19**, 746 (1925).
63. Momose, T., and Inaba, A.: Chem. Pharm. Bull. **7**, 541 (1959).
64. Weygand, F., and Hofmann, H.: Chem. Ber. **83**, 405 (1950).
65. Borel, E., and Deuel, H.: Helv. Chim. Acta **36**, 801 (1953).
66. Ekkert, L.: Pharm. Zentralhalle **75**, 515 (1934).
67. Bose, P. K.: Zeit. anal. Chem. **87**, 110 (1932).
68. Bowness, J. M.: Biochem. J. **70**, 107 (1958).

69. Vasseur, E.: Acta Chem. Scand. **2**, 693 (1948).
70. Turney, T. A.: Analyst **84**, 193 (1959).
71. Thies, H., and Kallinich, G.: Chem. Ber. **85**, 438 (1952).
72. Dreywood, R.: Ind. Eng. Chem., Anal. Ed. **18**, 499 (1946).
73. Fairbairn, N. J.: Chem. and Ind. **1953**, 86.
74. Helbert, J. R., and Brown, K. D.: Anal. Chem. **27**, 1791 (1955); **29**, 1464 (1957).
75. Momose, T., Ueda, Y., and Sugi, A.: Pharm. Bull. (Japan) **3**, 323 (1955); C. A. **50**, 12 954 (1956).
76. Link, K. P., et al.: J. Biol. Chem. **150**, 345 (1943); **133**, 293 (1940); J. Org. Chem. **5**, 639 (1940).
77. Browne, C. A., and Zerban, F. W.: Physical and Chemical Methods of Sugar Analysis. 3rd Ed., Wiley, New York, 1941.
78. Micheel, F.: Chemie der Zucker- und Polysaccharide. Akademische Verlagsgesellschaft, Leipzig, 1939.
79. Compton, J.: J. Am. Chem. Soc. **60**, 395 (1938).
80. Reynolds, D. D., and Evans, W. L.: J. Am. Chem. Soc. **60**, 2559 (1938).
81. Secor, G. E., and White, L. M.: Anal. Chem. **27**, 1998 (1955).
82. White, L. M., and Secor, G. E.: Anal. Chem. **27**, 1016 (1955).
83. Stahl, E.: Zeit. Anal. Chem. **181**, 303 (1961).
84. Stahl, E., and Kaltenbach, V.: J. Chromatog. **5**, 351 (1961).
85. Pastuska, G.: Zeit. Anal. Chem. **179**, 427 (1961).
86. Hais, I. M., and Macek, K. (Editors): Paper Chromatography. Publ. House Czechoslov. Acad. Sci, Prague, 1963, p. 289.

CHAPTER XV

AMINES

Amines are usually more soluble in dilute acids than in water. They can be detected by a precipitation reaction with sodium salt of tetraphenylboron or by using the test for nitrogen. It is further necessary to decide whether the substance to be identified is a primary, secondary, or tertiary amine, and whether it belongs to the aliphatic or to the aromatic series. To detect the primary amino group, isonitrile reaction, a spot-test reaction with ninhydrin, or color reactions with bindone (anhydro-bis-indanedione), sodium 1,2-naphthoquinone-4-sulfonate, or o-diacetylbenzene are usually applied. For the differentiation of aliphatic and aromatic amines the diazotization and coupling test or the reaction with glutaconaldehyde or chloranil is suitable. The detection of secondary amines can be carried out with sodium nitroprusside and acetaldehyde, or with carbon disulfide and a nickel (II) salt. For tertiary amines the reaction with Dragendorff reagent is characteristic.

A list of reactions used is given in Table 12.

For the identification of amines acylation methods (acetylation, benzoylation, 3,5-dinitrobezoylation), reaction with p-toluenesulfonyl chloride, preparation of substituted thioureas, diazotization and coupling (for aromatic primary amines), and the preparation of salts (picrates, tetraphenylboron salts) are most commonly used.

For the separation of amines preparative methods with p-toluenesulfonyl chloride or 3-nitrophthalic anhydride are employed. Paper, thin-layer, column, and gas chromatograpy are also very useful.

Reactions for the Detection of Amines

Table 12

Reaction	Aliphatic amines[a]			Aromatic amines[a]			Remarks
	I	*II*	*III*	*I*	*II*	*III*	
Diazotization and coupling	−	−	−	+	−	−	Certain NH_2 groups in heterocycles
with bindone	+	−	−	+	−	−	
with 1,2-naphthoquinone sulfonate	+	−	−	+	−	−	Some substances with active methylene group
with nitroprusside and acetaldehyde	−	+	−	−	−	−	
with fluorescein chloride	+	+	−	+	+	b	[b] Aromatic tert-amines react weakly
with aromatic aldehydes	+[c]	−	−	+	−	−	[c] See p. 322
with glutaconaldehyde	−	−	−	+	−	−	
with ninhydrin	+	d	−	−	−	−	[d] Sec-amines react only weakly
with Dragendorff reagent	−	−	+	−	−	+[e]	[e] Dialkylanilines or quaternary salts and other bases
with diazonium salts	−	−	−	+	+	+[f]	[f] Dialkylanilines
isonitrile reaction	+	−	−	+	−	−	
with benzotrichloride and $ZnCl_2$	−	−	−	−	−	+[g]	[g] Dialkylanilines, diphenly amines, carbazoles
with chloranil	−	−	−	+	−	−	At higher temperatures secondary and tertiary amines also react
with o-diacetylbenzene	+	−	−	+	−	−	
with 5-nitroisatin	−	−	−	−	−	+[h]	[h] Dialkylanilines, diphenylamines, carbazoles
with carbon disulfide + Ni^{2+}	−	+	−	−	+	−	

[a] The Roman numbers designate the primary, secondary, and tertiary amines, respectively.

1. Detection and Differentiation of Primary, Secondary, and Tertiary Amines

Isonitrile Reaction

When heated in alcoholic solution with alkali hydroxide and chloroform primary amines give offensive-smelling isonitriles (carbylamines):

$$RNH_2 + CHCl_3 + 3\,KOH \;\rightarrow\; RNC + 3\,KCl + 3\,H_2O$$

Secondary amines do not react in this way.

Reagents: alcoholic alkali hydroxide solution, alcohol, chloroform.

Procedure: A small amount of the amine is dissolved in alcoholic alkali, a few drops of chloroform are added, and the mixture is gently heated. Primary amine manifests itself by the characteristic odor of isonitrile.

Reaction with Ninhydrin

Similar to amino acids, primary aliphatic amines give a very sensitive color reaction with ninhydrin (see p. 278). In view of its sensitivity, it is advisable to combine the test with paper chromatography (p. 347). For the procedure see p. 278 (amino acids).

Reaction with Bindone

Anhydro-bis-indanedione (bindone) reacts with primary aliphatic amines forming a violet color, and with aromatic amines forming blue, and in dilute solutions green, colors. With aromatic amines the sensitivity is decreased by the presence of negative groups of the benzene nucleus; in the presence of more groups the reaction fails. On the other hand, some secondary amines also give a weak reaction. If the amine contains a sulfonic group, sodium acetate has to be added to the reaction mixture. For more details on the

course of this reaction and of its use for the detection of primary aliphatic amines see (1) and for the detection of hydrazine see (2).

Reagents: bindone, acetic acid or alcohol, light petroleum or benzene.

Procedure: The tested substance is dissolved or suspended in acetic acid, alcohol, ether, light petroleum, or benzene, a small amount of bindone is added, and the mixture is boiled for several minutes. Depending on the concentration of the amine, a color is formed either during the heating or only after prolonged standing. In the case of aromatic amines the amount of bindone should be enough to dissolve in hot solution, while in the case of aliphatic amines more is used, so that a part remains undissolved.

Reaction with Sodium 1,2-Naphthoquinone-4-sulfonate

1,2-Naphthoquinone-4-sulfonic acid reacts in alkaline media with compounds carrying a primary amino group with the formation of intensely colored quinoid condensation products.

The reactivity of the amino group can be strongly influenced by the presence of negative substituents in the amine molecule. For example, trinitroaniline and trichloroaniline do not react. On the other hand, in addition to primary aliphatic amines, aromatic amines, and amino acids, substances with an active methylene group also react (3). A positive reaction was also observed with substances such as guanidine, semicarbazide, phenylhydrazine, ethyleneimine, piperidine, piperazine, benzylcyanide, cyanoacetylacetone, diethyl malonate, ethyl acetoacetate, dibenzoylmethane, rhodanin, indole, methylphenylpyrazolone, resorcinol, phloroglucinol, and quaternary heterocyclic bases and oxonium salts.

Reagents: saturated sodium 1,2-naphthoquinone-4-sulfonate solution in 50% ethanol; 0.5 N NaOH, 2 N acetic acid.

Procedure: To a few drops of the tested amine solution two drops of the reagent solution are added and the mixture is alkalized with two drops of alkali. In a positive case a strong color is formed which changes after acidification with acetic acid; a precipitate may also be formed.

Reaction with Glutaconaldehyde

Glutaconaldehyde gives red to violet salts of Schiff's bases with primary aromatic amines in hydrochloric acid solution. However, as glutaconaldehyde is unstable, compounds which liberate it easily should be used for the reaction; such compounds are, for example, pyridine plus cyanogen bromide (4), pyridine irradiated with ultraviolet light (5), or 4-pyridyl-pyridinium dichloride (6), which in an alkaline medium sets free the enolate of the dialdehyde of glutaconic acid.

$$
\begin{array}{c}
\overset{\oplus}{N}\!\!-\!\!\!-\!\!\!-\!\!\!N \cdot HCl + 3\ NaOH \;\rightarrow \\
Cl^{\ominus}
\end{array}
$$

$$
\begin{array}{c}
H \\
| \\
C \\
\diagup\ \diagdown \\
CH\quad CH \\
|\qquad\ \| \\
CHO\ \ CHONa
\end{array}
\quad + \quad
\begin{array}{c}
NH_2 \\
\\
N
\end{array}
\quad + 2\ NaCl
$$

Simultaneously, 4-aminopyridine is formed, which does not react with glutaconaldehyde.

Detection with Cyanogen Bromide

Reagents: To a solution (40 ml) containing 4 g of KCN 5 ml of acetic acid are added followed by 40 ml of a solution containing 3 g of potassium bromate and 4 g of KBr. Then 4 ml of conc. sulfuric acid are added dropwise and under cooling and the mixture is made up with water to 100 ml. 1% pyridine in water; acetic acid.

Procedure: A dilute aqueous amine solution (5 ml) is mixed with 1 ml of pyridine solution and 2 ml of cyanogen bromide solution. A yellow to orange color and, eventually, an orange to brick-red precipitate are formed. If, after 10 min standing, 2 ml of acetic acid are added to the solution, the precipitate dissolves and the solution becomes red.

Detection with 4-Pyridylpyridinium Dichloride

Reagents: Pyridine (16 g) and 50 g of thionyl chloride are mixed, and after three days standing, the excess thionyl chloride is eliminated by evaporation under reduced pressure at 100 °C. Alcohol (15 ml) is added to the residue and evaporated. The residue is dissolved in dilute HCl, then decolorized by boiling it briefly with charcoal; it is eventually concentrated for crystallization. It is then diluted with alcohol, filtered off under suction, and washed with alcohol. Yield, 12 g; mp, 173 °C. 1% reagent solution in water; 2 N HCl; 2 N NaOH.

Procedure: The reagent (0.5 ml) is added to a dilute neutral solution of the amine (0.5 ml), and then alkali hydroxide solution is added dropwise until the solution is alkaline. The mixture is then immediately acidified with 2 N HCl; a red to violet color is observed; sometimes a precipitate is formed.

Reaction with Chloranil

Primary aromatic amines condense even in the cold or on gentle heating with chloranil in dioxane solution according to the equation (7)

$$ArNH_2 + O=\underset{\underset{Cl \quad Cl}{}}{\overset{\overset{Cl \quad Cl}{}}{\bigcirc}}=O \rightarrow Ar-N=\underset{\underset{Cl \quad Cl}{}}{\overset{\overset{Cl \quad Cl}{}}{\bigcirc}}=O + H_2O$$

If boiled in alcoholic solution, secondary and tertiary amines react as well. They can be differentiated from primary amines and from each other on the basis of their colors. Primary amines give a red color, secondary amines a violet color, and tertiary amines an emerald green color (8, 9).

Detection in Dioxane

Reagent: saturated chloranil solution in dioxane.

Procedure: Several milligrams of a primary arylamine are mixed with a few drops of the reagent. Violet to blue color appears.

Detection in Alcohol

Reagent: Chloranil (1 g) and 5 ml of glacial acetic acid are dissolved in 100 ml of ethanol.

Procedure: A few milliliters of the reagent are mixed with an equal amount of an aqueous or slightly acidified sample solution with acetic acid (free of ammonium salts), and the mixture is heated to boiling and then allowed to cool. Colors appear, depending on the type of amine present (see above).

Reactions with Aromatic Aldehydes

Aromatic aldehydes react with primary aromatic amines in the presence of acids with the formation of pronouncedly colored salts of Schiff's bases (see p. 215). To carry out the reaction, *p*-dimethylaminobenzaldehyde, vanillin, or fural, and hydrochloric acid are most commonly used. Sometimes secondary aromatic amines react as well, but these give a weaker color (10). Tertiary amines are nonreactive. In the case of fural, the reaction is probably more complicated (11). A positive reaction can also be obtained with hydrazine, urea and its derivatives, derivatives of pyrrole and indole, a number of aliphatic amino compounds, and, under certain conditions, even phenols.

If the reaction is carried out in a nonaqueous medium, it can be used much more extensively (12).

Reagent: 1% *p*-dimethylaminobenzaldehyde in a mixture of 95 parts of ethanol and five parts of conc. hydrochloric acid.

Procedure: One drop of the tested solution is mixed in a test tube with one drop of the reagent. In the presence of aromatic amines a yellow, orange, or red color is observed.

Reaction with Sodium Nitroprusside and Acetaldehyde

Secondary aliphatic amines, as well as cyclic amines and amino acids with an NH group, give a blue to violet color (13) with nitroprusside and acetaldehyde solution in an alkaline medium (Na_2CO_3). Neither primary nor tertiary amines react. However, it should be mentioned that in the presence of acetone sodium nitroprusside reacts with primary amines with the formation of a violet color (14). A solution of nitroprusside irradiated with a mercury lamp (formation of sodium pentacyanoferroate) gives a green color with primary aromatic amines (15).

Reagents: fresh 1% sodium nitroprusside solution; 2% sodium carbonate solution; acetaldehyde (an aqueous solution is also suitable).

Procedure: A drop of the nitroprusside solution to which acetaldehyde has been added (one tenth of the volume of the nitroprusside solution) is added to a drop of the aqueous amine solution and the mixture is alkalized with sodium carbonate solution. A blue to violet color is immediately formed.

Reaction with Carbon Disulfide, Nickel (II) Chloride, and Ammonia

All secondary amines react with carbon disulfide in ammoniacal solution with the formation of ammonium salts of dithiocarbamic acids, which do not decompose in the presence of metal salts (in contrast to similar compounds derived from primary amines), but give, for example, an insoluble salt with a nickel salt (16):

$$R_2N-C{\overset{\displaystyle S}{\underset{\displaystyle S}{\Big\langle}}}\ \frac{Ni}{2}$$

Reagent: $NiCl_2 \cdot 6H_2O$ (0.5 g) is dissolved in 100 ml of water and the solution is mixed under shaking with enough carbon disulfide to saturate the solution so that the droplets settle on the bottom of the test tube. In a well-stoppered bottle the reagent is quite stable.

Procedure: 1 ml of conc. ammonia and 1 ml of an aqueous solution of the amine are added to the reagent (1 ml) in a test tube (one droplet of the amine is dissolved in water; if necessary, a drop of conc. HCl may be added). After a while a precipitate is formed (if the reaction is positive) which can be extracted with chloroform to obtain a yellow-green extract.

Reaction with Dragendorff Reagent

Dragendorff's reagent is a solution of iodobismuthate. It gives orange to red precipitates with tertiary amines, quaternary nitrogen, or sulfonium salts and alkaloids. Among aliphatic tertiary amines, triethanolamine, for example, does not react.

Reagents: Solution I: Basic bismuth nitrate (850 mg) is dissolved in 10 ml of acetic acid and 40 ml of water. Solution II: Potassium iodide (8 g) is dissolved in 20 ml of water. The two solutions are mixed and can be kept in brown bottles for several months. Before use 10 ml of this stock solution is diluted with 20 ml of acetic acid and 100 ml of water.

Procedure: One drop of the tested solution of a tertiary amine or of a quaternary nitrogen base is spotted on filter paper and a drop of the reagent is spotted over it. An orange to red color appears, sometimes a precipitate.

Coupling with Diazonium Salts

Primary aromatic amines and their N-mono- and N,N-dialkyl derivatives can be coupled with diazonium salts in a slightly acid medium. The reaction can be carried out using stabilized diazonium salts (which do not contain free nitrous acid), for example, with *p*-nitrobenzenediazonium fluoroborate (17) or with a salt with 1-naphthalenesulfonate. The reaction takes place in dilute acetic acid in the presence of sodium acetate. Recently, 4-azo-benzenediazonium fluoroborate was proposed as the reagent, and dimethyl sulfoxide or dimethylformamide (18, 19) as the reaction medium in which intensely colored di-cations of bis-azo dyes can be formed.

Reagents: 0.2% aqueous *p*-nitrobenzenediazonium-1,5-naphthalene-disulfonate acidified with a drop of HCl; 10% sodium acetate.

Procedure: 1 ml of the reagent is added to a solution of amine in water, acidified with dilute hydrochloric acid, and, after a while, 5 ml of the sodium acetate solution; red-violet color.

Reaction with Fluorescein Chloride

Primary and secondary aliphatic and aromatic amines and tertiary aromatic amines with an N-methyl group give rhodamine dyes on fusion with fluorescein chloride and anhydrous zinc chloride. Different groups of these derivatives have characteristic fluorescences or colors making the differentiation of different types of amines possible: primary aliphatic amines give red colors with orange-red fluorescence, aromatic amines give red-violet colors without fluorescence, and pyrrole derivatives yellow-brown colors with a blue fluorescence. Acid amides and nitriles behave like primary aliphatic amines (20).

Reagents: HCl, 10% alcoholic HCl, fused $ZnCl_2$, fluorescein chloride [for preparation see (21)].

Procedure: One drop of the acidified (dilute HCl) amine solution is evaporated to dryness in a micro test tube, a small amount of fluoroscein chloride and a double amount of fused $ZnCl_2$ are added, and the mixture is heated at $250 - 260$ °C (air bath) until all the $ZnCl_2$ is melted. After cooling, the melt is extracted with ethanolic HCl, and the color and fluorescence of the solution are observed.

Reaction with Benzotrichloride and Zinc Chloride

Dialkyl derivatives of aniline, diphenylamine, and carbazole, when melted with benzotrichloride and fused zinc chloride, yield triphenylmethane dyes of the malachite green type. Phenols, diphenyl sulfide, and indole react as well, but the dyes obtained are red. This test can be completed by identifying the triphenylmethane dye formed by paper chromatography (22) to differentiate, for example, dimethyl- and diethylaniline. A similar reaction results from melting the sample with oxalic acid. This reaction leads to the formation of blue dyes, characteristic of the derivatives of diphenylamine and carbazole (23).

Reagents: benzotrichloride, fused $ZnCl_2$.

Procedure: A few milligrams of the tested substance are mixed in a micro test tube with approximately the same amount of $ZnCl_2$ and $1 - 2$ drops of benzotrichloride. The mixture is heated over a microflame until the whole mixture becomes a clear melt. After cooling, 1 ml of ethanol is added and the mixture is boiled briefly. In the presence of dialkylanilines, diphenylamine, carbazole, and thiodiphenylamine a green melt is formed which dissolves in ethanol with a green to green-blue color.

Other Color Reactions

In addition to the above reactions, a number of additional tests for amino compounds, important in special cases, have been worked out. For primary aliphatic and aromatic amines the reaction with o-diacetylbenzene (24) is very suitable. Instead of diazonium salts, hydrazone of 3-methyl-2-benzothiazole and ferric chloride can be used for the detection of aromatic amines

(25). Dialkylanilines and diphenylamines can also be detected by reaction with 5-nitroisatin (26). Many color reactions are based on oxidation with inorganic reagents: $FeCl_3$ (27, 28), potassium dichromate, and sulfuric acid (28−30), and hydrogen peroxide in the presence of V_2O_5 (31), or a number of other reactions (35−53).

2. Identification of Amines

A Review of Derivatives
Used for the Identification of Amines

1. Salts of inorganic and organic acids.
2. Acyl derivatives, $RNH_2 + R'COCl \rightarrow HCl + RNHCOR'$,

$$RNH_2 + \underset{CO}{\overset{CO}{\bigcirc}}O \rightarrow \underset{COOH}{\overset{CONHR}{\bigcirc}}$$

3. Sulfonamides, $RNH_2 + ArSO_2Cl \rightarrow HCl + ArSO_2NHR$.
4. Substituted ureas and thioureas,

$$RNH_2 + ArNCO \rightarrow ArNHCONHR$$
$$RNH_2 + ArNCS \rightarrow ArNHCSNHR$$
$$RNH_2 + ArCON_3 \rightarrow ArNHCONHR + N_2$$

5. Diazotization and coupling.
6. Alkylation and arylation, $R_3N + CH_3I \rightarrow R_3CH_3N^+I^-$.
7. Separation of amines with (a) p-toluenesulfonyl chloride, 3-nitrophthal-anhydride; (b) paper, thin-layer, gas, and column chromatography.

Procedure 1 is suitable for the identification of all types of amines, procedures 2−4 are designed for the identification of primary and secondary amines, while procedure 6 is meant primarily for the identification of tertiary amines. For the identification of primary amines procedure 5 is convenient.

An indirect identification of amines consists in catalytic denitrogenation and subsequent identification of the resulting hydrocarbon by gas chromatography (32). Aung et al. (33) recommend the preparation of Schiff's bases with picolinaldehyde and the measurement of the rate of dissociation by reaction with ferrous salt. The half-time of the reaction is characteristic of each amine. Differences in rates of acetylation were used for the identification of amines in petroleum fractions (34).

Amine Salts

Salts with Inorganic Acids

The basicity of amines can be made use of for their identification by converting them to salts which are not very soluble. The basicity of amines is dependent of the substituents on the nitrogen atom. Aliphatic amines are mostly strong bases, making the preparation of salts quite easy. No hydrolysis takes place during their crystallization. Aromatic residues bound to nitrogen decrease the basicity strongly, so that, for example, diphenylamine is such a very weak base that the preparation of salts practically cannot be utilized for identification purposes. Negative groups on the aromatic nucleus also greatly decrease the basicity of the amino nitrogen. Salts of amines with mineral acids can be prepared simply by neutralizing the solutions of amines with the corresponding mineral acid. Salts of less-basic amines must be prepared in an anhydrous medium (for example, by introducing dry HCl into a solution of the amine in benzene).

The following acids are commonly used: HCl, HBr, $HClO_4$, HNO_3, H_2SO_4, and also phosphotungstic acid (47), silicotungstic acid (48), Reinecke acids (49), and, sometimes, chloroplatinic and chloroauric acids (50−53).

It is impossible to give here a general recipe for when and how to use given salts for the identification of amines. Chlorides, bromides, or perchlorates are very suitable for identification in certain instances, while in other cases they are too soluble, hygroscopic, or have no sharp melting points (mostly derivatives of lower-molecular-weight amines). Salts with mineral acids can also be utilized for the identification of tertiary amines where the choice of other derivatives is restricted. Salts of phosphotungstic, silicotungstic, and Reinecke acids are especially suitable for the identification of amino acids, alkaloids, and tertiary amines.

If the base (undilute or in ethanolic solution, or also an aqueous solution of its hydrochloride) is added to an acidified (HCl) solution of chloroplatinic acid, poorly soluble and well-crystallizing salts of the composition $2 B . H_2PtCl_6$ (B = base) are formed. Chloroauric acid gives salts of the composition $B . HAuCl_4$. Both types of salt are prepared in the following manner: the solution of the amine (0.5%) in 10% HCl is added with shaking to a slight excess of chloroplatinic or chloroauric acid (the solutions contain ∼1% of Pt or Au), and the solution is heated to dissolution and then allowed to crystallize. Recrystallization is carried out from water or ethanol, best by acidifying with a droplet of HCl. Chloroplatinates often crystallize with a molecule of water or alcohol; chloroaurates easily lose hydrogen chloride. The derivatives mentioned usually partly decompose on heating, but the decomposition points are well reproducible. The advantage of these derivatives consists in the fact that they can be analyzed simply by ignition, which leaves metallic Pt or Au.

Salts with Organic Acids

A great number of organic acids yielding crystalline salts with amines have been proposed. The best known acids are 3,5-dinitrobenzoic (I) (54); 2,4-dinitrobenzoic (II) (55); 3,5-dinitro-o-toluic (III) (56); 3,5-dinitro-p-toluic (IV) (57); picric (V), picrolonic (VI), styphnic (VII), imidazoledicarboxylic

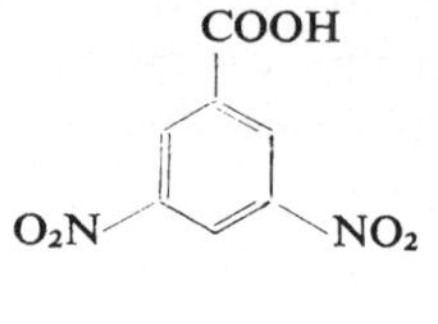

(I) 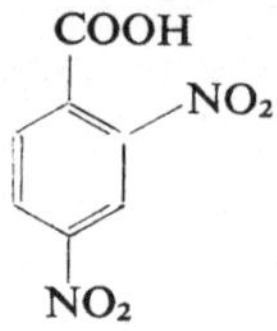   (II) 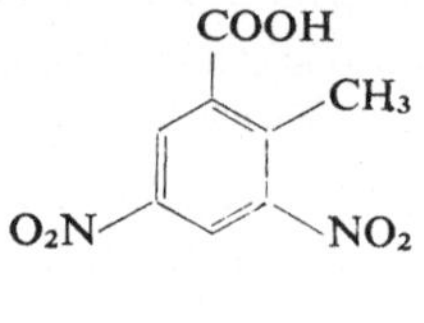   (III)

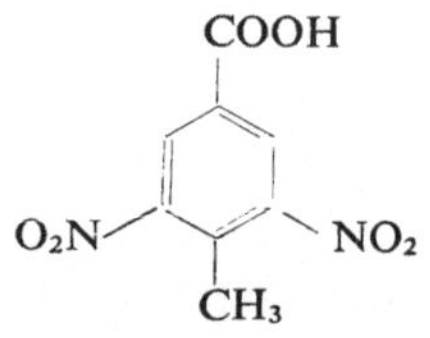

(IV) 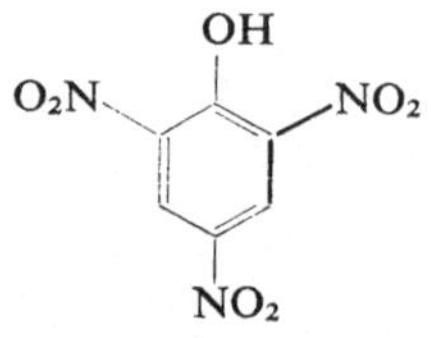   (V)

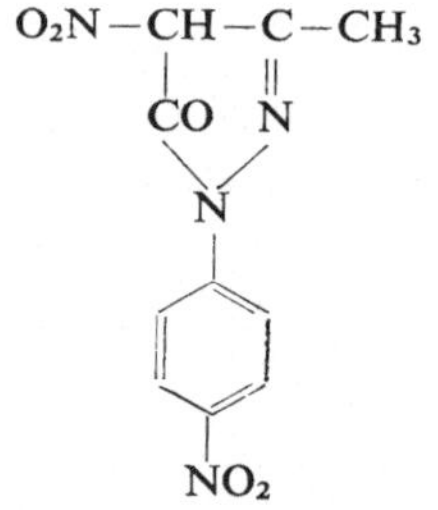

(VI) 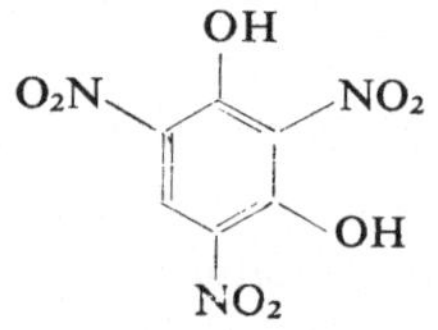   (VII)

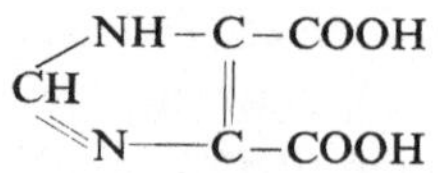

(VIII) 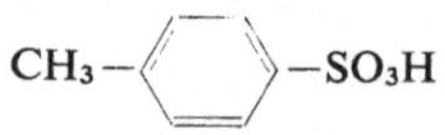   (IX)

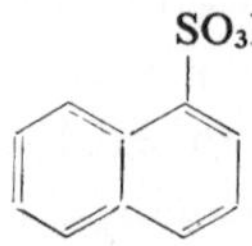

(X) 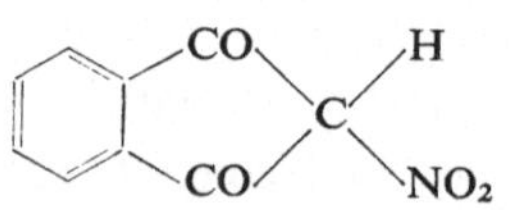   (XI)

(VIII) (58); *p*-toluenesulfonic (IX) (59, 60); and 1-naphthalenesulfonic (X) (61). Nitroindanedione (XI) (62, 63), dilituric acid (XII) (64), and disulfimide salts (XIII) are also used (65).

General Method of Preparation of Salts

The preparation of salts of organic acids with amines is identical in principle for all organic acids. Suitable amounts of a base and of the acid are mixed in a suitable solvent, and the mixture is allowed to crystallize. As for the choice of solvent, the basicity of the amine should be respected. Therefore, the salts of less-basic amines are prepared in less-polar solvents. The solvents used are water, ethanol, methanol, ethyl acetate (for picrates), ether, chloroform, and benzene. It is evident that when choosing the solvent the solubilities of the amine and the acid, and of the salt as well, should be taken into consideration.

Another method of preparing salts consists in using an aqueous, mostly saturated, solution of the sodium salt of the acid, which is then mixed with an aqueous solution of the hydrochloride or sulfate of the base. It is important to keep the solutions neutral; otherwise, the poorly soluble organic acid may precipitate in an acid medium (this should be kept in mind especially when picrates and dinitrobenzoates are prepared). This method of preparation should be used, for example, when isolating an amine from the reaction mixture by steam distillation. The distillate can be titrated with a normalized acid so that a measured amount of an organic acid salt can be added to the tested solution. Often, especially during isolation procedures, impure bases are obtained which give salts that do not crystallize well and precipitate (in the case of higher-molecular bases) in the form of oils. Methods which lead even in such cases to solid, crystalline derivatives are the following: (1) The base is liberated from the salt (see below) and then purified by distillation, chromatographically, or by means of ion exchangers. (2) The base is liberated and another salt is prepared. (3) The solutions of the base (or hydrochloride) and of the acid are mixed, drop by drop, with thorough stirring and at low temperature. (4) So-called fractional precipitation is carried out. Only a part of the reagent is added to the solution and the oil is separated by decantation. A new portion of the reagent is added, etc. (5) The precipitated oil is separated by decantation, washed several times by decantation (with water), and allowed to crystallize in a desiccator over a drying agent. It is also possible to smear the oil on an unglazed tile. (6) The precipitation is carried out at room temperature, the oil is separated by decantation, and the decanted solution is cooled in a refrigerator. The separated oil is dissolved again in hot water and the solution is allowed to cool in a refrigerator. The separated oil is separated again by decantation, and the decanted solution is allowed to cool again in the refrigerator, etc. (7) We can check the purity of the product or the composition of mother liquors by chromatography.

The solvents mentioned can be used for crystallization. It should be remembered that the melting points are not always sharp and that they depend on the method used for determination and on the rate of heating. Melting points determined on a Kofler block are generally higher than those determined in a capillary.

Among the above-mentioned salts, picrates, picronolates, and diliturates are the most commonly used. The use of disulfimides (XIII), proposed only recently (65), has several advantages: the derivatives generally have a sharp melting point, they crystallize well, and are not hygroscopic. They are mostly poorly soluble, and therefore suitable for the isolation of basic substances from mixtures.

Picrate of Dimethylaniline

Reagents: 10% picric acid solution in ethyl acetate; ethanol.

Procedure: Picric acid solution (2.5 ml; 10%) is added to 0.1 ml of dimethylaniline in a micro test tube under vigorous shaking. After 30 min standing in a refrigerator the picrate is filtered off on a filtration crucible and washed with 0.5 ml of ethyl acetate. Yield, 0.33 g; mp, 156 °C. After crystallization from 25 ml of ethyl acetate the weight of the product is 0.28 g, mp 156 °C.

Picrate of Quinoline

Reagents: saturated aqueous picric acid solution; methanol; 50% aqueous methanol.

Procedure: Quinoline (20 mg), 0.4 ml of the saturated picric acid solution, and 0.6 ml of methanol are heated to incipient boiling and then allowed to stand for 20 min. The picrate is filtered off and washed twice with 0.1 ml of methanol. Yield, 6 mg; mp, 200–201 °C. Crystallization from 1 ml of 50% methanol gave 3 mg, mp 199–200 °C.

Picrate of Aniline

Reagents: 5% aqueous sodium picrate solution; ethanol.

Procedure: Aniline hydrochloride (100 mg) is dissolved in a 10-ml test tube in 2 ml of water, and 5 ml of sodium picrate solution is added and heated until dissolved. After cooling, the picrate is filtered off and washed twice with 0.5 ml of water. Yield, 0.18 g; mp, 169 − 172 °C. After crystallization from 4 ml of ethanol the product weighed 0.11 g, mp 168 − 170 °C (decomposition).

Tetraphenylboron Salts

The detection and the identification of nitrogen compounds can be carried out by converting them to tetraphenylboron salts (66, 67):

$$R_1R_2R_3R_4N^\oplus + B(C_6H_5)_4^\ominus \rightarrow R_1R_2R_3R_4N \cdot B(C_6H_5)_4$$

where the R are alkyls, aryls, hydrogen atoms; the nitrogen atom may be a member of a ring. The precipitation reaction is very sensitive, and makes it possible to detect 0.3 − 40 µg in a dilution 1 : 1000 − 280,000 (68). The solution must not contain ions of K, Rb, Cs, NH$_4$, and Ag, which also give insoluble salts with the reagent. The precipitation is carried out from a weakly acid solution (pH ∼3; for example, from a solution of the base in 1% acetic acid), by adding the reagent solution in water (a saturated aqueous solution of the sodium salt of tetraphenylboron is ∼0.3 M).

General Method of Detection of Amino Compounds with Sodium Tetraphenylboranate

On a spot-test plate one drop of the tested solution (pH ∼ 3) is mixed with one drop of the reagent. The formation of the precipitate is observed on a black background.

The precipitation of salts is also suitable for the identification of bases. The salts are prepared by precipitation of aqueous, slightly acid solutions of bases with a saturated aqueous reagent solution. The precipitate is filtered off, washed with water, and dried at a suitable temperature. The salts are not crystallized, but they are sufficiently pure for identification purposes (precipitation from dilute solutions). Identification is carried out by melting-point determination. The values of melting points depend on the rate of heating. In view of the fact that the melting points of low-molecular amines are not sharp (they melt within a range of 10 − 15 °C), these salts are more suitable for the identification of higher-molecular amines, the melting points of which are sharper (range, 3° C) (69). The advantage of the salts of amines with tetraphenylboron consists in the fact that they can be isolated from dilute solutions. Determination of the equivalent weight may serve as an additional criterion for identification. For more details see the review article by Barnard and Wendlandt (70), which contains a list of melting points of 145 derivatives.

Triethanolamine Tetraphenylboranate

Reagents: 0.6% aqueous sodium tetraphenylboranate; N HCl.

Procedure: To an aqueous solution (10 ml) of 0.1 ml of triethanol-amine adjusted to pH 2−3 (test with indicator paper) 35 ml of the reagent solution are added slowly and under stirring. The separated precipitate is filtered off, washed with 30 ml of water, and dried at 60 °C. Yield, 240 mg; mp, 142−145 °C.

Regeneration of Amines from Salts

In certain cases the solution of the salt may be alkalized and the liberated amine can be separated, for example, in a separatory funnel or by steam distillation, or by extraction with ether or benzene. In other cases a more complicated procedure needs to be chosen: the bases are liberated from picrates by triturating the picrate with conc. HCl, and the liberated picric acid is filtered off on a sintered glass filter. The filtrate is evaporated, preferably in vacuo, and the hydrochloride obtained is converted to the free base according to the procedure given above.

The use of ion exchangers for the liberation of bases represents an elegant procedure (71). By filtering a solution of an amine salt through an anion exchanger column, the free base is obtained in the eluate. Eventually, the anion of an amine salt can be eliminated in the form of an insoluble salt, for example, from hydrochloride by shaking it with Ag_2O, or from sulfate by the addition of a calculated amount of barium hydroxide.

Acylation

Primary and secondary amines give substituted amides on acylation, while tertiary amines do not react, because they do not contain a hydrogen atom capable of substitution. Acylation can be carried out with acids (especially formic and acetic), acid chlorides, and acid anhydrides. Acetylation and, especially, formylation are carried out by heating the base with the acid under reflux; for acetylation it is more advantageous to use acetyl chloride or acetic anhydride.

$$RCOOH + ArNH_2 \rightarrow ArNHCOR$$
$$RCOCl + RR'NH \rightarrow RR'NCOR$$

General Method of Acetylation

Acetylation with acetyl chloride can take place without a solvent or in an inert solvent. An anhydrous solvent is necessary because the hydrolysis of acetyl chloride is faster than acetylation. The disadvantage of such acetylation is that only one half of the base is acetylated, because the second half is neutralized by the HCl formed. However, if the reaction is carried out in the presence of pyridine, the latter assumes the role of the acceptor of the HCl formed and all of the amine is acetylated. Acetylation is most commonly carried out with acetic anhydride, which is soluble in water and relatively stable in it at lower temperatures, so that acetylation can be carried out in water or in suspension. A 1.5 fold excess of acetic anhydride is used. Instead of water, dilute acetic acid can also be used. When a hydrochloride of a base is to be acetylated sodium acetate is added to the medium. Acetylation in aqueous media is advantageous because the formation of a diacetyl derivative is much diminished. The latter is usually formed if undiluted acetic anhydride is used. Alcohol may also serve as a solvent (as it is not acetylated at low temperatures). The use of alcohol makes possible the elimination of excess acetic anhydride by a double evaporation of the mixture with the alcohol. Amines which are acetylated reluctantly should be heated with a 1.5 − 3 fold excess of acetic anhydride in the presence of pyridine as solvent. Pyridine enhances the reaction and serves as a good solvent for amines. The presence of anhydrous zinc chloride also increases the reaction rate, especially of negatively substituted amines (72) (nitrodiphenylamines). Acetyl derivatives are most easily isolated by pouring the reaction mixture into water. If the derivative does not precipitate immediately, the solution should be cooled or even neutralized. Of course, the derivative can also be extracted with ether.

Formyl and acety derivatives of aliphatic amines are, as a rule, oils or low-melting substances and they are therefore not very suitable for identification. On the contrary, acetylated aromatic amines are very suitable for identification because they crystallize well and have sharp melting points distributed over a broad temperature range. Identification constants of practically all acetyl derivatives are known, which is an additional advantage. Acetyl derivatives are usually crystallized from water or alcohol, or a mixture of both, as well as from benzene, cyclohexane (or a mixture of these), or ethyl acetate.

Acet-*p*-toluide

Reagents: acetic anhydride; 10% HCl; methanol.

Procedure: *p*-Toluidine (20 mg) and 0.1 ml of acetic anhydride are introduced into a 5-ml micro test tube and the mixture is boiled for 5 min. After cooling, 0.3 ml water is added and the precipitated product is stirred with a glass rod. When the precipitate settles, the supernatant is taken off with a micropipette, the residue is stirred with 0.2 ml of 10% HCl and allowed to settle, and the upper phase is eliminated with a pipette. The procedure is repeated twice more with 0.1 ml of water. The crude product is dissolved in 1.5 ml of hot methanol, diluted with 0.5 ml of water, and heated again. After cooling, the mixture is allowed to stand in a refrigerator for 30 min and the acetate is eventually filtered off and washed twice with 0.5 ml of water. Yield, 5 mg; mp, 151 °C. The second crystallization is carried out in the same manner as the first, giving 3 mg of product, mp 151 °C.

Benzoylation

Benzoylation is carried out with benzoyl chloride, but substituted benzoyl chlorides are also useful. The most common method is that of Schotten (73) and Baumann (74).

General Method of Benzoylation

A solution or suspension of an amine in 10% NaOH is shaken under constant cooling (the temperature should remain below 25 °C) and a 1.5 − 2 fold amount of benzoyl chloride is added gradually. The solution must remain alkaline to the very end of the reaction, which is attained when the odor of benzoyl chloride has completely disappeared (the excess benzoyl chloride hydrolyzes to benzoic acid). The separated amine is filtered off under suction (after cooling, if necessary); if no product has separated (lower aliphatic amines), the mixture is extracted with ether. Benzoylation can also take place in a homogeneous solution, best by heating the amine solution in benzene and pyridine (2 : 1) with a 1.5 − 2 fold excess of benzoyl chloride at 60 − 70 °C for $\frac{1}{2}$ − 1 hr. The product is isolated by pouring the benzene solution into water, separating the benzene layer, washing it with 5% sodium carbonate, drying, and concentrating to a small volume. The corresponding benzamide crystallizes or separates after addition of an excess of hexane. Benzoyl derivatives are crystallized from aqueous ethanol, ethanol, benzene, benzene-cyclohexane, etc.

In order to prepare benzoyl derivatives and substituted benzoyl derivatives on semimicro and micro scales, a cold suspension of an alkaline (5% NaOH) amine solution is shaken with a 2 − 3 fold excess of the chloride dissolved in a minimum amount of benzene. The derivative is isolated by evaporating the benzene. If this procedure is unsuccessful, the reaction is carried out by heating the components in pyridine. If the micropreparation is carried out by the Schotten-Baumann method, the reaction mixture should be neutralized after the reaction has stopped, in order to completely separate the benzoyl derivative of the type RCONHR. In view of the presence of a hydrogen atom on the nitrogen, the latter is partly soluble in the aqueous phase; however, the pH of the solution should not drop below 8 (to prevent the precipitation of benzoic acid).

The advantages of benzoyl derivatives for identification are the same as those of acetyl derivatives. Benzoyl derivatives of lower aliphatic amines melt at low temperatures, and we therefore prefer the use of *p*-nitro- or 3,5-dinitrobenzoyl chloride [see also (75)]. The utilization of 3-nitrophthalic anhydride for the identification of amines is described in the discussion on the separation of primary, secondary, and tertiary amines (p. 345). The identification of amines in the form of their 4-nitrobenzeneazo-4-carboxamides is advantageous because the derivatives are colored and suitable for chromatography, and their molar extinction coefficient is not dependent on the type of substituent in the amine (76). The separation of the mentioned derivatives by thin-layer chromatography and their identification by IR and mass spectrography have also been described (77).

Benzoyl-*p*-chloroaniline

Reagents: benzoyl chloride; 15% HCl; 10% NaCO$_3$; methanol.

Procedure: A micro test tube fitted with a rubber stopper is filled with 0.3 g of *p*-chloroaniline, 0.6 ml of benzoyl chloride, and 0.9 ml of pyridine, and the mixture is heated at 80 °C in a water bath for 20 min. After the immersion of the tube into the water bath the rubber stopper is taken off very briefly (to diminish the pressure in the tube). The solvent is then evaporated to dryness on a boiling water bath (the vapors are sucked off with a water pump) and the residue is mixed with 1 ml of 10% HCl. When the mixture is settled the supernatant is eliminated with a micropipette and the residue is washed with 0.5 ml of water in the same manner. The washing is continued with two portions (0.5 ml each) of 10% sodium carbonate solution and two 0.5 ml portions of water. The residue is filtered off using a filtration tube and is washed with 0.5 ml of water. Yield, 0.48 g; mp (unsharp), from 45 °C. For crystallization the product is dissolved in 13 ml of boiling ethanol and the solution is diluted with water to incipient turbidity (1 ml). This is dissolved by heating again, and the solution is allowed to cool slowly. The crystals that formed weighed 0.33 g, mp 102 − 103 °C. After a second crystallization from 10 ml of methanol and 2 ml of water the yield was 0.20 g, mp 104 °C.

3,5-Dinitrobenzamide

Reagents: pyridine; 3,5-dinitrobenzoyl chloride; 1% sulfuric acid; potassium carbonate; anhydrous sodium sulfate; ethanol-free ether; benzene.

Procedure: Ammonium chloride (90 mg) is dissolved in a separatory funnel containing 10 ml of water, and 30 ml of ether are added, followed by 0.5 ml of pyridine and 0.5 g of 3,5-dinitrobenzoyl chloride dissolved in 2 ml of benzene. Potassium carbonate (11 g) is then added gradually in small portions to the mixture under constant cooling and shaking. After 20 min of standing under occasional shaking the aqueous layer is removed and the ethereal layer is gradually extracted twice with 10 ml of 1 % sulfuric acid and twice with 10 ml of water. The ether solution is then dried over 2 g of sodium sulfate, filtered into a distillation flask, and evaporated. The residue is dissolved in a small amount of boiling ethanol and water is added to incipient turbidity. The crystals separated after a prolonged standing in a refrigerator are filtered off using a filtration tube. Yield, 30 mg; mp, 172 – 177 °C (decomp.)

N-Ethylamide of 3,5-Dinitrobenzoic Acid

Ethylamine (100 mg) or its hydrochloride (140 mg) can be treated in the same manner as above. Yield, 100 mg; mp, 125 – 128 °C; after crystallization from 50% aqueous ethanol: 70 mg; mp, 125 – 126 °C.

N,N-Diethylamide of 3,5-Dinitrobenzoic Acid

When 110 mg of diethylamine or 165 mg of its hydrochloride are treated as above, the yield is 100 mg; mp, 90 °C. After crystallization from 50% aqueous ethanol the yield of crystals is 73 mg; mp, 90 °C.

Addition Compound of N,N-Diethyl-3,5-dinitrobenzamide and 1-Naphthylamine

Reagent: 10% 1-naphthylamine solution in 80% ethanol.

Procedure: N,N-diethyl-3,5-dinitrobenzamide (30 mg) is dissolved in 1 ml of ether, and 1 ml of 1-naphthylamine solution is added. The separated product is allowed to stand for 15 min in a refrigerator and it is then filtered off, washed with 0.5 ml of ice-cold 80% ethanol and twice with 0.5 ml of ice-cold water, and dried in a stream of air. Yield, 15 mg; mp, 155 – 156.5 °C.

Paper Chromatography of 3,5-Dinitrobenzamides

3,5-Dinitrobenzamides are very suitable derivatives for paper chromatography. The derivatives of primary and secondary aliphatic amines behave quite differently on chromatograms. As tertiary amines do not react with the reagent, the method offers a good opportunity for the differentiation of primary, secondary, and tertiary amines. In the solvent system 25% dimethylformamide/hexane the primary amines from methylamine to amylamine remain at the start, while higher amines, up to C_{14}, separate well (Fig. 34). Lower secondary amines also remain at the start. Only di-n-propylamine and di-n-butylamine move. The formamide/hexane system is suitable for the separation of dimethyl- to di-*n*-butylamine derivatives; primary amines up to C_6 remain at the start. For the separation of primary amines the systems formamide/hexane-benzene in various proportions are suitable. Secondary amines move with the solvent front. For the separation of the series of primary amines from hexalymine to octadecylamine the system 1-bromonaphthalene/90% acetic acid was found of value. Among aromatic amines, aniline, N-methyl, and N-ethylaniline can also be separated by this method in the system formamide/cyclohexane. Other aromatic amines do not give satisfactory results. 3,5-Dinitrobenzamides are dissolved in benzene, chloroform, or, preferably, acetone before application on the chromatogram. The amount applied is $10-40$ µg of the derivative. Detection is carried out as described for alcohols (p. 154). Dinitrobenzamides appear as yellow spots on a white background (78). Preparative separation of 3,5-dinitrobenzoyl derivatives of primary and secondary amines can be carried out on a column of alumina (79).

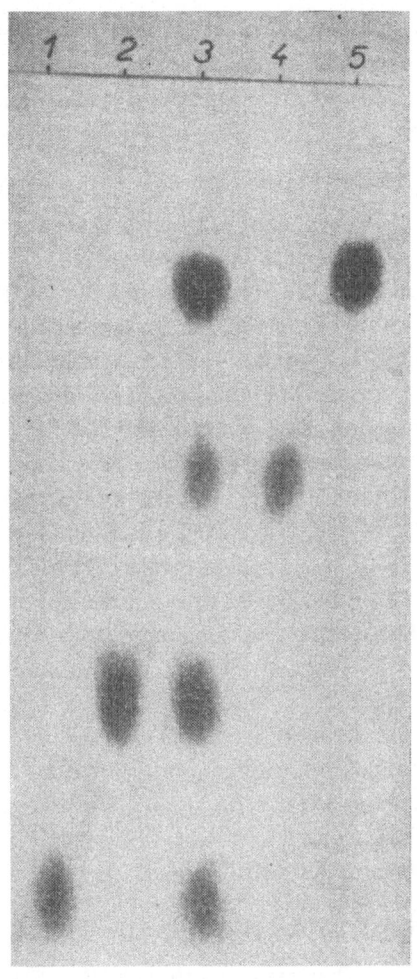

Fig. 34. Chromatogram of higher aliphatic amines (converted to 3,5-dinitrobenzamides) in dimethylformamide/hexane: (1) myristylamine, (2) laurylamine, (3) mixture, (4) decylamine, (5) octylamine.

Sulfonamides

Arylsulfonyl chlorides react with primary and secondary amines with the formation of sulfonamides, which are used both for the separation of primary, secondary, and tertiary amines, and for the identification of the former two types of compounds. For identification benzenesulfonyl chloride (80, 81) and *p*-toluenesulfonyl chloride (82) are especially suitable.

$$RNH_2 + \left\langle\!\!\!\bigcirc\!\!\!\right\rangle - SO_2Cl \rightarrow \left\langle\!\!\!\bigcirc\!\!\!\right\rangle - SO_2NHR + HCl$$

$$R, R'NH + \left\langle\!\!\!\bigcirc\!\!\!\right\rangle - SO_2Cl \rightarrow \left\langle\!\!\!\bigcirc\!\!\!\right\rangle - SO_2NR, R' + HCl$$

However, other substituted reagents were also proposed, especially for the identification of those amines which give low-melting derivatives with benzene- and p-toluenesulfonyl chloride: p-bromobenzenesulfonyl chloride (83) and m-nitrobenzenesulfonyl chloride (84).

General Method for the Preparation of Sulfonamides

Sulfonamides are prepared roughly by the same procedure as used for benzoylation according to Schotten and Baumann. A fourfold excess of alkali (in the form of a 12% solution) and a $1.5-2$ fold excess of the sulfonyl chloride are used per 1 mmole of the base. The reaction is carried out in an aqueous suspension, and the sulfonyl chloride is added in the form of an acetone solution. The mixture is usually shaken for 10 min and the reaction is completed by brief heating in a water bath, which is especially important when sulfonamides of secondary amines are prepared (excess sulfonyl chloride is also hydrolyzed in this way). When a primary amine is reacted, the mixture is filtered and sulfonylamide is separated from the filtrate by acidification. In the case of secondary amines the reaction mixture is simply poured into water. The reaction can also take place in pyridine. The isolation of sulfonamide is carried out in the same manner.

The advantages of sulfonamides as derivatives consist in the facts that a large number of them have been described, and that they have suitable melting points and can be purified well by crystallization. Sulfonamides of primary amines can also be purified by dissolution in alkali and precipitation with acids. However, these derivatives also possess certain disadvantages: lower amines give low-melting sulfonamides. Other disadvantages will be mentioned in the discussion on the separation of amines.

p-Toluenesulfonamide of p-Chloroaniline

Reagents: p-toluenesulfonyl chloride, 10% NaOH, HCl 1 : 1, 75% ethanol.

Procedure: In a micro test tube (5 ml volume) a mixture of 20 mg of p-chloroaniline, 50 mg of p-toluenesulfonyl chloride, and 1 ml of 10% NaOH is shaken for 5 min (the tube should be closed with a rubber stopper). After opening the test tube, the contents are heated briefly in a water bath at $50-60$ °C and shaken for an additional 2 min. After cooling, the mixture is neutralized with hydrochloric acid (test with litmus) and after 20 min standing in a refrigerator the supernatant is discarded with a micropipette and the settled crystals are washed by stirring them with 0.2 ml of cold water and filtering off the liquid. 75% Methanol (2 ml) is added to the crystals and the mixture is heated until dissolved. After 20 min standing in a refrigerator the product is filtered off, washed with 0.2 ml of water, and dried in a stream of dry air. Yield, 9 mg; mp, 92 °C. Crystallization from 0.5 ml of 75% methanol gave 2 mg of a product melting at 92 – 92.5 °C.

Substituted Ureas and Thioureas

These are prepared by reacting isocyanates, acid azides, or isothiocyanates with primary and secondary amines:

$$RNCO + R'NH_2 \rightarrow RNHCONHR'$$
$$ArCON_3 + RNHR' \rightarrow ArNHCONRR' + N_2$$
$$ArNCS + Ar'NH_2 \rightarrow ArNHCSNHAr'$$

The reaction of amines with isocyanates is analogous to the reaction of alcohols with isocyanates, and the same rules should be observed. However, the reaction with amines takes place at a much higher rate than with alcohols, so that it can also be carried out in an alcoholic medium. This is not recommended, however, in the case of phenylisocyanate. Therefore, although the reagent gives derivatives with suitable properties, the use of α-naphthylisocyanate should be given preference, as it is less sensitive toward water, so that an aqueous medium can also be used for the preparation of derivatives.

General Method of Preparation of Derivatives

The derivatives are best prepared by boiling both components, in an equimolar ratio, in light petroleum. After filtration (insoluble diphenyl- or dinaphthylurea is formed as a by-product) the derivative crystallizes out on cooling. Instead of isocyanate, acid azides can also be used with equal success. The reaction is carried out by boiling the components in toluene for 1 – 4 hr and the derivative is isolated after cooling.

Sah and co-workers proposed a large number of substituted benzazides (nitro-, chloro-, bromo-, iodobenzazides, naphthazides) [see, for example, (85)].

Among the most convenient derivatives of primary and secondary amines are substituted thioureas obtained by the reaction of amines with arylisothiocyanates. As isothiocyanates do not react under normal conditions either with water or with alcohol, it is not necessary to utilize anhydrous amines. This is the great advantage of this method. By-products are formed only to a small extent and can be eliminated easily by crystallization. If the melting point of the product is several degrees below the expected value, it is usually sufficient to extract it twice with 0.5 – 1 ml of light petroleum. A minor inconvenience of the method consists in the fact that the derivatives often separate in the form of oils which crystallize only reluctantly. Usually, thorough cooling and scratching with a glass rod induce crystallization. It is also advisable to keep a few crystals from an earlier batch for seeding the material. The reaction is carried out by 1 – 10 min heating of the components in an equimolar ratio either in the absence of solvent or in alcohol. In the second case, especially for lower-molecular amines, water must be added to precipitate the derivative. Crystallization is generally carried out from 50% aqueous ethanol. The most convenient derivatives are obtained from 1-naphthylisocyanate (86). The following isothiocyanates have also been proposed: phenylisothiocyanate and its alkylated, chlorinated, brominated, nitrated, and methoxylated derivatives (87 – 89).

N-1-Naphthyl-N'-cyclohexylthiourea

Reagents: 1-naphthylisothiocyanate, HCl 1 : 1, 90% ethanol, cyclohexane.

Procedure: A mixture of 0.05 ml of cyclohexylamine and 50 mg of 1-naphthylisothiocyanate in a 5-ml micro test tube is carefully heated over a microflame for 3 min. After cooling, the solidified melt is ground with a glass rod and mixed with 3 ml of HCl 1 : 1. When the solid material is settled, the acid is sucked off with a micropipette and the residue is washed in the same manner twice with 0.5 ml of 90% ethanol. Cyclohexane (1 ml) is then added to the crystals and the mixture is briefly boiled and then cooled. The supernatant is again separated from the sediment with a micropipette and the latter is washed in the same manner with 1 ml of 90% cold ethanol. The crude product is dissolved in approx. 2 ml of hot ethanol and diluted with 2 ml of water. After cooling, the mixture is allowed to stand in a refrigerator for 30 min and the crystals formed are filtered off and washed twice with 0.5 ml of 90% ethanol. Yield, 19 mg; mp, 182 – 183 °C. After crystallization from 1 ml of ethanol with the addition of three drops of water the product (14 mg) melted at 183 °C.

The derivative of ethylaniline prepared in the same manner gave 23 mg of product, mp 129 – 130 °C, which, after crystallization from 1.5 ml of ethanol and 0.5 ml of water, weighed 13 mg and had a mp 130 – 131 °C.

Note: If, after the reaction of the amine with 1-naphthylisothiocyanate, the product does not solidify even after scratching with a glass rod or after standing in a refrigerator, 1 ml of ethanol should be added to the oil and the mixture heated until complete dissolution. Water is then added to incipient turbidity and the sides of the test tube are scratched with a glass rod. If this intervention does not bring a positive result either, the oil formed is extracted with 90% ethanol and cyclohexane.

Diazotization and Coupling

The most important reaction of primary aromatic amines is diazotization, i.e. their reaction with nitrous acid with the formation of diazonium salts which can be coupled with phenols or with aromatic amines to give intensely colored azo dyes

$$ArNH_2 \xrightarrow[H^{\oplus}]{HNO_2} ArN_2^{\oplus} \xrightarrow[OH^{\ominus}]{ArOH} Ar-N=N-ArOH$$

This reaction is suitable both for the detection of the primary aromatic amino group and for the identification of amines by means of the preparation of their derivatives or by paper chromatography (90). The result of the diazotization is dependent on the basicity of the tested amine, and the pH value of the medium during the reaction should be adjusted in accordance with the amine's basicity. Usually, diazotization is carried out in the presence of 2.5 equivalents of hydrochloric acid, but in the case of negatively substituted, and hence less-basic, amines a $1-3$ fold excess of acid must be used. If even this excess does not prevent the hydrolysis of the salt, the diazotization should be carried out in suspension. In the case of dinitrochloro- and nitrochloroaniline the diazotization is best carried out in sulfuric acid. With phenols the coupling takes place in an alkaline medium, and the excess of nitrite from the preceding reaction does not impair the reaction. The most convenient passive component is 2-naphthol, which yields orange or red azodyes with the azo group unequivocally at position 1. These are water-insoluble and can be identified either by isolation and melting-point determination or by extraction with chloroform and identification by paper chromatography. Dyes soluble in water can be obtained by coupling with R salt. Coupling with amines is carried out in acid media, and the excess sodium nitrate must be eliminated with sulfamic acid. Suitable passive components are 1-naphthylamine, N-1-naphthylethylenediamine, and N-ethyl-1-naphthylamine (91, 92).

Some amino groups bound on a heterocycle are also diazotizable — for example, 4-aminotriazole, 4-aminothiazole, 4-aminoantipyrine, 3-aminoquinoline, etc.

Aliphatic primary amines also give diazonium salts with nitrous acid. However, in this case the diazonium ion is not stabilized by resonance structures as in the case of arylamine, and decomposition to nitrogen and carbonium ions takes place. During the performance of the reaction the evolution of nitrogen takes place and a conveion to the corresponding alcohol. In addition to primary amines, polymethylenediamines, alkylarylamines with the amino group in the side chain, and amino acids also react in this manner. Among aliphatic compounds, diazo compounds are only given by esters of amino acids and by certain aminoketones.

All secondary amines, i.e., aliphatic, aromatic, and mixed, give, in the presence of nitrous acid, nitrosoamines which can be steam distilled (for the detection of N-nitroso compounds see p. 361).

Tertiary aliphatic amines do not react in the cold with nitrous acid, and they can be regenerated from the mixture. Only aromatic dialkylamines with a free para position give strongly yellow *p*-nitroso derivatives in many cases.

Test by Coupling with N-1-Naphthylethylenediamine

Reagents: 4 N HCl, 0.1% $NaNO_2$, 0.5% ammonium sulfamate, 0.1% N-1-naphthylethylenediamine dihydrochloride.

Procedure: To a few milliliters of an aqueous solution containing 50 – 100 µg of amine, 0.5 ml of 4 N HCl and 1 ml of $NaNO_2$ solution are added. After 5 min 1 ml of ammonium sulfamate solution is added to the mixture, and after an additional 3 min the mixture is stirred and mixed with 1 ml of N-1-naphthylethylenediamine dihydrochloride solution. A purple color is formed.

Test by Coupling with 2-Naphthol

Reagents: 2 N HCl; N $NaNO_2$; 2% naphthol in 2 N NaOH; iodine-starch paper.

Procedure: Sodium nitrite solution is added dropwise and under cooling to a solution of a few milligrams of amine in 3 ml of 2 N HCl until one drop of the mixture causes an immediate blue-violet color when applied to the iodine-starch paper. The diazonium salt solution formed is poured under stirring into 50 ml of 2-naphthol solution, and the mixture is allowed to stand for 30 min in ice-cold water. An orange to red precipitate is formed.

Paper Chromatography of Arylazo-2-naphthols

The preparation of derivatives for paper chromatography is carried out as in the preceding case, and the precipitate formed is extracted with a small amount of chloroform. The solution formed is applied directly onto the paper. The following solvent systems may be used for chromatography: 50% dimethylformamide/hexane, paraffin oil/ethanol-ammonia (4 : 1), or ethanol-water (4 : 1), paraffin oil/85% acetic acid, and 1-bromonaphthalene/ /90% acetic acid. A suitable amount of the azodye for chromatography is 10 – 15 µg.

General Methods of Preparation of Azo-2-naphthols

Amine (1 mmole) is dissolved in 1 ml of conc. HCl, diluted with water (approx. 10 ml), and mixed, after cooling, with several portions of a freshly prepared solution (10−20%) of sodium nitrite. The end of the diazotization is controlled by means of iodine-starch paper. The temperature during the diazotization is kept below 5 °C. The cold solution of the diazo compound is added dropwise and under vigorous stirring to an alkaline solution (15 ml of 10% NaOH) of 2-naphthol (1.2−1.5 mmole). After 30 min standing the dye is isolated and crystallized, best from acetic acid. In the case of the product being precipitated during the coupling in the form of an oil, it is cooled in a refrigerator and brought to crystallization by scratching with a glass rod.

o-Methylbenzeneazo-2-naphthol

Reagents: 5% solution of 2-naphthol in 10% NaOH; 10% $NaNO_2$; 10% HCl; sulfamic acid; ethanol; KI-starch paper.

Procedure: In a 10-ml test tube 132 mg of o-toluidine hydrochloride are dissolved in 1.5 ml of 10% HCl, the solution is cooled with ice, and 10% NaOH is added dropwise and with shaking until the reaction mixture turns the KI-starch paper blue immediately (usually after 10 drops). The excess of nitrous acid is eliminated by the addition of several crystals of sulfamic acid (test with KI-starch paper) and the cold solution is added dropwise and under vigorous shaking into 3 ml of an alkaline 2-naphthol solution. The mixture is allowed to stand in a refrigerator for 30 min and the product is filtered off on a filtration crucible; it is washed three times with 2 ml of water and dried. Yield, 140 mg; mp, 126−128 °C. After crystallization from 7 ml of ethanol the product weighed 85 mg, mp 130 °C.

Alkylation and Arylation

Tertiary amines react with alkylating agents with the formation of quaternary ammonium salts. For alkylation, methyliodide, methyl p-toluenesulfonate, and benzoyl chloride are used. Methyl p-toluenesulfonates are suitable derivatives for the identification of nitrogenated cyclic compounds. The general method of preparation of quaternary ammonium compounds consists of mixing equimolar quantities of the tertiary amine and of the reagent and heating for several minutes. The reaction can also be carried out in an inert solvent (benzene, nitromethane, diisopropyl ether).

Tetramethylammonium Iodide

Reagents: nitromethane, ethanol, methyl iodide.

Procedure: Trimethylamine (0.3 ml) is dissolved in a micro test tube in 1 ml of nitromethane, and 0.2 ml of methyl iodide is added. The tube is closed with a rubber stopper and allowed to stand overnight. The separated crystals are filtered using a filtration tube and washed with 0.2 ml of nitromethane and 0.1 ml of ethanol. They are then dried in a stream of dry air and weighed. Yield, 0.19 g; mp, 290 − 295 °C (in capillary, with decomposition). After crystallization from 0.5 ml of ethanol the melting point of the product (25 mg) was 290 − 294 °C (decomp., capillary).

N-Substituted 2,4-Dinitroanilines

Primary and secondary amines react with 2,4-dinitrochlorobenzene with the formation of N-substituted 2,4-dinitroanilines; 0.01 mole of amine is dissolved in 10 ml of ethanol and mixed with 2 g of dinitrochlorobenzene and 1 g of anhydrous sodium acetate. The reaction mixture is boiled for 5 − 10 min under reflux. 2,4-Dinitroanilines usually crystallize after cooling. If no derivative crystallizes, the solution is concentrated. If, instead of free amine, its salt is used, the reaction should be carried out in the presence of 2 g of anhydrous sodium acetate (93).

3. Separation of Amines

Chemical Methods

Among chemical methods for the separation of amines, those of Hinsberg (reaction with p-toluenesulfonyl chloride) and of Alexander (94) (reaction with 3-nitrophthalic anhydride) are most commonly used. Tertiary amines do not react with the reagents mentioned and they can be separated after the reaction − for example, by extraction. Derivatives of primary amines with p-toluenesulfonyl chloride are soluble in alkali hydroxide solutions, in contrast to sulfonamides of secondary amines; this is utilized for their separation. When primary and secondary amines are separated by reacting them with 3-nitrophthalic anhydride, use is made of the fact that only phthalimine acids derived from primary amines can be cyclized. Practical utilization of both procedures is demonstrated by the separated of a mixture of aniline, ethylaniline, and diethylaniline. However, it should be mentioned that in a number of cases the procedures fail or do not lead to a sufficiently sharp separation. Negatively substituted amines which do not react with p-toluenesulfonyl chloride can be separated with 3-nitrophthalic anhydride. Some p-toluenesulfonamides of primary amines are poorly soluble in alkali. The derivative of primary amine with 3-nitrophthalic anhydride is cyclized merely by boiling in benzene, and the phthalimide formed is soluble in benzene and can be isolated together with the tertiary amine.

Separation of a Mixture of Aniline, Ethylaniline, and Diethylaniline with p-Toluenesulfonyl Chloride

Reagents: p-toluenesulfonyl chloride, 10% NaOH, ether, HCl 1 : 1, sodium picrate, ethanol, light petroleum.

Procedure: A mixture (150 mg) of aniline, ethylaniline, and diethylaniline (1 : 1 : 1) is mixed in a 10-ml test tube with 5 ml of 10% NaOH and 0.3 g of p-toluenesulfonyl chloride, and the tube is closed (rubber stopper) and shaken for 3 min. The test tube is immersed for several minutes into a 60 °C water bath and shaken again for 3 min. After 5 min standing the solution is cooled and extracted three times with 5 ml of ether. The ethereal layer is reextracted twice with 5 ml of 5% HCl and the acid aqueous extract is evaporated on a water bath to dryness. The residue is dissolved in a minimum amount of water, 100 mg of sodium picrate are added, and the mixture is heated until a clear solution is obtained (water may be added if necessary) and then allowed to cool. The separated picrate is filtered off and crystallized from ethanol. Yield, 60 mg; mp, 134−135 °C. The combined ethereal extracts are washed twice with 2 ml of water and ether is distilled off. Three milliliters of 10% NaOH are added to the oily residue and the oil is mixed (scratching the sides of the test tube) with a glass rod until it solidifies. It is then filtered off using a filtration tube and washed twice with 1 ml of water, and the crude N-ethyl-N-phenyl-p-toluenesulfonamide is crystallized from aqueous ethanol. Yield, 15 mg; mp, 85 °C.

After extraction with ether the alkaline solution is acidified with hydrochloric acid 1 : 1 and the separated derivative is filtered off. If the precipitated product is oily, it should ba cooled in a refrigerator and brought to crystallization by scratching with a glass rod. The solid product is washed twice with 1 ml of water and, after drying in a stream of dry air, with 1 ml of light petroleum. Yield, 57 mg; mp, 100−101 °C.

Separation of a Mixture of Aniline, Ethylaniline, and Diethylaniline with 3-Nitrophthalic Anhydride

Reagents: 3-nitrophthalic anhydride, benzene, ether, ethanol, acetone, 10% NaOH, picric acid, HCl 1 : 1, saturated NaHCO₃ solution.

Procedure: In a 200 ml-flask provided with a reflux condenser a mixture of aniline (1 g), ethylaniline (1 g), and diethylaniline (1 g) dissolved in 50 ml of benzene is boiled with 8 g of 3-nitrophthalic anhydride for 1 hr. After cooling, the benzene solution is extracted twice with 20 ml of 10% NaOH and the alkaline extract is extracted with 30 ml of ether. The combined benzene and ethereal layers are distilled off. From the residue a portion of N-phenyl-3-nitrophthalimide separates out. It is filtered off under suction and washed with a small amount of 0 °C cold benzene. Picric acid (2.5 g) is then added to the filtrate, followed by ethanol until the mixture becomes a clear solution (heating necessary). The mixture is then cooled and the separated diethylaniline picrate is filtered off and crystallized from ethanol (mp, 136–138 °C). The alkaline layer (40 ml of 10% NaOH) is mixed with HCl 1 : 1 (gradually) until the mixture is distinctly acid (test with Congo paper). The precipitated mixture of phthalimine acids is filtered off on a small Büchner funnel and washed with water. The material on the filter is dried well and transferred into a 50-ml Erlenmeyer flask. The flask is immersed into a 170 °C hot paraffin oil bath for 10 min and 0.5 g of 3-nitrophthalic anhydride are then added. The heating acids is continued for an additional 30 min at 160 °C. After cooling, the melt is broken and triturated with a glass rod and extracted systematically with three portions (25 ml each) of saturated NaHCO₃ solution. The extract is filtered each time and the solid material is washed with 10 ml of water and then combined with the extract from the benzene layer. It is crystallized from aqueous acetone (mp of N-phenyl-3-nitrophthalimide is 138.5–139.5 °C). Hydrochloric acid 1 : 1 is added carefully to the NaOH₃ solution in a 500 ml Erlenmeyer flask until acid (test with Congo paper) and the precipitate is filtered off and washed with 5 ml of water. It is crystallized from aqueous ethanol (mp of the amide of N-phenyl-N-ethyl-3-nitrophthalic acid is 194–201 °C).

Separation of Amines in the Form of 3,5-Dinitrobenzamides by Column Chromatography on Alumina (79)

3,5-Dinitrobenzamides of primary and secondary amines can be separated by column chromatography on alumina using a mixture of chloroform and benzene (for example, in the ratio 1 : 3) as eluent. The separation and the purity of fractions are controlled by paper or thin-layer chromatography. As primary and secondary amines react quantitatively with 3,5-dinitrobenzoyl chloride, while tertiary amines do not, the separation of amines and their purification can be easily performed on this basis. The preparation of derivatives is easy (see p. 334), both starting from free

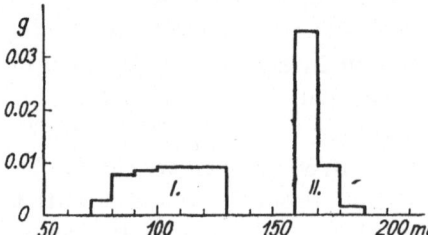

Fig. 35. Separation of a mixture of 50 mg dimethyl-3,5-dinitrobenzamide (I) and 50 mg of methyl-3,5-dinitrobenzamide (II) on alumina (50 g), activity III with a mixture of chloroform-benzene (1 : 3) as eluent; 10 ml fractions.

bases and from their hydrochlorides and aqueous solutions. Tertiary amines can be isolated after the reaction from the alkaline medium by extraction or steam distillation. A typical record of a separation is shown in Fig. 35. The procedure can also be carried out on a nonadhering thin layer of alumina.

Paper and Thin-Layer Chromatography

Aliphatic Amines. Paper chromatography of aliphatic monoamines and their salts requires special conditions. These substances must be chromatographed in a manner that excludes the hydrolysis of salts and the influence of the presence of various anions (95). Figure 36 shows the results of the chromatography of amines with alkyl groups up to C_4 in butanol saturated with a KCl solution. The chromatographic paper is impregnated by dipping it in and drawing it through a 5% aqueous KCl solution and drying in an oven at 105 °C, or allowing it to hang freely in the air at room temperature overnight. The solutions of amine salts ($2-6$ µl of a $0.5-1\%$ solution) are applied on a paper treated in such a way. n-Butanol saturated with a KCl solution serves as the mobile phase. After the development the chromatogram is dried either in the air or at 105 °C in a drying oven. Primary amines are detected by spraying with a 0.2% alcoholic ninhydrin solution acidified with a few drops of acetic acid, with heating in the oven at 100 °C for a few minutes. Primary amines appear as red-violet spots, which can be stabilized by spraying the chromatogram with a 1% Cu^{2+} salt solution in aqueous alcohol. Secondary amines can be detected by a spray consisting of a 5% sodium nitroprusside solution in 10% aqueous acetaldehyde containing 2% of soda. Blue spots appear. Tertiary amines can be detected with Dragendorff reagent (p. 324), appearing as red-orange spots. The difficulties encountered during the chromatography of amines can be avoided by working in a strongly acid medium [for example, using the solvent system n-butanol–conc. HCl (4 : 1) for development] (96).

Higher aliphatic primary amines up to C_{18} can be separated in the solvent system lauryl alcohol/ethanol–N HCl (1 : 1). Their detection is carried out with a 0.2% alcoholic ninhydrin solution containing 1% of pyridine, by heating at 100 °C for several minutes. Usually, $40-60$ µg of amines dissolved in alcohol are applied on the start. For the chromatography of aliphatic amines in the form of 3,5-dinitrobenzamides see p. 336.

Ethanolamines. Mono-, di-, and triethanolamines are best separated in an ethyl acetate-methanol-ammonia (12 : 4 : 1) mixture. The R_F values are strongly dependent on the temperature, increasing with the latter, but they can be regulated by changing the methanol content in the mobile phase. The addition of methanol increases these values, while a decrease in methanol concentration causes them to drop. Ethanolamines are applied as 0.5−1% alcoholic or aqueous solutions in quantities of 10−50 µg. The detection is carried out by spraying with 0.1% alcoholic alizarine solution. Bases appear as violet spots on a yellow background. Other reagents may also be applied

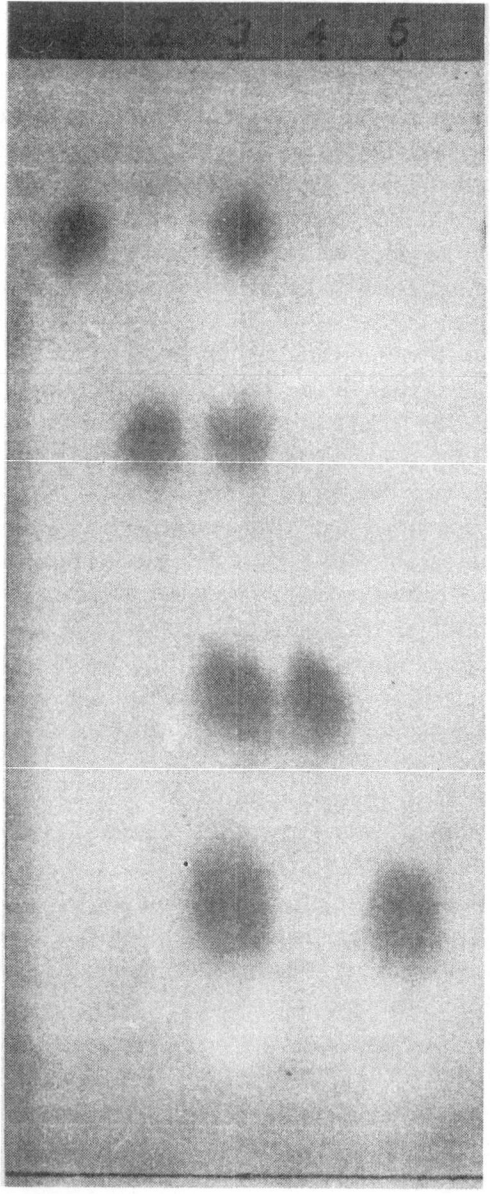

for amines; only triethanolamine does not react with Dragendorff reagent (97). The solvent system ethyl acetate-dimethylformamide (2 : 1) is also suitable (96).

Aliphatic Diamines (98). These are best chromatographed on papers impregnated with a solution of the composition 0.1 N HCl−0.1 N KCl−water (1 : 1 : 1.5). After drying at room temperature, chromatography is carried out in a mixture of one part of the impregnation solution and four parts of phenol. The developed chromatogram is dried at room temperature and sprayed with ninhydrin, as in the case of higher aliphatic amines. For the utilization of strongly acid solvent systems see (96).

Fig. 36. Chromatogram of salts of aliphatic amines on potassium chloride-impregnated paper; mobile phase *n*-butanol saturated with conc. KCl. (1) Methylamine, (2) ethylamine, (3) mixture of methyl-to *n*-butylamine, (4) *n*-propylamine (5) *n*-butylamine.

Aromatic Amines. Aromatic amines can be chromatographed in the form of their salts in *n*-butanol-2,5 N HCl or ethyl acetate-acetic acid-water (12 : 4 : 1). If free amines (bases) do not undergo oxidation easily or if they are not volatile, solvent systems consisting of formamide as the stationary phase and of hexane, benzene, chloroform (also mixtures of these), or carbon tetrachloride as the mobile phase can be used (99 − 102). Halogenated amines can be chromatographed with advantage in the solvent system 1-bromonaphthalene/pyridine − water (2 : 1). In this case they can be applied on the paper either as free bases or as salts. Detection is carried out by spraying with 1% *p*-dimethylaminobenzaldehyde or *p*-dimethylaminocinnamaldehyde solution in a mixture of 95 parts of ethanol and five parts of conc. HCl. Primary aromatic amines can also be detected by hanging the chromatogram for a few minutes in a glass cylinder on the bottom of which sodium nitrite solution acidified with hydrochloric acid is placed, and then spraying with a solution containing 1% of resorcinol and 5% of soda. Amines appear as yellow to orange spots. N,N-Dialkylanilines can be detected with Dragendorff reagent and mono and dialkylanilines; primary aromatic amines with free para position can be detected by spraying with a solution of a stabilized diazonium salt, as, for example, 1% solution of *p*-nitrobenzenediazonium fluoroborate. For N-alkylated benzidines a spray composed of one part of a 0.1 N cerium (IV) sulfate solution in 4 N sulfuric acid and three parts of water was found suitable (103).

Amino derivatives of anthraquinone should be chromatographed in the following systems: 1-bromonaphthalene/pyridine-water (2 : 1 and 1 : 1) (104) and 1-bromonaphthalene/80% acetic acid.

Aromatic amines can sometimes be chromatographed successfuly after their conversion to derivatives. For example, isomeric toluidines and aniline, which are poorly separated as salts, can be clearly separated after their conversion to bromo derivatives (105); a series of primary amines can be separated after conversion to arylazo-2-naphthole (p. 342), and for the separation of isomers the products of the reaction of amines with diazonium salts (p. 324) can also be employed. They can be chromatographed in a formamide/hexane system. Thin-layer chromatography of free bases can be carried out on nonadhering layers of alumina (106, 107), or on silica gel G layers using the mobile phase cyclohexane − carbon tetrachloride-ethyl ecetate (10 : 70 : 20):

Quaternary Ammonium Salts. For lower members of the series the system n-butanol saturated with N HCl is the most suitable; for higher members of the homologous series reversed phases must be used. One of the best is the solvent system lauryl alcohol/ethanol − N HCl (1 : 1). Detection is carried out with Dragendorff reagent or potassium iodoplatinate (5 ml of 5% PtCl$_4$ in N HCl is mixed with 45 ml of a 10% aqueous KI solution and 100 ml of water). Red-orange spots are formed with the first reagent; with the second the spots are blue-green on a pink background.

Identification of N-Alkyls by Paper Chromatography

The cleavage is carried out with hydriodic acid using the procedure for the quantitative determination of N-alkyls (21), and the split off alkyl iodide is converted to the corresponding ester of 3,5-dinitrobenzoic acid in a manner similar to that described for the identification of O-alkyls on p. 202. Chromatographic analysis is also carried out in the same manner.

Gas Chromatography

A sharp separation of ammonia, methylamine, dimethylamine, and tri-methylamine can be effected by chromatography on triethanolamine as the stationary phase (108). The following stationary phases have also been used for the separation of amines: diethanolamine (109), mineral oils and paraffins (110), polyethylene glycol (111), silicone oil DC550 (112), and undecanol (109). For the separation of pyridine bases, glycerol, paraffin, polyethylene glycol, silicone oils, triethanolamine, and trixylyl phosphate have been employed (113). A thorough study of the separation, identification, and determination of amines of various types (long-chain aliphatic amines, alicyclic monoamines and diamines, aromatic amines) by gas chromatography has been published by Vanden Heuvel et al. (114). For the application of Kováts indices enabling the differentiation among primary, secondary, and tertiary amines see (115).

References

1. Singewald, A.: Zeit. anal. Chem. **164**, 219 (1958).
2. Vanag, G. Ya., and Zhagat, R. A.: Dokl. Akad. Nauk SSSR **133**, 362 (1960).
3. Sachs, F., and Graveri, M.: Ber. **38**, 3685 (1905).
4. Pesez, M.: Ann. Chim. Anal. **25**, 37 (1943).
5. Feigl, F., and Anger, V.: J. prakt. Chem. **139**, 180 (1934).
6. Feigl, F., Anger, V., and Zappert, R.: Mikrochemie **16**, 67 (1934).
7. Frehden, O., and Goldschmidt, L.: Mikrochim. Acta **1**, 338 (1937).
8. Sivadjian, J.: Bull. Soc. Chim. France **2**, 623 (1935).
9. Sass, S., Kaufman, J. J., Cardenas, A. A., and Martin, J. J.: Anal. Chem. **30**, 529 (1958).
10. Nicksic, S. W., and Judd, S. H.: Anal. Chem. **30**, 2002 (1958).
11. McGowan, J. C.: J. Chem. Soc. **1949**, 777.
12. Menzie, C.: Anal. Chem. **28**, 1321 (1956).
13. Feigl, F., and Anger, V.: Mikrochim. Acta **2**, 107 (1937).
14. Rimini, E.: Ann. Farmacotherap. Chim. **27—28**, 193 (1898); Zeit. anal. Chem. **41** 438 (1902).

15. Anger, V.: Mikrochim. Acta **2**, 3 (1937).
16. Duke, F. R.: Ind. Eng. Chem., Anal. Ed. **17**, 196 (1945).
17. Hanot, C.: Bull. Soc. Chim. Belge **66**, 76 (1957).
18. Sawicki, E., Stanley, T. W., and Hauser, T. R.: Chemist-Analyst **48**, 30 (1959).
19. Sawicki, E., Noe, J. L., and Fox, F. T.: Talanta **8**, 257 (1961).
20. Feigl, F., Anger, V., and Zappert, R.: Mikrochemie **16**, 67 (1934).
21. Jureček, M.: Organická analysa I, p. 163, Nakladatelství ČSAV, Prague, 1955.
22. Gasparič, J., and Matrka, M.: Collection Czech. Chem. Commun. **24**, 1943 (1959).
23. Feigl, F., and Goldstein, D.: Anal. Chem. **32**, 861 (1960).
24. Weygand, F., Weber, H., Maekawa, E., and Eberhardt, G.: Chem. Ber. **89**, 1994 (1956).
25. Sawicki, E., Stanley, T. W., Hauser, R., Elbert, W., and Noe, J. L.: Anal. Chem. **33**, 722 (1961).
26. Sawicki, E., Stanley, T. W., and Elbert, W.: Mikrochim. Acta **1961**, 505.
27. Noelting, E., and Thesmar, G.: Ber. **35**, 628 (1902).
28. Bertetti, J.: Ann. Chim. (Roma) **44**, 495 (1954).
29. Agulhon, H., and Thomas, P.: Bull. Soc. Chim. France **11**, 69 (1912).
30. Murray, H. D.: Chem. News **130**, 23 (1925).
31. Kozlov, N. S., and Andreeva, A. A.: Sb. Nauch. Tr. Perm. Politekh. Inst. No. **18**, 165 (1965).
32. Thompson, C. J., Coleman, H. J., Ward, C. C., and Rall, H. T.: Anal. Chem. **34**, 151 (1962).
33. Aung, T., Healy, E. E., and Murmann, R. K.: Chem. Analyst **49**, 73 (1960).
34. Nicksic, S. W., and Judd, S. H.: Anal. Chem. **32**, 998 (1960).
35. Candea, C., and Macovski, E.: Bull. Soc. Chim. France **4**, 1398 (1937).
36. Hearn, W. E., and Kinghorn, R.: Analyst **85**, 766 (1960).
37. Gasparič, J., and Matrka, M.: Collection Czech. Chem. Commun. **24**, 643 (1959).
38. Afanas'ev, B. N., and Kruzhevnikova, A. I.: Zhur. Anal. Khim. **12**, 143 (1957).
39. Kehrmann, F., and Roy, G.: Ber. **55**, 156 (1922).
40. Frieser, R. G., and Scardaville, P. A.: Anal. Chem. **32**, 196 (1960).
41. Lellmann, E.: Ann. **228**, 248 (1885).
42. Thiele, J., and Steimmig, G.: Ber. **40**, 955 (1907).
43. Albert, A.: J. Chem. Soc. **1939**, 920.
44. Heim, O.: Ind. Eng. Chem., Anal. Ed. **7**, 146 (1935).
45. Merz, K. W., and Kammerer, A.: Arch. Pharm. **256−8**, 198 (1953).
46. Massmann, W., and Menge, G.: Zeit. Anal. Chem. **148**, 100 (1955−56).
47. Fischer, E., and Bergmann, M.: Ann. **398**, 96 (1913).
48. Bertrand, G.: Compt. Rend. **128**, 742 (1899).
49. Dausi, A., Mamoli, L., and Ciocca, B.: Ann. Chim. Applic. **22**, 561 (1932).
50. Meyer, H.: Monatsh, **15**, 164 (1894).
51. Fischer, E.: Ann. **199**, 281 (1879).
52. Stoehr, C.: J. prakt. Chem. (3) **45**, 20 (1892).
53. Salkowski, H.: Ber. **31**, 776 (1898).
54. Buehler, C. A., Curriee, E. J., and Lawrence, R.: Ind. Eng. Chem., Anal. Ed. **5**, 277 (1933).
55. Buehler, C. A., and Calfer, J. D.: Ind. Eng. Chem., Anal. Ed. **6**, 351 (1934).
56. Sah, P. P. T., and Tien, C. H.: J. Chin. Chem. Soc. **4**, 490 (1936).

57. Sah, P. P. T., and Yuin, K. H.: J. Chin. Chem. Soc. **5**, 129 (1937).

58. Pauly, H., and Ludwig, E.: Hoppe-Sayler's Zeit. physiol. Chem. **121**, 165 (1922).

59. Keyworth, C. M.: J. Soc. Chem. Ind. London **46**, 20 (1924).

60. Noller, C. R., and Liang, P.: J. Am. Chem. Soc. **54**, 670 (1932).

61. Forster, R. B., and Keyworth, C. M.: J. Soc. Chem. Ind. London, 43, 165, 299 (1924).

62. Wanag, G.: Ber. **69**, 1066 (1936).

63. Wanag, G., and Lode, A.: Ber. **70**, 547 (1937).

64. Francis, C. V.: Microchem. J., **6**, 606 (1962).

65. Runge, F., Engelbrecht, H. J., and Franke, H.: Ber. **88**, 533 (1955).

66. Wittig, G.: Angew. Chem. **62**, 231 (1950).

67. Wittig, G., Keicher, G., Rückert, A., and Raff, P.: Ann. **563**, 110 (1949).

68. Crane, F. E., Jr.: Anal. Chem. **30**, 1426 (1958).

69. Wendlandt, W. W., and Dunham, R.: Anal. Chim. Acta **19**, 505 (1958).

70. Barnard, A. I., Jr., and Wendlandt, W. W.: Revista de la sociedad química de México **3**, 269 (1959).

71. Večeřa, M., and Friedrich, K.: Chem. Listy **51**, 283 (1957).

72. Kehrmann, F., and Baumgartner, E.: Helv. Chim. Acta **9**, 673 (1926).

73. Schotten, C.: Ber. **17**, 2544 (1884).

74. Baumann, E.: Ber. **19**, 3218 (1886).

75. Voláková, B.: Collection Czech. Chem. Commun. **26**, 1332 (1961).

76. Neurath, G., and Doerk, E.: Chem. Ber, **97**, 172 (1964).

77. Heyns, K., Harke, H. P., Scharmann, H., and Gruetzmacher, H. F.: Zeit. anal. Chem. **230**, 118 (1967).

78. Gasparič, J., and Borecký, J.: J. Chromatog. **5**, 466 (1961).

79. Večeřa, M., and Gasparič, J.: Chem. and Ind. **1957**, 263.

80. Hinsberg, O.: Ber. **23**, 2962 (1890); **33**, 3526 (1900).

81. Hinsberg, O., and Kessler, J.: Ber. **38**, 906 (1905).

82. Fischer, E., and Bergmann M.: Ann. **398**, 96 (1913).

83. Marvel, C. S., and Smith, F. E.: J. Am. Chem. Soc. **45**, 2696 (1923).

84. Marvel, C. S., Kingsbury, F. L., and Smith, F. E.: J. Am. Chem. Soc. **47**, 166 (1925).

85. Sah, P. P. T.: J. Chin. Chem. Soc. **5**, 100 (1937).

86. Suter, C. M., and Moffett, E. W.: J. Am. Chem. Soc. **55**, 2497 (1933).

87. Sah, P. P. T., and Lei, H. H.: J. Chin. Chem. Soc. **2**, 153 (1934).

88. Campbell, K. N., Campbell, B. K., and Patelski, S. J.: Proc. Indiana Acad. Sci. **53**, 119 (1943).

89. Brown, E. L., and Campbell, K. N.: J. Chem. Soc. **1937**, 1699.

90. Gasparič, J., Novotná, M., and Jureček, M.: Collection Czech. Chem. Commun. **25**, 2757 (1960).

91. Bratton, A. C., and Marshall, E. K.: J. Biol. Chem., **128**, 537 (1939).

92. Averell, P. R., and Norris, M. V.: Anal. Chem. **20**, 753 (1948).

93. van der Kam, E. J.: Rec. Trav. Chim. **45**, 722, 734 (1926).

94. Alexander, J. W., and McElvain, S. M.: J. Am. Chem. Soc. **60**, 2285 (1938).

95. Večeřa, M., and Gasparič, J.: Chem. Listy **52**, 611 (1958).

96. Borecký, J.: Mikrochim. Acta **1966**, 279.

97. Gasparič, J, Borecký, J., Obruba, K., and Hanzlík, J.: Collection Czech. Chem. Commun. **26**, 2950 (1961).

98. Obruba, K., Gasparič, J., and Borecký, J.: Collection Czech. Chem. Commun. **27**, 1498 (1961).

99. Cee, A., and Gasparič, J.: Mikrochim. Acta **1966**, 295.

100. Gasparič, J., Petránek, J., and Večeřa, M.: Mikrochim. Acta **1955**, 1026.

101. Latinák, J.: Collection Czech Chem. Commun. **24**, 2939 (1959).

102. Gemzová, I., and Gasparič, J.: Mikrochim. Acta **1966**, 310.

103. Gasparič, J., and Matrka, M.: Chem. Listy **52**, 749 (1958); Collection Czech. Chem. Commun. **24**, 643 (1959).

104. Gasparič, J.: J. Chromatog. **4**, 75 (1960).

105. Latinák, J.: Chem. Listy **51**, 1493 (1957).

106. Gemzová, I., and Gasparič, J.: Collection Czech. Chem. Commun. **31**, 2525 (1966).

107. Gasparič, J.: Zeit. anal. Chem. **218**, 113 (1966).

108. Issoire, J., and Chaput, L.: Chim. anal. **43**, 313 (1961).

109. James, A. T., and Martin, A. J. P.: Biochem. J. **48**, vii (1951).

110. James, A. T.: Anal. Chem. **28**, 1564 (1956).

111. Landault, C., and Guiochon, G.: J. Chromatog. **13**, 327 (1964).

112. Kaiser, R.: Gas-Chromatographie, p. 188. Akademische Verlagsgesellschaft, Leipzig, 1960.

113. Brooks, V. T., and Collins, G. A.: Chem. and Ind. **1956**, 1021.

114. Vanden Heuvel, W. J. A., Gardiner, W. L., and Horning, E. C.: Anal. Chem. **36**, 1550 (1964).

115. Golovya, R. V., Mironov, G. A., and Schuravljova, I. P.: Abh. Deutsch. Akad. Wiss., Berlin, Kl. Chem., Geol. Biol. **1966**, 555.

CHAPTER XVI

SOME NITROGEN-CONTAINING COMPOUNDS

1. Nitro Compounds

For the detection and identification of aliphatic and aromatic nitro compounds reduction is used (catalytic; with hydrogen or with the system: metal plus HCl); amines formed on reduction can be detected and identified by methods described on p. 317.

Aromatic nitro compounds can also be identified by the introduction of an additional nitro group or by the oxidation of the alkyl group — if present on the nucleus — and, finally, in the case of di- and polynitro compounds by the preparation of addition compounds (1), preferably with aromatic hydrocarbons.

The choice of procedure is determined by general rules and depends on the character of the nitro compound identified and also on the amount of substance available.

Hydroxamate Test

Aliphatic nitro compounds are reduced with zinc to corresponding hydroxylamines, which, under the influence of benzoyl chloride, change to hydroxamic acids. These can be detected by the reaction with ferric chloride (2).

Reagents: 50% ethanol, NH_4Cl, powdered zinc, HCl, 10% $FeCl_3$ solution.

Procedure: A solution of the nitroso compound in 50% ethanol (1 ml) is boiled with 50 mg of ammonium chloride and 50 mg of powdered zinc. After decantation one drop of benzoyl chloride is added to the clear solution, which is then acidified with hydrochloric acid. Eventually, a drop of $FeCl_3$ solution is added, causing a red color.

Hydrolysis

Aliphatic nitro compounds are hydrolyzed, with the formation of nitrous acid, which can be detected easily. The reaction is also given with aromatic *o*- and *p*-dinitro compounds containing a labile nitro group (3).

Reagents: Metanilic acid (0.5 g) is dissolved in 150 ml of 2 N CH_3COOH and mixed with 0.1 g of l-naphthylamine dissolved in 20 ml of boiling water and 150 ml of 2 N CH_3COOH; a solution of 10 g potassium carbonate in 5 ml of water; dilute CH_3COOH (1 : 1).

Procedure: A few milligrams of an aliphatic nitro compound are boiled with 1 ml of potassium carbonate solution for 2 min. After cooling, 1 ml of water is added. A few drops of the resulting solution are acidified with 1 ml of dilute acetic acid, and 0.05 ml of the reagent is added; a pink to red color is formed.

Coupling

Primary aliphatic nitro compounds can be coupled with diazonium salts, for example, with diazotized metanilic acid, with the formation of an orange to red color. Secondary aliphatic nitro compounds do not react (4, 5).

Reagents: The reagent is prepared by mixing equal parts of 0.9% $NaNO_2$ solution with 0.72% metanilic acid solution in 18% hydrochloric acid; 10% methanol; phosphate buffer, pH 4.3.

Procedure: To 1 ml of a solution of a primary aliphatic nitro compound in 10% methanol 2 ml of the buffer solution (pH 4.3) and 5 ml of the reagent are added, giving rise to color ranging from orange to red.

Reaction with Sodium 1,2-Naphthoquinone-4-sulfonate

Nitromethane reacts with this reagent with the formation of a violet to blue color, as with amines (p. 320) (6,7). Nitroethane does not react. This makes this reaction suitable for the detection of nitromethane in the presence of higher nitroparaffins (8).

Detection of Nitromethane

Reagents: 0.5% aqueous sodium 1,2-naphthoquinone-4-sulfonate; CaO, or 5% Na_2CO_3 solution.

Procedure: One drop of an alcoholic solution of the tested substance is mixed with one drop of the reagent and several mg of CaO or two drops of sodium carbonate solution. The presence of nitromethane is revealed by the formation of violet to blue color.

Other Color Reactions of Aliphatic Nitro Compounds

To detect aliphatic nitro compounds, especially primary ones, use can be made of their reaction with nitrous acid (9); for the differentiation of nitromethane and nitroethane their degradation to formaldehyde and acetaldehyde, respectively, is suitable (10).

A sensitive color reaction of aliphatic nitro compounds (down to 0.02 μg) is based on their decomposition with H_2O_2 in alkaline media and the color reactions of the nitrite ion formed (11).

Janovsky Reaction

In the literature the name of Janovsky is connected with the reaction of aromatic polynitro compounds, especially those of *m*-dinitro compounds with alkalies in the presence of compounds with an active methylene group $(-CH_2-)$ – for example, methyl ketones. An orange, red, or violet to blue color appears, depending on the nature of the polynitro compound (see also p. 234) (12, 13).

Various hypotheses have been put forward concerning the mechanism of this reaction. It seems that the reaction is rather complicated, and that the reaction of polynitro compounds with alkalies in alcohol (14, 15, 16), as well as the reaction of dinitro compounds with aliphatic amines and diamines, most recently carried out in dimethylformamide or dimethyl sulfoxide, or the reaction with alkali cyanides, also belong to this type of color reactions.

As for the question of which nitro compounds give a positive reaction, the data in the literature are not consistent. Generally, *m*-dinitro compounds are considered as active, but exceptions are known. Among papers giving lists of reacting compounds, the following should be mentioned (21 – 23).

Colored reaction products were isolated by Aketsuda (24).

Reagents: 10% NaOH, acetone.

Procedure: The tested nitro compounds are dissolved in acetone, and sodium hydroxide solution is added dropwise. The appearance of an orange, red, violet, brown-red, or even blue color means a positive reaction.

Reaction with Sugars

o-Dinitro compounds, which do not give a positive Janovsky reaction, give a strong color in alcoholic solution in the presence of reducing sugars (see p. 306).

Reagents: 25% sodium carbonate solution, ethanol, 1% aqueous glucose solution.

Procedure: To one drop of an alcoholic *o*-dinitro compound solution a drop of glucose solution and five drops of Na_2CO_3 solution are added, and the mixture is heated over a microflame. If the reaction is positive, a red to violet color appears.

Color Reactions after the Reduction
of Nitro Compounds

Aromatic nitro compounds can be reduced in several ways, either to nitroso compounds only, or up to amines. The usual color tests can then be carried out with the products or the reduction. Electrolytic reduction or reduction with zinc in $CaCl_2$ solution leads to nitroso compounds which can be detected by the reaction with sodium pentacyanoammoferroate (25); reduction with zinc in glacial acetic acid or in NH_4Cl, as well as reduction with stannous chloride, give rise to primary aromatic amines which can be detected either with p-dimethylaminobenzaldehyde, or, after diazotization, by coupling with a suitable passive component (p. 341).

Reduction to Nitroso Compound

Reagents: ethanol, 10% aqueous $CaCl_2$ solution, zinc dust, 1% aqueous sodium pentacyanoammoferroate solution.

Procedure: A small amount of a mononitro compound is dissolved in 3 ml of hot ethanol, and to this solution 7 − 10 drops of $CaCl_2$ solution are added, followed by 50 mg of zinc. The mixture is boiled for 1 min on a water bath, then cooled and filtered. Four drops of the filtrate (if necessary, after dilution) are mixed with one drop of the reagent. The formation of a red to violet color means a positive reaction.

Reduction to Amine

Reagents: powdered Zn, N HCl; or Zn dust, glacial acetic acid; or a solution of 0.7 g of $SnCl_2 . 2 H_2O$ in 15 ml of conc. HCl and 100 ml of water.

Procedure: A sample of the analyzed material is heated over a boiling water bath with several drops of stannous chloride solution or with a small amount of zinc and several milliliters of acid for several minutes. Some times standing for an additional 10 min is useful. After cooling, the mixture is filtered and the color reactions are carried out.

Test for Aromatic Amines

Reagents: 1% solution of p-dimethylaminobenzaldehyde in a mixture of 95 parts of ethanol and five parts of HCl; or 1% aqueous $NaNO_2$ solution and 1% solution of resorcinol in 10% Na_2CO_3; or 1% aqueous $NaNO_2$ solution, 1% ammonium sulfamate, and 0.1% N-l-naphthylethylenediamine dihydrochloride in water.

Procedure: (a) A small amount of the filtrate from the reduction is spotted on a piece of filter paper, followed by a drop of p-dimethylaminobenzaldehyde solution. In a positive case a yellow, orange, or red color is observed.

(b) To a small amount of the filtrate cooled with icy water a drop of dilute HCl and several drops of $NaNO_2$ solution are added. After a while one drop of this mixture is spotted on filter paper, followed by one drop of resorcinol solution; the formation of an orange to orange-brown color shows a positive reaction.

A small amount of the cooled filtrate (cooled with ice water) is mixed in a test tube with a drop of dilute HCl and several drops of $NaNO_2$ solution, and the mixture is shaken for a few minutes. An excess of sulfamate is then added dropwise to the mixture (until no nitrogen gas bubbles are liberated after shaking), followed by a few drops of the N-l-naphthylethylene-diamine solution. The formation of a red to blue-violet color indicates a positive reaction.

Reduction of Aliphatic Nitro Compounds (Identification)

Reduction with Tin

The substance (4 mmole) is mixed in a small 10-ml flask wih 3 ml of conc. HCl and 0.25 g of tin (in the form of small granules). After 10 min an additional 0.25 g of tin are added and the mixture is boiled under reflux for 30 min. After cooling, the reaction mixture is diluted with 6 ml of 25% NaOH, the reflux condenser is substituted by a descending one, and 4 ml are distilled over. The flask for the condensate contains a few drops of methyl red, and the pH value of the condensate is kept low by the addition of 2 N HCl. Benzoyl chloride (0.4 ml) is then added to the distillate, and the mixture is cooled and mixed with 8 ml of 25% NaOH. The flask is then stoppered and shaken vigorously for 10 min. The separated benzamide is filtered off and crystallized as described on p. 334.

Note: Higher-boiling aliphatic nitro compounds yield higher-boiling amines. In such cases the isolation of the amine by extracting it with ether is recommended. If benz-amide is not suitable as a derivative, *p*-toluenesulfonylamide (p. 337) or picrate can be prepared.

Reduction with Zinc

In a 200-ml beaker 4 mmole of nitro compound are dissolved (or suspended) in 30 ml of water and acidified with 20 ml of conc. HCl. The beaker is placed in a vessel containing crushed ice, and as soon as the temperature drops to +5 °C, 3 g of zinc dust are added, with continuous stirring. After 30 min an additional 2 g of zinc dust are added and the stirring is continued for another 30 min. The mixture is filtered and the filtrate is worked up by one of the following procedures: (1) diazotization and coupling with 2-naphthol (see p. 341); (2) coupling with diazotized *p*-toluidine (see p. 324); (3) isolation of the base by steam distillation and acetylation.

Note: If it is not certain that the reduction can take place at room temperature, the solution should be cooled to prevent the formation of by-products.

Reduction of Nitrobenzene with Zinc

Reagents: zinc, conc. HCl, 6 N NaOH, 2 N NaOH, sodium sulfate, acetic anhydride, methanol, ether.

Procedure: In a 10-ml flask provided with a reflux condenser 0.2 ml of nitrobenzene is dissolved in 1 ml of methanol, and 0.5 g of Zn is added, followed by 2 ml of conc. HCl. The mixture is gently boiled until all zinc is dissolved. After cooling, the mixture is diluted with 5 ml of water and then with 15 ml of 6 N NaOH. The well-cooled solution is then extracted three times with 5 ml of ether. The ethereal extract is dried over sodium sulfate (1 hr standing with occasional shaking) and filtered through a cotton-wool plug into a distillation flask. Acetic anhydride (0.3 ml) is added to the filtrate and the ether is evaporated. The residue is heated in a boiling water bath for 5 min, 1 ml of water is added, and the mixture is neutralized with 2 N NaOH (test with indicator paper). The separated acetyl derivative is filtered off under suction and washed twice with 0.5 ml of water. Yield, 0.10 g; mp, 97 – 99 °C. Crystallization from 7 ml of water gave 30 mg of pure product, mp, 112 – 113 °C.

Catalytic Reduction of Nitro Compounds

This is carried out by reduction with hydrogen in the presence of PtO_2 (Adams catalyst) or palladium on a carrier (26) in ethanol. After filtering off the catalyst, the solution is worked up best by evaporating alcohol after the addition of acetic acid. The residue can be worked up immediately by acetylating it with acetic anhydride. The reduction of nitro groups takes place easily even in the case of dinitro compounds and nitrophenols.

Reduction of Nitrobenzene with Hydrogen

Reagents: ethanol, PtO_2, electrolytic hydrogen from a hydrogen cylinder; sodium picrate.

Procedure: Nitrobenzene (30 mg) is dissolved in a test tube in 10 ml of ethanol, and 50 mg of PtO_2 are added to the solution. Hydrogen is then introduced into the solution through a capillary from the hydrogen cylinder. The operation is carried out in the fume-cupboard and the rate of hydrogen introduction should be such as to cause a rapid stream of bubbles to whirl the catalyst around. The test tube is also shaken occasionally. The mixture is filtered through a small folded filter paper and the platinum is washed with 3 ml of ethanol. The filtrate is acidified with 0.3 ml of conc. HCl and alcohol is distilled off. The excess HCl is eliminated by heating on a water bath. The residue is dissolved in a minimum amount of water, and 2 ml of saturated sodium picrate solution are added. The mixture is heated until a clear solution results, and is then cooled for crystallization. The picrate is filtered off using a filtration tube. Yield, 30 mg; mp, 180 °C.

Nitration of Nitro Compounds

This method comes into consideration mainly for the preparation of suitable derivatives of mononitro compounds. Procedures are given on p. 124. It must be remembered that the introduction of a second nitro group can lead in certain cases to one dinitro compound from two isomeric mononitro compounds: *o*- and *p*-nitrotoluene both give 2,4-dinitrotoluene on nitration.

Oxidation of Alkyl Groups of Aromatic Nitro Compounds (General Method)

To finely ground $KMnO_4$ (1 g) in 50-ml flask 0.5 ml of 25% NaOH, 20 ml of water, and $300 - 400$ mg of the nitro compound are added, and the mixture is refluxed (boiling stone) until the pink color of permanganate ion disappears $(1-2$ hr). After cooling, the reaction mixture is acidified with 10% H_2SO_4 (test with Congo paper), and heated to the boiling point, and, if the solution contains MnO_2, a small amount of sodium hydrogen sulfite is added.

After cooling, the reaction mixture is filtered under suction and purified by crystallization from ethanol.

Paper Chromatography

Paper chromatography of nitrated aromatic hydrocarbons is usually carried out on papers impregnated with a 5% paraffin oil solution with ethanol-water-acetic acid mixture (20 : 14 : 1) as solvent (27, 28), or on formamide-impregnated papers with hexane as the mobile phase. Nitro derivatives of hydrocarbons appear under UV light as dark spots.

For nitrophenols the combination l-bromonaphthalene/80% acetic acid (29) is most suitable; *o*-mono- and *o,o*-dinitrophenols are best separated on papers impregnated with formamide and in hexane as mobile phase. Mononitrophenol is almost always near the front; dinitrophenol forms a streak near the start. A suitable quantity for application is $10-20$ µg. The samples are dissolved in benzene.

Nitrated acids are chromatographed in current ammonia containing solvent systems common for the chromatography of acids (n-propanol − ammonia 2 : 1).

Nitro compounds can be detected on chromatograms with p-dimethylaminobenzaldehyde after previous reduction with stannous chloride as described on p. 154. Nitrophenols are detected by simple spraying with 1% alcoholic alkali hydroxide.

2. Nitro Esters

Organic esters of nitric acid give a color reaction with ferrous sulfate and conc. sulfuric acid [this reaction is also given by nitramines (30, 31)] or with phenoldisulfonic acid (32), and also with diphenylamine in sulfuric acid [see also nitramines (33, 34)].

3. Nitrous Acid Esters

Nitrous acid esters give with conc. H_2SO_4 the same reaction as nitro compounds.

Nitrous acid esters react in an acid medium with sulfanilic acid with the formation of corresponding diazonium salts, which can be coupled with N-1-naphthylethylenediamine (35).

Reagents: Sulfanilic acid (5 g) is dissolved in 1 liter of water containing 140 ml of glacial acetic acid and mixed with 20 ml of a 0.1% N-1-naphthylethylenediamine dihydrochloride solution; dilute acetic acid.

Procedure: A few drops of the reagent are added to a few drops of nitrite (ester) solution in dilute acetic acid. A red-violet color appears.

4. Nitroso Compounds

Nitroso compounds are characterized by the presence of a $-\overset{\displaystyle |}{\underset{\displaystyle |}{C}}-N=O$ group, and they are met predominantly in the aromatic series of compounds; in the aliphatic series they occur only in those cases where the carbon atom carrying the nitroso group does not carry a hydrogen atom. If the nitroso group is bound to a nitrogen atom, the compounds are called N-nitroso compounds (nitrosoamines).

Reaction with Sodium Pentacyanoammoferroate

With this reagent, $Na_3[Fe(CN)_5NH_3]$, aromatic nitroso compounds give strongly colored complexes in which the molecule of ammonia is substituted by a molecule of a nitroso compound (36). A positive reaction is also given by hydrazines, thioaldehydes, and thioketones.

Reagent: Sodium nitroprusside is dissolved in a treble amount of conc. ammonia and allowed to stand at 0 °C for 24 hr. Alcohol is then added and the separated crystals are filtered off and washed with ethanol. To carry out the color reaction, a 1% aqueous solution of the product is prepared.

Procedure: To one drop of the tested solution a few drops of freshly prepared reagent solution are added, giving rise to a strongly green, sometimes violet, color.

Color Reactions after Reduction

Similar to nitro compounds, aromatic nitroso compounds can also be reduced to amines, which can then be detected (see p. 357).

Reaction with Concentrated Sulfuric Acid

Aromatic tertiary nitroso compounds and compounds containing an −NO group bound to oxygen or nitrogen liberate nitrous acid under the influence of conc. sulfuric acid. In the presence of phenols Liebermann's reaction takes place (p. 195). Free nitrous acid can also be detected on the basis of its reaction with diphenylamine. The latter reaction is also given by nitroesters.

Detection by Liebermann Reaction

Reagents: conc. sulfuric acid, phenol.

Procedure: A few milligrams of the tested substance are heated with a granule (crystal) of phenol and 1 ml of conc. H_2SO_4 on a boiling water bath. In a positive case a red color appears after a few moments.

Detection with Diphenylamine

Reagents: conc. H_2SO_4, diphenylamine.

Procedure: A sample of the nitroso compound is added to conc. sulfuric acid containing traces of diphenylamine; a dark blue color is formed.

Reaction with Diphenylbenzidine

Nitroso compounds oxidize diphenylbenzidine in conc. H_2SO_4 to diphenylbenzidine blue (37). The reaction is analogous to the preceding one.

Reagent: 10 mg of diphenylbenzidine are dissolved in 10 ml of 85% H_2SO_4.

Procedure: A crystal of the tested substance is added to a solution of diphenylbenzidine in H_2SO_4. A violet-blue color appears.

Identification of C-Nitroso Compounds

A general method for the identification of C-nitroso compounds consists in their reduction (Sn + HCl, catalytic) to corresponding amines, which are then identified by one of the methods described on p. 326. The reaction is carried out as with nitro compounds.

Nitrosodimethylaniline

Reagents: PtO_2, ethanol, conc. HCl, sodium picrate, hydrogen gas from a hydrogen cylinder.

Procedure: 30 mg of *p*-nitrosodimethylaniline are placed in a test tube and 50 mg of PtO_2 are added, followed by 5 ml of ethanol. A hydrogen stream is then introduced into the solution through a capillary tube; the whole operation is carried out in the fume cupboard until the solution is decolorized. This takes approx. 40 min, and after the reduction the solution is filtered through a folded filter paper. Platinum on the filter is washed twice with ethanol (caution, the mixture is self-inflammable!). Concentrated HCl (0.3 ml) is added to the filtrate, and ethanol is distilled off. Excess HCl is eliminated from the residue by heating it over a water bath. Sodium picrate (50 mg) is then added to the residue, followed by just enough boiling water to make it a clear solution. After cooling, the separated picrate is filtered off under suction, yielding 30 mg of product, mp 130 °C. After crystallization from ethanol with addition of water to incipient turbidity and standing in a refrigerator the yield of the separated picrate was 10 mg, mp 130 °C.

$LiAlH_4$ reduces nitroso compounds to azo compounds (38); this can be used for the identification of small amounts of nitroso compounds.

Condensation of aromatic nitroso compounds with arylamines [for identification *p*-bromoaniline is most suitable (39)] leads to azo compounds,

$$ArNO + Ar'NH_2 \rightarrow H_2O + ArN=NAr'$$

Identification of N-Nitroso Compounds

Depending on the reaction conditions and the reagents used, the reduction of N-nitroso compounds gives either substituted hydrazines or amines. However, reduction to hydrazine is not quite unambiguous and generally takes place in small yields. In addition, substituted hydrazines are not described to such an extent that they can be used for identification. Therefore, the reduction to the corresponding amine (which can be converted to a suitable derivative) is more useful. The reduction is carried out with zinc in hydrochloric acid as was described for the reduction of nitro compounds.

5. Azo Compounds

Azo compounds are industrially important derivatives, since they represent a prominent group of dyes. The systematic identification of azo compounds is beyond the scope of this book, and therefore reference should be made to special monographs (40) and original papers.

The basic reaction for identification is the energetic cleavage of the azo group with the formation of two aromatic amines. After cleavage the original coupling component contains an additional amino group in the position of the former azo group. A characteristic accompanying feature of this reaction is the decoloration of the original, generally strongly colored, solution.

$$Ar-N=N-Ar' \xrightarrow[\text{HCl}]{\text{Zn}} Ar-NH_2 + Ar'-NH_2$$

Amines obtained by reduction can be separated either on the basis of their chemical reactions or can be identified by paper or thin-layer chromatography (41, 42, 43).

Azo groups are split either with $SnCl_2$ (44) or dithionite (45). In some cases cleavage by catalytic hydrogenation is recommended (46), because solutions of the liberated bases can be obtained free of inorganic salts. Azo compounds insoluble in water can be cleaved in glacial acetic acid with powdered zinc (43).

As an example of identification, the cleavage of methyl orange and the identification of components will be described.

Cleavage of Methyl Orange

Reagents: stannous chloride, conc. HCl, 20% NaOH, acetic anhydride, ether, 5% $NaHCO_3$.

Procedure: One gram of methyl orange placed in an Erlenmeyer flask is dissolved in a minimum amount of hot water, and 4 g of stannous chloride in 10 ml conc. HCl is added to this solution. The mixture is heated on a water bath until decolorized and is then thoroughly cooled in a refrigerator. The precipitated sulfanilic acid is filtered off under suction. The precipitation of the product can be enhanced by scratching with a glass rod, or the solution may be partly evaporated on a porcelain dish.

Identification is carried out on the basis of color reactions or by chromatographic or spectral methods. The filtrate is additioned with 20% sodium hydroxide until the dissolution of the precipitate is complete. The solution is then extracted three times with ether, and the extract is dried over potassium carbonate and filtered rapidly through a cotton-wool plug. Acetic anhydride (0.5 ml) is added, and after 20 min standing the ether is evaporated. Water (5 ml) is added, followed by 5 ml of 5 % $NaHCO_3$ and charcoal. The mixture is heated and filtered through a folded filter and the filrate is allowed to cool in a refrigerator. The separated acetyl derivative is filtered off under suction. Yield, 100 mg; mp, 131−132 °C.

Paper and Thin-Layer Chromatography

On paper chromatography, water-soluble azo compounds (i.e., those containing a sulfonic group) have a profoundly different behavior from those insoluble in water (lacking both −SO₃H and −COOH groups).

For azo compounds, i.e., azo dyes soluble in water, common solvent systems and untreated paper are used. As the componds have their own strong color, detection is no problem. However, a new problem is met, the substantivity of the dyes, i.e., their affinity toward cellulose. For simple azo dyes a sufficient number of suitable solvent systems can be found, but a number of dyes are only very slowly eluted from the start, so that for their chromatographic separation the solvent must be allowed to overrun, sometimes for several days. We also know that the quality of the paper has an influence on the separation, and in our case the paper Schleicher-Schüll 2045/B gl gave very good results. Among many suitable solvent systems, the following may be mentioned: 5% ammonia with the addition of 2% of sodium citrate; pyridine−water (8 : 2); n-butanol−ethanol−water (1 : 1 : 1); isoamyl alcohol−pyridine−ammonia (1 : 1 : 1); n-propanol−ammonia (2 : 1); n-butanol−formic acid (1 : 1); n-butyl acetate−pyridine−water (3 : 4 : 3); dimethylformamide−water (4 : 1) (or other ratios); and dimethylformamide−ammonia (4 : 1).

For the chromatography of substantive dyes it is much more convenient to apply thin-layer chromatography on silica gel G (47), for example, in the following solvent systems: n-propanol−ammonia (2 : 1), pyridine−n--amyl alcohol−ammonia (1 : 1 : 1), or n-butyl acetate−pyridine−water (2 : 2 : 1).

Water-insoluble azo dyes soluble in fats (Sudan dyes), or disperse dyes can usually be easily chromatographed on papers impregnated with organic solvents. For example, Sudan dyes are well separated in systems with paraffin oil as stationary phase (impregnation) and ethanol-water mixture (8 : 2) or ethanol-ammonia (8 : 2) as the mobile phase; for obtaining rapid information papers impregnated with 50% dimethylformamide and hexane as the moving phase may be used. Dyes can be applied onto the paper in acetone solution in an amount up to 20 μg (48).

For azo compounds utilized as disperse dyes the solvent system l-bromonaphthalene/pyridine-water (1 : 1 or 2 : 1) is suitable. The dyes are applied on the chromatograms after dissolution in pyridine, up to 100 μg per spot. In addition to the above-mentioned, most-suitable solvent systems, the following solvent mixtures may be used (49): lauryl alcohol/ethanol−ammonia (1 : 1); lauryl alcohol/ethanol−N HCl (1 : 1); 1-bromonaphthalene/ 90% acetic acid; or formamide/ hexane (or benzene, or chloroform, or their mixture).

For the separation and identification of azo dyes insoluble in water but soluble in organic solvents thin-layer chromatography on silica gel G can be employed. Hexane, benzene, chloroform, or acetone, or the mixtures thereof, may be used as the mobile phase.

6. Azoxy Compounds

In addition to the detection of the azoxy group in the molecule by the re-arrangement of the substance to hydroxyazobenzenes, the $N = N$ bond can be cleaved, as in azo compounds, by an energetic reduction with Zn or $SnCl_2$ in an acid medium (50).

In an alkaline medium reduction to an azo compound or hydrazo compound can take place

$$\underset{\overset{|}{O}}{ArN=NAr} \xrightarrow[C_2H_5OH]{Zn,\ NaOH} ArN=NAr \xrightarrow[C_2H_5OH]{Zn,\ NaOH} ArNHNHAr$$

In the case of azoxybenzene, sublimation with a threefold amount of iron dust suffices for conversion to azobenzene (51).

7. Hydrazo Compounds

This class of compounds is characterized by the grouping $-NHNH-$ and it represents an important group of intermediates of the production of aromatic diamines. The basic aromatic compound of this class is hydrazobenzene, which, under the influence of acids, undergoes the so-called benzidine rearrangement, which can also be used for identification.

Rearrangement of Hydrazobenzene

Reagents: zinc dust, 3% HCl, 30% KOH.

Procedure: Hydrazobenzene (100 mg), 3% HCl, (5 ml), and Zn (0.2 g) are mixed and shaken in a stoppered (rubber stopper) test tube for 20 min. The mixture is then heated to 50 °C and shaken at this temperature for an additional 10 min. After cooling, 7 ml of 30% KOH are added and the mixture is cooled in a refrigerator. The separated benzidine is filtered off under suction and washed twice with 0.5 ml of water. Yield, 30 mg; mp, 124−125 °C.

However, we have to point out that the reaction is not always unambiguous, and that by-products are formed, both isomers of aromatic *p, p'*-diamines and derivatives of diphenylamine, so-called semidines. In addition, in a number of cases disproportionation can take place instead of the benzidine rearrangement, leading to azo compound and the liberated base. The original literature should be consulted for details of this reaction (52—54).

Aliphatic and aliphatic-aromatic hydrazo compounds of the $RNHNHR_2$ and $ArNHNHR_1$ types also give amines on energetic reduction (R_1NH_2 and R_2NH_2, or $ArNH_2$ and R_1NH_2, respectively). The reduction is carried out with zinc in an acid medium or with sodium amalgam in acetic acid (55, 56). A mild oxidation of aromatic hydrazo compounds leads to azo compounds. The bubbling of a stream of air through a suspension of the hydrazo compound to which ammonia was added is fully sufficient to carry out the oxidation.

For the identification of simple aromatic hydrazo compounds paper chromatography can be used successfully on papers impregnated with dimethylformamide (cyclohexane as the mobile phase) or formamide (cyclohexane or benzene as the mobile phase). Hydrazo compounds can be spotted on the paper in the form of an alcoholic or ethereal solution in concentrations from 0.5 to 200 µg. Detection is carried out by spraying with p-dimethylaminobenzaldehyde (p. 349). Under the influence of the hydrochloric acid present in the reagent, a rearrangement to benzidines or semidines takes place, and they then react with the reagent, with the formation of orange spots (53).

8. Diazo Compounds

The most important color reaction of diazonium salts is their coupling with a suitable passive component, i.e., with a phenol in alkaline medium, or with an aromatic amine in acid medium (see p. 341). Among phenols, resorcinol and 2-naphthol or R-salt are suitable passive components; among amines, 1-naphthylamine or N-1-naphthylethylenediamine are suitable.

In dyes obtained on coupling with resorcinol or R-salt the amine from which the diazonium salt has been prepared can be identified by determining the spectrum in sulfuric acid (57).

For identification by paper chromatography, coupling with 2-naphthol and chromatography along the lines given on p. 342 may be used, or, if the diazonium salt contained a sulfo group, the procedure on p. 365 can be followed. Commercial products of stabilized diazonium salts can be chromatographed directly in n-butanol saturated with 2.5 N HCl; detection is carried out by spraying with an alkaline R-acid solution.

The reactions with HCl, ArOH, or with carboxylic acids may serve for the identification of aliphatic diazo compounds:

$$RCHN_2 + HCl \rightarrow RCH_2Cl + N_2$$
$$RCHN_2 + ArOH \rightarrow RCH_2OAr + N_2$$
$$RCOCHN_2 + R'COOH \rightarrow RCOCH_2OOCR' + N_2$$

The reaction products, unless solid crystalline compounds, are identified by procedures given in the discussion of the identification of alkyl halogenides (p. 136) and phenol ethers (p. 201).

Aromatic diazo compounds are identified by coupling them with 2-naphthol or 1-phenyl-3-methyl-5-pyrazolone in an alkaline medium, or with an aromatic amine in a weakly acid medium (see the identification of amines on p. 341).

9. Nitriles

Detection by Conversion to Thiocyanic Acid

The detection of aromatic and aliphatic cyanides can be carried out by heating the solid sample or a dry residue in a micro test tube in admixture with elemental sulfur. Thiocyanic acid is formed, and this can be detected on the basis of its reaction with ferric salts (58).

Reagents: powdered sulfur, 1% aqueous ferric nitrate solution.

Procedure: To a small amount of the sample or of a dry residue of its evaporated solution in a test tube 10 mg of sulfur are added gradually, and the test tube is placed on an asbestos plate and heated. The mouth of the tube is covered with a filter paper on which a drop of the acidified ferric nitrate solution has been spotted. The escaping HCNS is shown by the appearance of red color on the paper.

Detection by Conversion to Hydrogen Cyanide

A specific test for aliphatic cyanides consists, according to Feigl et al. (58), in the heating of the substance with a mixture of CaO and $CaCO_3$ at 250 °C, during which hydrogen cyanide is liberated. It can be detected by the sensitive reaction with copper (II) acetate and benzidine.

Reagent: This is prepared by mixing equal parts of solutions I and II: Solution I: copper (II) acetate (2.86 g) in 1 liter of water. Solution II: a mixture of 475 ml of a saturated solution (at room temperature) of benzidine acetate and 525 ml of water.

Procedure: A small amount of the tested substance or a dry residue of a drop of the tested solution is mixed in a micro test tube with a mixture of CaO and $CaCO_3$ (1 : 1) and heated in a bath at 250 °C. The mouth of the micro test tube is covered with a piece of filter paper wetted with the reagent. The escaping hydrogen cyanide produces a blue color.

Nitriles also react with fluorescein chloride (p. 324). Another detection scheme for nitriles is based on their pyrolysis with soda lime. The escaping ammonia or alkylamine is detected (59).

Identification of Nitriles

For the identification of nitriles use can be made of their hydrolysis to carboxylic acid or amide:

$$R-C{\equiv}N \;\rightarrow\; RCONH_2 \;\rightarrow\; RCOOH$$

This can be carried out either with acids or with alkalies.

Another method of identification consists in the reduction to amine either with sodium in alcohol, or, preferably, with $LiAlH_4$ in ether.

The addition of thioglycolic acid to a nitrile group in the presence of hydrogen chloride,

$$RCN + HSCH_2COOH \xrightarrow{\;HCl\;} R-\underset{\substack{\| \\ NH \cdot HCl}}{C}-S-CH_2COOH$$

has a limited use (60) because the iminoalkylmercaptoacetic acids formed, though well crystalline and in a high yield, melt with decomposition and the melting points depend on the rate of heating. In addition, the reaction fails with ortho-substituted aromatic nitriles.

Finally, the reaction of nitriles with Grignard reagent was also recommended, leading eventually to ketones which can be easily identified. The method is especially suitable for the identification of low-molecular nitriles; phenylmagnesium bromide (61) is used for the Grignard reaction

$$RCN + ArMgBr \;\rightarrow\; R-\underset{\substack{| \\ Ar}}{C}{=}NHMgBr \xrightarrow[\;H_2O\;]{\;HCl\;} RCOAR + NH_4Cl + MgBrCl$$

Alkaline hydrolysis is carried out by heating the nitrile under reflux with an alcoholic KOH solution (62, 63), sometimes in a sealed tube, or — in order to increase the temperature — in amyl alcohol (64), ethylene glycol, or glycerol (65).

Those nitriles which are hydrolyzable with great difficulty are saponified by melting them with alkali hydroxide (66).

Acid hydrolysis of a nitrile is carried out in conc. HCl, 75% H_2SO_4, or in its mixture with 85% phosphoric acid, at temperatures from 150 to 190 °C. In the case of volatile nitriles the reaction should be carried out in a sealed tube.

For isolation and identification a suitable procedure depends on the properties of the carboxylic acid formed by hydrolysis (see p. 265).

Partial hydrolysis of nitriles to amides is carried out with an acid or with an alkaline solution of hydrogen peroxide if the formed amide has properties suitable for identification. The reaction conditions (temperature, concentration, and duration of hydrolysis) differ in different cases, and the reaction is usually carried out in 95% H_2SO_4 at 25–80 °C.

In hydrolysis with alkaline hydrogen peroxide (67) the nitrile is mixed with 3% hydrogen peroxide and alcohol or acetone (the amount necessary for complete dissolution should be used) and the mixture is alkalized with 6 N NaOH (test with litmus) and heated at 60 °C for 4 hr. The solution should be alkaline throughout the experiment. After hydrolysis the mixture is neutralized with dilute sulfuric acid and the solvent is distilled off. The amide is isolated either by filtration under suction or by extraction with chloroform. The method is not universal and fails with sterically hindered nitriles (68, 69).

Acid Hydrolysis

In a 15-ml flask fitted with a reflux condenser 10 mmole of nitrile are boiled with 4 ml of 85% H_3PO_4 and 2 ml of 95% H_2SO_4 for 10 min. After cooling, the mixture is diluted with 5 ml of water and distilled. Four milliliters of the distillate are collected and the volatile acid in the distillate is identified in the form of its S-1-naphthylmethylthiuronium salt. If a nonvolatile acid is formed during hydrolysis, it is extracted with ether after partial neutralization of the react on mixture with 6 N NaOH.

Alkaline Hydrolysis

Nitrile (2–3 mmole) is mixed in a 15-ml flask with 4 ml of glycerol and 2 g of KOH and the mixture is refluxed for 1 hr. After dilution with 1 ml of water the solution is cooled and extracted with 2 ml of ether. The alkaline solution is neutralized with 6 N HCl, weakly acidified, and extracted three times with 4–5 ml portions of ether. The ethereal extracts containing the acid are worked up in the usual manner. If the acid formed by hydrolysis is volatile with steam, it can be isolated by distillation.

Hydrolysis of Nitriles of Aromatic Acids to Benzamides

Aromatic acid nitrile (1 mmole) is placed on the bottom of a 10-ml test tube (by rotating the tube), and after the addition of five drops of 95% H_2SO_4 the nitrile is stirred with a rodlike thermometer. If the reaction takes place, the temperature rises 40–50 °C and the mixture solidifies to a glassy mass.

The tube is then immersed into a $60-70$ °C bath for $2-3$ min while its contents are continuously stirred. After the addition of 1 ml of water the solution is cooled, 2 ml of 10% Na_2CO_3 are added, and the mixture is thoroughly shaken (to dissolve the formed acid, if any). The amide is filtered off under suction and washed with water.

Hydrolysis of Benzyl Cyanide
(S-1-Naphthylmethylthiuronium Salt of Phenylacetic Acid)

Reagents: sulfuric acid (1 : 1), 5% NaOH, saturated aqueous solution of S-1-naphthylmethylthiuronium chloride, ethanol.

Procedure: In a 10-ml flask provided with a reflux condenser 0.05 ml of benzyl cyanide is mixed with 3 ml of sulfuric acid (1 : 1) and the mixture is refluxed for 2 hr. The cooled solution is diluted with 5 ml of water and the separated acid is filtered off and washed twice with 0.2 ml of water. The crude acid is dissolved in 2 ml of alcohol and neutralized with 5% NaOH (to phenolphthalein). Saturated solution of S-1-naphthylmethylthiuronium chloride (7 ml) is added and the mixture is heated until entirely dissolved. It is then filtered through a folded filter paper and the filtrate is allowed to cool slowly. Yield, 25 mg; mp, $160-161$ °C.

Hydrolysis of Benzyl Cyanide to Amide

Reagents: conc. sulfuric acid, 5% $NaHCO_3$.

Procedure: In a test tube with a conical bottom 50 mg of benzyl cyanide are mixed with two drops of conc. sulfuric acid and the mixture is allowed to stand for 5 min under occasional stirring with a glass rod. The micro test tube is then immersed into a 70 °C bath for 5 min under continuous stirring (the solid phase is triturated to a fine powder). Water is added (0.8 ml), and, after cooling, the amide is filtered off on a filtration tube. The material on the filter is stirred thoroughly with 0.8 ml of a 5% sodium hydrogen carbonate solution and submitted to suction. It is then washed three times with 0.3 ml of water (applying suction) and dried in a stream of dry air. Yield, 40 mg; mp, $135-154$ °C.

Reduction of Nitriles

When sodium in ethanol is used as a reducing agent a minimum of 0.4 ml of nitrile should be worked up. The reaction is carried out in 10 ml of ethanol with the addition of 0.7 g of sodium cut into small pieces. When the reduction is complete ($10-15$ min) the reaction mixture is solidified, the solvent is distilled off, the solution is alkilized again, and the formed amine is steam distilled. The yields of this reaction are low. It is more advantageous to carry out the reduction with $LiAlH_4$ in ether.

Reduction of Acetonitrile with Lithium Aluminum Hydride
(N-Ethylamide of 3,5-Dinitrobenzoic Acid)

Reagents: see pp. 276, 336.

Procedure: Acetonitrile (0.1 ml) is worked up in the same manner as described on p. 276 for the reduction of acetamide. The amine obtained on reduction is distilled off and converted to 3,5-dinitrobenzamide using the procedure described on p. 336. The crude amide is crystallized from 10 ml of cyclohexane and 1 ml of ethanol. Yield, 70 mg; mp, 124 °C.

10. Isonitriles

On hydrolysis with dilute acids isonitriles are converted to amines even in the cold. During the reaction formic acid is formed:

$$RNC + 2\,H_2O \xrightarrow{\ H^{\oplus}\ } RNH_2 + HCOOH$$

Identification of the amine in aqueous media is carried out by procedures described in the discussion on the identification of amines (p. 326).

On reduction with sodium in amyl alcohol isonitriles give secondary amines (70):

$$RNC + 2\,H_2 \rightarrow RNHCH_3$$

which can also be used for identification.

11. Nitramines

Nitramines are obtained from primary or secondary amines by substitution of the nitrogen-bound hydrogen by a $-NO_2$ group.

Several color reactions can be used to detect nitramines: reaction with aromatic amines in acetic acid (71), reaction with ferrous sulfate in sulfuric acid (72, 73), diazotization with nitrous acid in 50% acetic acid, and coupling with a suitable component (74) (for aromatic nitramines):

$$ArNHNO_2 + HNO_2 \rightarrow Ar-N_2^{\oplus} + NO_3^{\ominus} + H_2O$$

The Liebermann reaction can also be used, as in the case of nitroso compounds. This reaction is also given by nitramines—compounds with the groupings

$$\begin{matrix} R\diagdown \\ \diagup \\ R \end{matrix}\!\!N-NO_2 \quad \text{or} \quad \begin{matrix} R\diagdown \\ \diagup \\ R \end{matrix}\!\!C=N-NO_2$$

Nitramines also give a positive test with diphenylamine in conc. sulfuric acid, similar to nitrates, but they can be recognized on the basis of the fact that the reaction does not take place in phosphoric acid (75).

On reduction nitramines give corresponding amines (76), which can be identified by some of the methods described on p. 326.

$$RNHNO_2 \xrightarrow[HCl]{FeCl_2} RNH_2 + NO + H_2O$$

12. Hydrazines

Color Reactions

Hydrazines reduce Fehling's reagent. The course of the reaction depends, however, on the type of hydrazine used: hydrazines of the $RNHNH_2$ and $RNHNHR'$ types react in the cold, those of $ArNHNH_2$, Ar_2NNH_2, and $ArRNNH_2$ react easily in boiling solutions, while hydrazines of the type R_2NNH_2 do so with difficulty even when boiled, and compounds of the ArNHNHAr type (hydrazo compounds) do not reduce at all. For procedure see p. 210.

Hydrazines substituted only on one nitrogen atom with an aliphatic alkyl give a strong isonitrile reaction.

Reagents: alcoholic alkali, alcohol, chloroform.

Procedure: A sample of hydrazine is dissolved in alcohol and mixed with a few ml of alcoholic alkali and several drops of chloroform, and the mixture is gently heated. The odor of isonitrile can be detected very soon.

Asymmetrically substituted hydrazines can be converted to N-nitrosoamines by reaction with nitrous acid. They can be detected by using the Liebermann reaction (p. 195).

Hydrazines with one amino group give a positive reaction with sodium 1,2-naphthoquinone-4-sulfonate (p. 320) as well as with *p*-dimethylaminobenzaldehyde (see p. 322) (78).

Similarly to nitroso compounds, aliphatic and aromatic hydrazines, as well as hydrazides, react with sodium pentacyanoammoferroate (79).

Reagents: 1% aqueous sodium pentacyanoammoferroate (p. 361); 2 N NaOH.

Procedure: A few drops of the reagent are added to an aqueous or alcoholic solution of the sample. After a while a red to violet color is formed which changes to yellow on alkalization if the above-mentioned substances are present in the sample.

Under the influence of selenous acid (selenium dioxide) arylhydrazines are oxidized (80) to diazonium salts, which can be detected by coupling them with 1-naphthylamine (81)

Reagents: hydrochloric acid (1 : 5), SeO_2, solid sodium acetate, a solution of 1-naphthylamine. Naphthylamine (0.3 g) is dissolved in a boiling mixture of 70 ml of water and 30 ml of glacial acetic acid and the solution is filtered and kept in a dark place.

Procedure: To one drop of the tested solution one drop of hydrochloric acid and a crystal of SeO_2 are added. Red selenium separates, which is filtered off after the addition of water and $1-2$ min standing. A few drops of 1-naphthylamine solution are added to the filtrate, followed by a few crystals of sodium acetate. The test is positive if a red color appears which becomes more pronounced on addition of hydrochloric acid.

Identification

Hydrazines, if they have free hydrogens, react with chlorides of organic acids to give hydrazides; they also react in a similar manner with acid anhydrides and with isocyanates (see p. 326).

However, the most suitable method for the identification of hydrazines of the $R_1R_2NNH_2$ type is their condensation with carbonyl compounds. They are usually condensed with benzaldehyde.

General Method of Condensation

(a) Fifty per cent acetic acid is added to hydrazine (1 g) suspended in 10 ml of water with continuous shaking until the hydrazine is completely dissolved. Benzaldehyde (0.5 ml) is then added to the solution, as is alcohol to obtain a homogeneous solution. If hydrazone is not formed on standing, the solution should be heated on a water bath. It is then diluted with $2-3$ ml of water and the separated hydrazone is filtered off under suction and recrystallized (ethanol, benzene).

(b) Hydrazine (0.5 g) is dissolved in $3-5$ ml of acetic acid and mixed with 0.3 ml of benzaldehyde and the solution is set aside. After several hours of standing the hydrazone separates.

If the hydrazone is not separated in a crystalline form, it is recommended to dissolve the oil in pyridine (or a mixture of ethanol and pyridine) and to precipitate the hydrazone by addition of benzene, light petroleum, ether, or, sometimes, water (82).

On condensation with aldehydes hydrazine gives azines (ArCH = NN = CHAr); if the hydrazones prepared by condensation with benzaldehyde are not crystalline, they can be condensed with a higher-molecular carbonyl compound, as, for example, o- or p-nitrobenzaldehyde, p-hydroxybenzaldehyde, etc.

13. Hydrazides

1. Hydrazides reduce an ammoniacal silver nitrate solution (p. 210).

2. They react with a solution of p-dimethylaminobenzaldehyde in an acid medium (p. 322), splitting off an acyl group and giving rise to azines:

$$R-CO-NH-NH_2 + 2\,HOC-\langle\bigcirc\rangle-N(CH_3)_2 \xrightarrow{\text{HCl}}$$

$$RCOOH + (CH_3)_2N-\langle\bigcirc\rangle-CH=N-N=CH-\langle\bigcirc\rangle-N(CH_3)_2$$

The reaction is also given by symetrically substituted diacylhydrazines.

3. Lower members of the series react with ninhydrin (83) (p. 278).

4. Monoacylhydrazines react easily with one molecule of a diazonium salt with the formation of a corresponding diazohydrazide, which gives, in alkaline media with the elimination of water the corresponding 1,2,3,4-tetrazole (84,85),

$$
\begin{array}{ccccccc}
NH_2 & & HN-N=N-Ar & & N=N-NH-Ar & & N=N \\
| & \rightarrow & | & \rightarrow & | & \rightarrow & | \quad\rangle N-Ar \\
NH-OC-R & & HN-OC-R & & N=C-OH & & N=C \\
& & & & \quad| & & \quad| \\
& & & & R & & R
\end{array}
$$

Reaction with diacylhydrazine takes place less easily, because in alkaline medium one acyl group is split off.

Reagents: 0.2% aqueous solution of p-nitrobenzenediazonium-1,5-naphthalenedisulfonate; 5% NaOH.

Procedure: 1 ml of reagent solution is added to an aqueous solution of hydrazine and the mixture is alkalized with sodium hydroxide. A strong red color is obtained.

5. Paper chromatography of hydrazides can be carried out in isoamyl alcohol-collidine-water (10 : 2 : 1) (86), and one of the above-mentioned reactions can be used for detection.

14. Oximes

Color Reactions

Hydroxylamine, oximes, and hydroxamic acids can be oxidized with iodine quantitatively to nitrous acid. If this oxidation takes place in the presence of sulfanilic acid, the latter is diazotized so that it can be detected, after the excess of iodine had been eliminated with thiosulfate, by coupling it with 1-naphthylamine.

Reagents: 0.1 N solution of iodine in acetic acid; sulfanilic acid solution (10 g of sulfanilic acid in 750 ml of water and 250 ml of glacial acetic acid); solution of 1-naphthylamine (3 g of 1-naphthylamine in 700 ml of water and 300 ml of glacial acetic acid); conc. HCl.

Procedure: The tested substance is acidified with three drops of conc. HCl and the solution is heated until 4/5 of its volume is evaporated. Solid sodium acetate (on the tip of a spatula) is added, followed by 1−2 drops of sulfanilic acid solution and one drop of iodine solution in acetic acid. After 2−3 min the excess iodine is eliminated by thiosulfate, and 1-naphthylamine solution is added to the solution. If an oxime is present, a pink to red color appears (87).

Another method of detection of oximes is based on the oxidation with benzoyl peroxide: nitrous acid is liberated, which can be detected by reaction with sulfanilic acid and 1-naphthylamine. The reaction is also given by hydroxamic acids.

Reagents: 10% benzoyl peroxide solution in benzene; a solution prepared by mixing equal amounts of 1% sulfanilic acid solution in 30% acetic acid and 1% 1-naphthylamine solution in 30% acetic acid.

Procedure: The sample is placed into a micro test tube, one drop of benzoyl peroxide is added, and the mixture is evaporated to dryness. A piece of filter paper impregnated with the 1-naphthylamine reagent solution is held near the mouth of the test tube and the mixture is heated in a glycerol bath at 120−130 °C. Within 3−10 min a pink to red color appears on the filter paper.

The reaction is not given by diphenylglyoxime, salicylaldoxime, benzyl-α-monoxime, benzyl-α-dioxime, benzoyl-α-oxime, etc. (88).

Cleavage of Oximes and Hydrazones of Carbonyl Compounds

These are usually solid compounds which are used for the identification of carbonyl compounds. If a carbonyl compound has to be identified in a hydrazone or oxime, the cleavage is carried out by acid hydrolysis, by expelling the carbonyl compound by a more reactive carbonyl compound, and, in the case of oximes, sometimes by oxidative cleavage. Derivatives of ketones undergo cleavage most easily, followed by aromatic and unsaturated aldehydes, while derivatives of aliphatic aldehydes undergo cleavage less easily.

Dilute oxalic acid (89, 90) and phthalic acid (91) were used for the cleavage of hydrazones of α,β-unsaturated aldehydes, while for easily cleaved oximes sulfurous acid was used (92). More stable hydrazones require a mineral acid for their cleavage, often a very concentrated one, as well as a high temperature. The presence of organic solvents may also be necessary (93–95). In those cases where degradation of the formed components can be expected, the cleavage should be carried out in an inert atmosphere (96).

The carbonyl compounds are isolated by separation, extraction, or steam distillation, depending on their nature. They can also be cleaved by expelling the carbonyl compound with a more reactive aldehyde; benzaldehyde is most commonly used, but p-nitrobenzaldehyde, 2,4-dinitrobenzaldehyde, and formaldehyde can be used as well (97–99). The expulsion of 2,4-dinitrobenzaldehyde from its hydrazone can be carried out with glyoxal or diacetyl (100).

The exchange of carbonyl compounds is carried out by boiling the components in water or aqueous ethanol, and the regenerated carbonyl compound is isolated after filtering off the insoluble hydrazone either by the evaporation of the solvent or by extraction, depending on its nature. In those cases where the aldehyde liberated by cleavage is unstable under the conditions of hydrolysis (dioxime of succinic acid aldehyde), the oxime can be cleaved oxidatively in an acid medium. Nitrous acid or amyl nitrite (101–103), ferric chloride (104), or bromine in water (105) can be used as oxidative agents. Of course, this method of cleavage can be applied only for those compounds which are not easily oxidized (106).

15. N-Heterocycles

There are no general, standardized methods for the identification of nitrogen-containing heterocycles.

Liquid pyrrole and its derivatives can be cleaved by reaction with hydroxylamine in alkaline medium, leading to the oxime of succinic acid dialdehyde (107); this method was also used for the study of the structure of pyrrole homologs (108). Reduction of pyrrole and its homologs with metals in an acid medium often yields a mixture of bases [see, for example, (109)]. Catalytic hydrogenation is more advantageous (110).

If nitrogen heterocycles have basic properties, the preparation of picrates, chloroplatinates, or the quaternization of the heterocyclic nitrogen with methyl iodide (see p. 326) is recommended; this is true especially for pyridine-, pyrroline-, and quinoline-type bases. Tetraphenylboranates (111) are also suitable for the identification of nitrogen heterocycles. For their preparation see p. 331. The preparation of quinoline picrate is described on p. 330.

Tetraphenylboranates of many heterocyclic bases as well as 78 alkaloids, antibiotics, etc. have been described (111).

This class of compounds can be identified by spectral methods in the UV and infared regions; after the degradation of the molecule the fragments can also be identified.

In addition, characteristic color reactions are also employed. For example, for the identification of purine and pyrimidine a procedure has been elaborated based on the solubility determination (N NaOH, acetone, chloroform, carbon tetrachloride) and color reactions with HNO_2, $AgNO_3$, $CuSO_4$, $CoSO_4$, $Co(SCN)_2$, Fe^{2+}, picric acid, and $Ba(OH)_2$ of the substances to be identified (112).

For the identification of piperazine derivatives powder X-ray diffraction of their picrates has been proposed (113); for gas chromatography of heterocyclic nitrogen compounds on columns with inorganic salts see (114).

Color Reactions of Pyridine

Most of the color reactions of pyridine are based on the cleavage of the pyridine nucleus by suitable reagents and the detection of the cleavage products with sensitive reagents.

For example, when pyridine reacts in 2,4-dinitrochlorobenzene in an alkaline medium [for review see (115)], leading to a red solution, it is supposed that the corresponding pyridine salt is formed first, which is then cleaved with the formation of 2,4-dinitroanil of glutaconic dialdehyde, the sodium salt of which is red (116).

The cleavage with cyanogen bromide (117) or chloride (KCN plus chloramine T) (118), or with thionyl chloride, has a similar course. In all instances glutaconic dialdehyde is the cleavage product and it is detected by condensation either with an aromatic amine (see p. 215), for example, aniline, 2-naphthylamine (119) or benzidine, with the formation of dianils, or with compounds with an active methylene group, as, for example, barbituric acid or dimedone with the formation of pentamethine dyes (120, 121).

The reaction is often used for the detection and colorimetric determination of pyridine, for example, in air. As pyridine derivatives with a substituted α-position do not react, the reaction can be used for carrying out the detection of pyridine in α-picoline.

Detection by Cleavage with Cyanogen Bromide

Reagents: To 40 ml of a solution containing 4 g of KCN are added 5 ml of acetic acid and 40 ml of a solution containing 3 g of potassium bromate and 4 g of potassium bromide. Concentrated sulfuric acid (4 ml) is then added dropwise and the solution is diluted with water to make its volume 100 ml. 10% sodium acetate solution; acetone; 0.05% solution of benzidine in ethanol containing 2% of acetic acid.

Procedure: 10 ml of the sodium acetate solution are added to a solution of pyridine in water (1 ml), followed by 2 ml of the reagent and 5 ml of acetone. Eventually, 2 ml of benzidine solution are added to the mixture; an orange-red color is formed.

Detection by Clevage with Thionyl Chloride

Reagents: thionyl chloride, 4 N NaOH, 1% 1-naphthylamine in 2 N HCl.

Procedure: One drop of the sample solution is additioned with a few drops of thionyl chloride, and the mixture is heated and evaporated to dryness. One drop of water and one drop of sodium hydroxide are added to the residue, and the solution is then acidified by addition of the 1-naphthalamine solution. If pyridine is present, a red color is formed.

Among other color reactions of pyridine, the one given by Dragendorff reagent is often used; to distinguish pyridine derivatives substituted with alkyls, the reaction with pentacyanoammoferroate and *p*-nitrosomethylaniline (122) may serve, and for the detection of N-oxides the reaction with dimethylaniline and conc. HCl is indicated (123).

Pyridine-2-carboxylic acids give a red color with ferrous salts (124). The reaction is effected with a solution of acid in 50% methanol and with aqueous 0.174 M solution of ferrous ammonium sulfate hexahydrate.

Paper Chromatography of Pyridine
and Its Homologs

Pyridine and its homologs can be well chromatographed and separated in the solvent system n-butanol-N-HCl (4 : 1), or n-propanol-water-HCl (20 : 2 : 1); detection is done with Dragendorff reagent (p. 324). Isomeric alkylpyridines do not separate, however. The solutions for application on paper are prepared by mixing 0.2 ml of pyridine, 1 ml of conc. HCl, and 10 ml of ethanol. Then 2−6 µg of the solution are applied.

Color Reactions of Pyrrole, Indole, and Carbazole

The most common reaction of pyrrole, indole, and their derivatives is that with p-dimethylaminobenzaldehyde (see p. 322), which takes place with derivatives having a free position 1; a red color is produced (125, 126).

A pine splinter dampened with HCl is also colored red by these substances. When pyrrole or indole react with a 10% selenium dioxide solution and conc. nitric acid an intense violet color is formed (127).

With isatin in conc. sulfuric acid pyrroles produce the same blue color as thiophene.

Indole and its derivative with unsubstituted positions 2 or 3 give a violet color when boiled with xanthydrol in acetic acid (128).

Pyrrole, indole, and carbazole give a color reaction with 2-bromo-2-nitroindandione-1,3 in glacial acetic acid (129, 130),

When alkalized, a mixture of indole and sodium nitroprusside produces a violet color which changes to blue after acidification with acetic acid.

Carbazole gives a number of interesting color reactions. Its solution in conc. sulfuric acid is yellow, but the color changes to blue-green on addition of oxidants (nitrite, nitrate). The addition of formaldehyde also produces a blue color or even a precipitate. If a solution of an aromatic aldehyde or of the corresponding benzal chloride in ethanol or acetic acid is added to one drop of a carbazole solution in the same solvent, a red to violet-blue color is produced (131−133). If a drop of benzotrichloride is added to a very dilute solution of carbazole in conc. sulfuric acid and shaken, a brilliant red color is formed, caused by the formation of dicarbazylphenylmethane dye. Similarly, if carbazole is heated with a drop of benzotrichloride and a double amount of anhydrous zinc chloride until completely melted (over a microflame), the same dye is produced, which on dissolution of the melt in alcohol manifests itself by a blue-green color. Reactions in which triphenylmethane dyes are formed take place at positions 3 or 6 of the carbazole nucleus. If these positions are substituted, the reaction does not take place.

An intense green color is also produced with antimony pentachloride in carbon tetrachloride. The reaction with xanthydrol is also very sensitive (134).

Paper and Thin-Layer Chromatography of Indole and Carbazole Derivatives

Indole derivatives can be chromatographed on paper in various solvent systems; we choose them according to the substituents present on the indole nucleus: if they are hydrophilic ($-COOH$), ammoniacal solvent mixtures, as, for example, isopropanol – conc. ammonia – water (8 : 1 : 1) (135), or acid mixtures, such as chloroform – acetic acid (100 : 2) or benzene – propionic acid – water are suitable; for derivatives of more lipophilic character the systems light petroleum – methanol (100 : 4) or formamide/hexane can be employed (136).

For chromatography on thin layers of silica gel G various systems have been proposed. Chloroform with addition of methanol and benzene-ethanol in various proportions is among the simplest. For a review see (137).

Detection is carried out by dipping the chromatogram into a 20% methanolic p-toluenesulfonic acid solution and heating at 60 °C. Colored spots are formed if an amount of 2–100 µg is applied. Other detection reagents, such as, for example, p-dimethylaminobenzaldehyde (p. 349) or diazotized p-nitraniline (p. 192), also give beautiful colored spots.

Carbazole and its derivatives are best chromatographed on paper impregnated with 1-bromonaphthalene with 90% acetic acid. Detection is carried out by dipping the chromatograms in a 10% $SbCl_5$ solution in carbon tetrachloride. Green spots are formed. Picric acid (a saturated solution in benzene) has also been proposed for detection; brick-red spots result (138).

References

1. Dermer, O. C., and Smith, R. B.: J. Am. Chem. Soc. **61**, 748 (1939).
2. Davidson, D.: J. Chem. Educ. **17**, 81 (1940).
3. Bose, P. K.: Analyst **56**, 504 (1931).
4. Turba, F., Haul, R., and Uhlen, G.: Angew. Chem. **61**, 74 (1949).
5. Cohen, I. R., and Altschuler, A. P.: Anal. Chem. **31**, 1638 (1959).
6. Turba, F., Haul, R., and Uhlen, G.: Angew. Chem. **61**, 74 (1949).

7. Feigl, F., and Goldstein, D.: Anal. Chem. **29**, 1522 (1957).
8. Jones, L. R., and Riddick, J. A.: Anal. Chem. **28**, 1493 (1956).
9. Altshuller, A. P., and Cohen, I. R.: Anal. Chem. **32**, 802 (1960).
10. Feigl, F., and Goldstein, D.: Anal. Chem. **29**, 1521 (1957).
11. Meisel, T., and Erdey, L.: Mikrochim. Acta **1966**, 1148.
12. Janovsky, J. V., and Erb, L.: Ber. **19**, 2155 (1886).
13. Janovsky, J. V.: Ber. **24**, 971 (1891).
14. Farmer, R. C.: J. Chem. Soc. **1959**, 3425, 3430, 3433.
15. Abe, T.: Bull. Chem. Soc. Japan **32**, 887, 997 (1959); **33**, 41 (1960); **34**, 21 (1961).
16. Gitis, S. S.: Zhur. obshch. khim. **27**, 1894 (1957).
17. Anger, V.: Mikrochim. Acta **2**, 3 (1937).
18. Smith, G. N.: Anal. Chem. **32**, 32 (1960).
19. Smith, G. N., and Swank, M. G.: Anal. Chem. **32**, 978 (1960).
20. Heotis, J. P., and Cavett, J. W.: Anal. Chem. **31**, 1977 (1959).
21. Bost, R. W., and Nicholson, F.: Ind. Eng. Chem., Anal. Ed. **7**, 190 (1935).
22. Porter, C. C.: Anal. Chem. **27**, 805 (1955).
23. Newlands, M. J., and Wild, F.: J. Chem. Soc. **1956**, 3686.
24. Aketsuda, M.: Yakugaku Zasshi **80**, 389 (1960); C. A. **54**, 18 408 (1960).
25. Hackett, C. B., and Clark, R. M.: Analyst **85**, 683 (1960).
26. Cheronis, N. D., and Levin, N.: J. Chem. Educ. **21**, 603 (1944).
27. Franc, J.: Chem. Listy **49**, 872 (1955).
28. Franc, J., and Knížek, J.: Collection Czech. Chem. Commun. **24**, 2299 (1959).
29. Gasparič, J.: J. Chromatog. **13**, 459 (1964); **15**, 83 (1964).
30. Bandelin, F. J., and Pankratz, R. E.: Anal. Chem. **30**, 1435 (1958),
31. Laccetti, M. A., Semel, S., and Roth, M.: Anal. Chem. **31**, 1049 (1958).
32. Gardon, J. L., and Leopold, B.: Anal. Chem. **30**, 2057 (1958).
33. Whitman, C. L., and Fauth, M. I.: Anal. Chem. **30**, 1672 (1958).
34. Finnie, T. M., and Yallop, H. J.: Analyst **82**, 653 (1957).
35. Altshuller, A. P., and Schwab, C. A.: Anal. Chem. **31**, 314 (1959).
36. Feigl, F., Anger, V., and Frehden, O.: Mikrochemie **15**, 181 (1934).
37. Anger, V.: Mikrochim. Acta **1960**, 58.
38. Rudinger, J. and Ferles, M.: Hydrid lithnohlinitý (Lithium Aluminum Hydride). Nakladatelství ČSAV, Prague, 1956.
39. Levy, W. J., and Campbell, N.: J. Chem. Soc. **1939**, 1442.
40. Brunner, A.: Analyse der Azofarbstoffe. Springer, Berlin, 1929.
41. Panchártek, J., Allan, Z. J., and Mužík, F.: Collection Czech. Chem. Commun. **25**, 2783 (1960).
42. Cee, A., and Gasparič, J.: Mikrochim. Acta **1966**, 295.
43. Gemzová, I., and Gasparič, J.: Collection Czech. Chem. Commun. **32**, 2740 (1967).
44. Jacobson, P., and Hönigsberger, F.: Ber. **33**, 4069, 4093 (1903).
45. Grandmougin, E.: Ber. **39**, 2494 (1906).
46. Rosenmund, K. W., and Pfankuch, E.: Ber. **56**, 2258 (1923).
47. Gasparič, J., and Cee, A.: J. Chromatog. **14**, 484 (1964).
48. Gasparič, J., and Matrka, M.: Collection Czech. Chem. Commun. **25**, 1969 (1960).
49. Gasparič, J., and Gemzová-Táborská, I.: Collection Czech. Chem. Commun. **27**, 2996 (1962).
50. Müller, E.: Die Azoxyverbindungen. Enke, Stuttgart, 1936.

51. Schmidt, H., and Schultz, G.: Ann. **207**, 320 (1881).
52. Večeřa, M.: Chem. Listy **52**, 1373 (1958).
53. Večeřa, M., Petránek, J., and Gasparič, J.: Chem. Listy **21**, 1553 (1957). Collection Czech. Chem. Commun. **23**, 333 (1958).
54. Večeřa, M., and Petránek, J.: Chem. Listy **48**, 1351 (1954).
55. Fischer, E.: Ann. **199**, 325 (1879).
56. Schlenk, O.: J. prakt. Chem. **(2) 78**, 52 (1908).
57. Allan, Z. J., and Mužík, F.: Chem. Listy **47**, 380 (1953); Collection Czech. Chem. Commun. **18**, 663 (1953).
58. Feigl, F., Gentil, V., and Jungreis, E.: Mikrochim. Acta **1959**, 47.
59. Trofimenko, S., and Sease, J. W.: Anal. Chem. **30**, 1432 (1958).
60. Condo, F. E., Hinkel, E. T., Fassero, A., and Shriner, R. L.: J. Am. Chem. Soc. **59**, 230 (1937).
61. Shriner, R. L., and Turner, T. A.: J. Am. Chem. Soc. **52**, 1267 (1930).
62. Frankland, L., and Kolbe, H.: Ann. **65**, 288 (1848).
63. Shriner, R. L., Fulton, J. M., and Burks, D., Jr.: J. Am. Chem. Soc. **55**, 1496 (1933).
64. Cleve, P. T.: Ber. **25**, 2475 (1892).
65. Campbell, K. N., et al.: J. Am. Chem. Soc. **68**, 1844 (1946).
66. Friedländer, P., and Littner, S.: Ber. **48**, 328 (1915).
67. Radziszewski, B.: Ber. **18**, 355 (1885).
68. Deinert, J.: J. prakt. Chem. **(2) 52**, 431 (1895).
69. Kaufmann, A., and Albertini, A.: Ber. **44**, 2052 (1911).
70. Nef, J. U.: Ann. **270**, 267 (1892).
71. Franchimont, A. P. N.: Rec. Trav. Chim. **16**, 227 (1897); C. **1897**, II, 477.
72. Thiele, J., and Lachman, A.: Ann. **288**, 267 (1895).
73. Laccetti, M. A., Semel, S., and Roth, M.: Anal. Chem. **31**, 1049 (1959).
74. Bamberger, E.: Ber. **30**, 1248 (1897).
75. Whitman, C. L., and Fauth, M. I.: Anal. Chem. **30**, 1672 (1958).
76. Lehmstedt, K., and Zumstein, O.: Ber. **58**, 2024 (1925).
77. Lehmstedt, K.: Ber. **60**, 1910 (1927).
78. McKennis, H., and Yard, A. S.: Anal. Chem. **26**, 1960 (1954).
79. Feigl, F., Anger, V., and Frehden, O.: Mikrochemie **15**, 181 (1934).
80. Postowsky, J. J., Lugowkin, B. P., and Mandryk, G. Th.: Ber. **69**, 1913 (1936).
81. Feigl, F., and Demant, V.: Mikrochim. Acta **1**, 322 (1937).
82. Neuberg, C.: Ber. **32**, 3384 (1899).
83. Hinman, R. L.: Anal. Chim. Acta **15**, 125 (1956).
84. Dimroth, O., and de Montmollin, G.: Ber. **43**, 2904 (1910).
85. Cee, A.: J. Chromatog. **8**, 421 (1962).
86. Satake, K., and Seki, T.: Kagaku no Ryoiki **4**, 557 (1950); C. A. **45**, 4604 (1951).
87. Feigl, F., and Demant, V.: Mikrochim. Acta. **1**, 322 (1937).
88. Feigl, F., and Silva, E.: Analyst **82**, 582 (1957).
89. Kiliani, H.: Ber. **56**, 2016 (1923).
90. Wallach, O.: Ann. **347**, 316 (1906).
91. Tiemann, F.: Ber. **33**, 3713 (1900).
92. Gluud, W.: Ber. **48**, 420 (1915).
93. Fischer, E., and Hirschberger, J.: Ber. **22**, 365, 3218 (1889).
94. Auwers, V. K.: Ann. **439**, 132 (1924).
95. Liebermann, C., and Lindenbaum, S.: Ber. **40**, 3570 (1907).

96. Večeřa, M.: Chem. Listy **45**, 475 (1951).
97. Zemplén, G.: Ber. **59**, 2408 (1926); **60**, 1309 (1927).
98. Fischer, E.: Ber. **35**, 3144 (1902).
99. Sachs, F., and Kempf, R.: Ber. **35**, 1224 (1902).
100. Kaufmann, H.: Ber. **35**, 473 (1902).
101. Mamrich, C.: Arch. Pharm. **270**, 283 (1932).
102. Cope, A. C.: J. Am. Chem. Soc. **73**, 3416 (1951).
103. Henle, F., and Schupp, G.: Ber. **38**, 1372 (1905).
104. Gabriel, S.: Ber. **15**, 2004 (1882); **14**, 823, 2332 (1881).
105. Piloty, O.: Ber. **30**, 3161 (1897).
106. Goldschmidt, S., and Veer, W. L. C.: Rec. Trav. Chim. **65**, 796 (1946).
107. Willstätter, R., and Hübner, W.: Ber. **40**, 3869 (1907).
108. Fischer, H., and Zimmermann, W.: Zeit. physiol. Chem. **89**, 163 (1914).
109. Večeřa, M.: Dissertation, Prague, 1951 (On the Isomerism in Pyrroline Series).
110. Weygand, C.: Organisch-chemische Experimentierkunst, p. 170, 2nd Ed., Barth, Leipzig, 1948.
111. Barnard, A. J., Jr., and Wendlandt, W. W.: Revista de la Sociedad Quimica de Mexico **3**, 269 (1959).
112. Weiss, R.: Mikrochim. Acta **1961**, 11.
113. Quentin, R. J.: Appl. Spectroscopy **18**, 25 (1964).
114. Hanneman, W. W.: J. Gas Chromatography **1**, 18 (1963).
115. Freytag, H.: Zeit. anal. Chem. **152**, 86 (1956).
116. Wanag, G.: Zeit. anal. Chem. **126**, 21 (1943).
117. König, W.: J. prakt. Chem. **69**, 105 (1904); **70**, 19 (1904); **83**, 406 (1911).
118. Nielsch, W., and Giefer, L.: Zeit. anal. Chem. **171**, 401 (1959).
119. Barta, L.: Biochem. Z. **277**, 412 (1935).
120. Asmus, E., and Garschagen, H.: Zeit. anal. Chem. **139**, 81 (1953).
121. Asmus, E., and Kurandt, H. F.: Zeit. anal. Chem. **149**, 3 (1956).
122. Biddiscombe, D. P., and Herington, E. F. G.: Analyst **81**, 711 (1956).
123. Coats, N. A., and Katritzki, A. R.: J. Org. Chem. **24**, 1836 (1959).
124. Acheson, R. M., and Taylor, G. A.: J. Chem. Soc. **1959**, 4140.
125. Chernoff, L. H.: Ind. Eng. Chem., Anal. Ed. **12**, 273 (1940).
126. Muhs, M. A., and Weiss, F. T.: Anal. Chem. **30**, 259 (1958).
127. Montignie, E.: Bull. soc. chim. France **51**, 689 (1932).
128. Fearon, W. R.: Analyst **69**, 122 (1944).
129. Vanag, G.: Zhur. Anal. Khim. **9**, 217 (1954).
130. Vanag, G. Ya., and Mackanova, M. A.: Zhur. Anal. Chim. **13**, 485 (1958).
131. Sawicki, E.: Chemist-Analyst **46**, 67 (1957).
132. Sawicki, E., Stanley, T. W., and Hauser, T. R.: Chemist-Analyst **47**, 69 (1958).
133. Strafford, N., and Stubbings, W. V.: Rec. trav. chim. **67**, 918 (1947).
134. Gilbert, G., Stickel, R. M., and Morgan, H. H., Jr.: Anal. Chem. **31**, 1981 (1959).
135. Leemann, H. G., and Weller, H.: Helv. Chim. Acta **43**, 1359 (1960).
136. Procházka, Ž., Šanda, V., and Macek, K.: Collection Czech. Chem. Commun. **24**, 2928 (1959).
137. Stahl, E. (Editor): Dünnschicht-Chromatographie. Springer, Berlin-Heidelberg-New York, 1967.
138. Chakrabarty, D. P., Das, K. C., and Das, B. P.: Indian J. Chem. **4**, 416 (1966).

CHAPTER XVII

SULFUR COMPOUNDS

1. Mercaptans and Thiophenols

These compounds are characterized by a very unpleasant odor if they have a sufficiently high vapor pressure. Among color reactions used for their detection, those with sodium nitroprusside, phosphotungstic acid, and nitrous acid are most commonly used. The preparation of heavy-metal salts is important both for detection and identification; also used for identification are the reactions with 2,4-dinitrochlorobenzene, m-nitrobenzamide, or (for chromatography) 3,5-dinitrobenzoyl chloride and 2,4-dinitrophenylsulfenyl chloride (see (1)]. On mild oxidation mercaptans give disulfides — for example, with iodine,

$$2\,RSH \xrightarrow{I_2} RS-SR + 2\,HI$$

The disulfides formed can be detected or identified by gas chromatography (2).

With strong oxidants (potassium permanganate, nitric acid) sulfonic acids are formed,

$$RSH \xrightarrow{O} RSO_3H$$

Reactions with Nitroprusside
or with Grote's Reagent

Mercaptans give a red color with nitroprusside in an alkaline medium; Grote's reagent is generally used for the reaction (3).

According to Havíř et al. (4), fluorescein-1,3,6,8-tetramercuri-tetraacetate is an even more sensitive reagent than Grote's reagent.

Reagents: Sodium nitroprusside (0.5 g) is dissolved in 10 ml of water and mixed with 0.5 g of hydroxylamine hydrochloride and 1 g of sodium hydrogen carbonate. As soon as the evolution of gas subsides two drops of bromine are added. The excess bromine is eliminated by a stream of air and the solution obtained is filtered and made up to a final volume of 25 ml; $NaHCO_3$.

Procedure: A small sample of the tested substance is dissolved in 2–3 ml of water and the solution is additioned with excess $NaHCO_3$ and 0.5 ml of the reagent. In the presence of a mercapto compound a purple color is produced either immediately or after 10 min. In the presence of thioketones the color is blue.

Reaction with Phosphotungstic Acid

Mercapto compounds reduce a solution of phosphotungstic acid with the formation of a blue color (5, 6).

Reagents: 100 g of sodium tungstate are refluxed with 33 ml of 85% H_3PO_4 and 150 ml of water for 1 hr. A drop of bromine is then added and the excess is expelled by boiling. After cooling, the solution is made up to 500 ml. Acetate buffer of pH 5.2.

Procedure: To an aqueous solution (1 ml) of the sample are added 10 ml of the buffer and 4 ml of the reagent. In the presence of mercapto compounds a blue color is produced.

Reaction with Iodine and Sodium Acetate

The reaction between sodium azide and iodine takes place very slowly with the formation of iodide ions, but it can be accelerated by the presence of trace amounts of organic sulfur compounds, especially mercaptans or sulfides.

Reagents: 3% aqueous sodium azide; 0.1 N iodine.

Procedure: One drop of the tested solution or a few milligrams of the solid is dropped on a watch glass and a drop of azide solution and a drop of iodine solution are added. In the presence of both organic mercaptans and sulfides the evolution of nitrogen is observed as well as the decolorization of the iodine solution.

Identification of Mercaptans and Thiophenols with 2,4-Dinitrochlorobenzene

2,4-Dinitrochlorobenzene reacts with sodium salts of mercaptans and thiophenols with the formation of well-crystallized sulfides which can be further oxidized to corresponding sulfones, thus enabling a second series of derivatives to be obtained (7, 8):

$$RSNa + ClC_6H_3(NO_2)_2 \rightarrow RSC_6H_3(NO_2)_2 + NaCl$$

$$RSC_6H_3(NO_2)_2 \xrightarrow{\text{oxid.}} (NO_2)_2C_6H_3SO_2R$$

General Method of Preparation of Alkyl (Aryl) 2,4-Dinitrophenyl Sulfides

Mercaptan (0.01 mole) is dissolved in 30 ml of ethanol and mixed with 1 ml of 10 N NaOH and 2 g of 2,4-dinitrochlorobenzene in 10 ml of ethanol. The reaction usually takes place spontaneously and should be completed by heating the mixture on a water bath. The reaction mixture is filtered while hot, and the sulfide crystallizes out from the filtrate on cooling.

Benzyl 2,4-Dinitrophenyl Sulfide
(Identification of Benzyl Mercaptan)

Reagents: 10% ethanolic 2,4-dinitrochlorobenzene solution, N NaOH, methanol.

Procedure: In a 10-ml flask fitted with a reflux condenser are placed 0.1 ml of benzyl mercaptan, 2 ml of a 10% methanolic 2,4-dinitrochlorobenzene solution, 1 ml of N NaOH, and methanol until dissolution is complete. The solution is boiled for 20 min and then filtered, while hot, through a small folded filter. The filtrate is cooled in a refrigerator and the separated derivative is filtered off with suction. Yield, 78 mg; mp 129 to 130 °C.

Phenyl 2,4-Dinitrophenyl Sulfide
(Identification of Thiophenol)

Reagents: methanol, N NaOH, 10% methanolic 2,4-dinitrochlorobenzene, ethanol.

Procedure: In an ampoule of 4 ml volume, ready for sealing, 10 mg of thiophenol are mixed with 0.5 ml of methanol, 0.12 ml of N NaOH, and 1.5 ml of 10% methanolic 2,4-dinitrochlorobenzene solution. After sealing, the ampoule is immersed in a boiling water bath for 20 min. After opening the ampoule, its contents are filtered through a cotton-wool plug (in a funnel) and the filtrate is allowed to cool in a refrigerator. If necessary, a few drops of water can be added, and the mixture is scratched with a glass rod. The precipitate is filtered off on a filtration tube and washed with 1 ml of methanol. Yield, 30 mg; mp, 118—119 °C. After crystallization from 2 ml of ethanol the product weighed 18 mg, mp 121 °C.

Paper and Thin-Layer Chromatography

For chromatographic purposes mercaptans can be converted to 3,5-dinitrobenzoyl thioesters by the usual procedure, described in the chapter on alcohols. Chromatography is also carried out as in the case of 3,5-dinitrobenzoates of alcohols. On chromatograms mercaptans behave as alcohols with an additional carbon atom, i.e., an ethyl mercaptan derivative behaves as a derivative of n-propanol. For details see p. 154. Thin-layer chromatography can be used in a similar manner. For paper chromatographic identification alkyl 2,4-dinitrophenyl sulfide have also been used (9).

2. Sulfides

On oxidation with potassium permanganate or hydrogen peroxide, sulfides yield corresponding sulfones, some of which have properties suitable for identification (especially aromatic ones). Sulfones have also been identified by mass spectrometry (10). Another identification method makes use of the reaction of sulfides with alkyl halogenides (the formation of sulfonium salts):

$$R_2S + R'X \rightarrow R_2R'S^{\oplus}X^{\ominus}$$

p-Bromophenacyl bromide is employed as an alkylation agent (11, 12). The bromide formed can be converted to other, less-soluble salts, such as, for example, perchlorates or picrates.

However, sulfides with branched chains or with the alkyl group either do not form p-bromophenacylsulfonium salts at all, or do so only in low yields (12, 13).

Addition compounds of sulfides with mercuric chloride (14, 15) can also be used for identification. They are prepared by mixing the sulfide with a small molar excess of mercuric chloride in ethanol. Alkyl phenyl sulfides do not give these addition compounds (15).

A very suitable method of identification of sulfides consists in their conversion to sulfilimines on reaction with the sodium salt of sulfonylchloramine,

$$R_2S + ArSO_2NClNa \rightarrow R_2SNSO_2Ar + NaCl$$

Chloramine T and chloramine N (17) are most convenient for identification purposes.

General Method of Preparation of Sulfilimines

To prepare these derivatives 1.5 mmole of chloramine in 5 ml of methanol is mixed with 1.5 mmole of sulfide in 5 ml of methanol, and the mixture is allowed to react at room temperature for 15 min. The solvent is then evaporated and the crude derivative is washed with 5 ml of 2 N NaOH and water, and is then crystallized from cyclohexane, benzene, chloroform, or their mixture.

When alkyl benzyl, alkyl aryl, and diaryl sulfides are to be identified the reaction time should be prolonged to 12 hr; alternatively the reaction can be catalyzed by the addition of 1 ml of 10% HCOOH in methanol. S,S-Dialkyl-N-*p*-nitrobenzenesulfonylsulfilimines of 34 sulfides have been prepared (17).

On reaction with diallyl sulfide and other sulfides containing a reactive allyl group, rearrangement of the sulfilimine originally formed takes place to N-allylthio-N-allyl-*p*-nitrobenzenesulfonamide (18):

$$
\begin{array}{ccc}
CH_2{=}CH{-}CH_2 & & CH_2{=}CH{-}CH_2{-}S \\
& \diagup S^{\oplus}{-}N^{\ominus}SO_2C_6H_4NO_2 \;\rightarrow & \diagup NSO_2{-}C_6H_4NO_2 \\
CH_2{=}CH{-}CH_2 & & CH{=}CH_2{-}CH_2
\end{array}
$$

The main advantage of *p*-nitrobenzenesulfonylsulfilimines as derivatives suitable for identification consists in the possibility of also using them for the identification of a mixture of sulfides by paper or column chromatography.

Dibenzyl Sulfone

(Identification of Dibenzyl Sulfide)

Reagents: 10% H_2O_2, 50% aqueous alcohol, glacial acetic acid.

Procedure: Dibenzyl sulfide (50 mg) is dissolved in 1 ml of glacial acetic acid and mixed with 0.2 ml of 10% hydrogen peroxide. The mixture is heated and boiled gently for 3 min. The hot solution is diluted with water until it becomes turbid, the turbidity is dissolved by heating, and eventually the solution is allowed to cool slowly. The precipitated derivative is filtered off under suction and washed twice with 0.5 ml of water. Yield 48 mg, mp 134 °C, which drops after crystallization from 2 ml of 50% ethanol to 27 mg, mp 134 °C.

Benzyl 2,4-Dinitrophenyl Sulfone

Reagents: glacial acetic acid, 30% H_2O_2, methanol.

Procedure: In a test tube containing a solution of 20 mg of benzyl 2,4-dinitrophenyl sulfide in 2 ml of acetic acid, hydrogen peroxide (0.3 ml) is added and the mixture is boiled gently for 2 min. After partial cooling an additional 0.3 ml of hydrogen peroxide is added and the mixture is boiled for a further 2 min. Water is then added until a colloidal precipitate is formed, and the mixture is allowed to stand in a refrigerator for 30 min. The separated sulfone is filtered off and washed twice with 1 ml of water. Yield, 23 mg; mp, 168 – 170 °C. Crystallization from 2 ml of methanol yielded 14 mg of the product, mp 177 – 178 °C. A second crystallization from 1.5 ml of methanol left 8 mg of product, mp 181 – 182 °C.

Addition Compound of Dibutyl Sulfide and Mercuric Chloride

Reagents: mercuric chloride, ethanol.

Procedure: 8 g of mercuric chloride are dissolved in 30 ml of ethanol in an Erlenmeyer flask, and 1 ml of dibutyl sulfide is added. After 30 min standing in a refrigerator the formed addition compound is filtered off on a filtration crucible and dried. Yield, 4.2 g; mp, 106–107 °C. On crystallization from 11 ml of methanol the yield diminished to 3.4 g, mp 109–110 °C, and after an additional crystallization from 8 ml of ethanol the product weighed 2.8 g, mp 110 °C.

S,S-Dimethyl-N-*p*-tolylsulfonylsulfilimine
(Identification of Dimethyl Sulfide)

Reagents: chloramine T, methanol, 2 N NaOH, toluene, cyclohexane.

Procedure: Chloramine T (0.12 g) dissolved in 10 ml of methanol is mixed with a solution of 0.03 ml of dimethyl sulfide in 10 ml of methanol. The solvent is then distilled off on a water bath and the oily residue is stirred thoroughly with 5 ml of 2 N NaOH. The alkaline solution is then carefully poured out and the crude product remaining in the beaker is washed twice with 10 ml of water. The residue is then dissolved in 10 ml of toluene by heating it over a water bath, the solution is filtered through a folded filter paper, and cyclohexane is added to the boiling solution to incipient turbidity. The separated crystalline derivative is filtered off under suction after 30 min of standing. Yield, 45 mg; mp, 159 °C.

S,S-Dibenzyl-N-*p*-nitrobenzenesulfonylsulfilimine
(Identification of Dibenzyl Sulfide)

Reagents: chloramine N, methanol, 2 N NaOH, chloroform, cyclohexane.

Procedure: Chloramine N (200 mg) is dissolved in 5 ml of methanol and additioned with 150 mg of dibenzyl sulfide. After 15 min standing methanol is distilled off and the residue is washed by stirring it first with 5 ml of 2 N NaOH and then with 20 ml of water. The product is crystallized from a chloroform-cyclohexane mixture. Yield, 200 mg; mp, 232–233 °C.

Paper Chromatography of p-Nitrobenzenesulfonylsulfilimines

The separation of these derivatives by means of paper chromatography is carried out best on papers impregnated with formamide. The mobile phase for sulfides with C_1-C_4 alkyls is benzene or a mixture of benzene and cyclohexane (3 : 2); the latter mixture is also suitable for benzyl and arylalkyl sulfides with an alkyl group up to C_4.

Preparation of derivatives for chromatography: To a solution of 0.5 mmole of sulfide in 2 ml of methanol a solution of 200 mg of sodium p-nitrobenzenesulfochloroamide in 5 ml of methanol is added and the reaction mixture is suitably acidified with one drop of 1% formic acid in methanol (to make the reaction smoother). The reaction is accompanied by decolorization of the yellow reaction mixture. After 15 min the reaction mixture is diluted with 20 ml of water and additioned with 10 ml of 2 N NaOH. It is then extracted twice with 5 ml of chloroform. The chloroform solution is applied directly onto the paper, a suitable amount being 5 to 15 μg. Detection is carried out by spraying the chromatogram with a solution of p-dimethylaminobenzaldehyde after previous reduction with stannous chloride (p. 154). Yellow spots on a white background are produced.

Separation of Sulfides by Gas Chromatography

For the separation of C_1-C_4 dialkyl sulfides and alkyl benzyl sulfides the following stationary phases have been employed (20): silicone oil; 7,8-benzoquinoline; 5,6-benzoquinoline; 2,4-dimethyl-7,8-benzoquinoline; and phenanthrene; Apiezon M and polymethylphenylsiloxane have also been used (21).

Sulfides are separated on silicone oil according to their boiling points; the use of 5,6- and 7,8-benzoquinolines enables the separation of sulfides with branched chains from those with straight chains.

3. Disulfides

Detection after Conversion to Mercaptans

Disulfides can be easily reduced to corresponding mercaptans—for example, with alkali cyanide or with zinc and sulfuric acid in acetic acid, or also with lithium aluminum hydride—and the mercaptans can then be easily detected or identified by known procedures (p. 386).

Detection with Grote's Reagent

Reagents: as on p. 385; aqueous KCN solution; $NaHCO_3$.

Procedure: Excess $NaHCO_3$ is added to a solution of the tested substance in water, followed by 0.5 ml of the reagent and 3 ml of KCN solution; the appearance of a red color represents a positive reaction.

Cleavage of Dibenzyl Disulfide with Lithium Aluminum Hydride

Reagent: LiAlH$_4$ (0.5 g) is introduced quickly into a 100-ml Erlenmeyer flask containing 20 ml of dry ether (dried over sodium) and the flask is immediately closed with a cork provided with a protecting (drying) tube filled with layers of calcium chloride, solid NaOH, and again CaCl$_2$. The mixture is heated to incipient boiling and then stirred for 2−3 min. The operation is repeated several times. Eventually, after 15 min heating the solution is allowed to stand overnight.

Procedure: 150 mg of dibenzyl disulfide in 2 ml of ether is placed in a ground-joint, 20-ml flask provided with a dry reflux condenser and a protecting drying tube, and the mixture is cooled with ice and salt. When thoroughly cooled, 5 ml of LiAlH$_4$ solution are added at once to the solution through the reflux condenser, and the mixture is stirred for 2−3 min, keeping the flask in the coolant bath. After 5 min the coolant bath is substituted by a mixture of ice and water (0 °C), and the flask is kept in for an additional 5 min. Eventually, the mixture is allowed to stand at room temperature under the hood. Water is then added dropwise until the evolution of hydrogen ceases. Sulfuric acid (3 ml of 10% solution) is then added, followed by 5 ml of ether, and the mixture is transferred to a separatory funnel, where it is shaken and separated. The ethereal layer is washed with 2 ml of water and then mixed with 2 ml of N NaOH, 20 ml of methanol, and 0.5 g of 2,4-dinitrochlorobenzene. The mixture is placed in a flask and refluxed for 1 hr. Ether is then distilled off and the separated product is filtered off. Yield, 0.34 g; mp, 84−90 °C (unsharp). Crystallization from 20 ml of ethanol gives 0.16 g, mp 120−121 °C. The sulfide is converted oxidatively to sulfone (see p. 389).

Paper Chromatography of Disulfides (22)

Dibenzyl sulfide, disulfide, trisulfide, and tetrasulfide can be separated chromatographically on paper impregnated with 1-bromonaphthalene; 80% acetic acid can serve as the mobile phase which is allowed to overrun overnight. Three to five microliters of a 5−10% solution are applied on the paper. Detection is carried out with potassium iodoplatinate, giving light yellow spots of dibenzyl sulfide and yellow to brown-yellow spots of disulfides and polysulfides on a pink-red background.

4. Sulfonic Acids and Their Chlorides

Aromatic and aliphatic sulfonic acids are generally easily soluble in water. They are also hygroscopic and can be identified unchanged only with difficulty. In addition, their melting points are not sharp. Sulfonic acids containing other functional groups in the molecule that substantially influence their properties (for example, amino groups) are an exception.

To prove the identity of sulfonic acids, fusion with sodium hydroxide and the qualitative analysis of the products, or the reaction with thionyl chloride (see p. 394), is usually applied.

Detection of the Sulfonic Group by Fusion with Sodium Hydroxide

When fused with sodium hydroxide aromatic sulfonic acids are converted to corresponding phenols with the simultaneous formation of sodium sulfite. Sulfinic acids and sulfones react in a similar manner. The presence of these groups in the original sample can be detected on the basis of the evolution of sulfur dioxide after the acidification of the melt [this reaction is also given by aliphatic sulfonic acids (23)], or by detecting phenols in the melt, using some of the known reactions. The phenol may also be identified paper chromatographically (24).

$$C_6H_5SO_3H + 2 NaOH = C_6H_5OH + Na_2SO_3 + H_2O$$

Test Based on the Detection of SO_2

Reagents: solid NaOH; conc. HCl; $Ni(OH)_2$ freshly precipitated from a solution of $NiCl_2$ with alkali and washed until neutral; a solution of 0.05 g of benzidine or of its hydrochloride in 10 ml of 2 N acetic acid, made up to 100 ml with water.

Procedure: In a narrow test tube blown out to a small bubble at its end is placed a small amount of the tested substance and a pellet of NaOH. This is then heated over a small flame until melted. After cooling, two drops of HCl are added (control with litmus paper), and the sides of the tube near its mouth are carefully rinsed and cleaned. Filtration paper with the $Ni(OH)_2$ precipitate is then held to the mouth of the tube, which is heated in a hot water bath. The escaping SO_2 changes the light green precipitate to black. If the color is weak, the precipitate is dampened with drops of benzidine solution, which makes it turn blue.

Detection Based on the Identification of Phenol

Reagents: solid NaOH; 50% H_2SO_4; ether; solution of a diazonium salt.

Procedure: A sample of the tested substance is melted with a threefold amount of NaOH and a drop of water as in the preceding case. After cooling, the melt is dissolved in a few milliliters of hot water and the solution is transferred into a test tube and acidified with sulfuric acid. The liberated phenol is extracted with a few milliliters of ether and detected by coupling it with a diazonium salt (see p. 192).

Detection by Paper-Chromatographic Identification of Phenol

Reagents: solid KOH; benzene; conc. HCl.

Procedure: The tested substance (0.2 g) and 2.5 g of KOH are heated in a glass tube at 360−370 °C for 3−5 min. After cooling, 5 ml of water are added, and the mixture is stirred with a glass rod and extracted with an equal part of benzene. The aqueous phase is acidified with hydrochloric acid (test with Congo red) and extracted with 5 ml of hexane. The hexane solution is applied directly on the chromatogram, which is then run in the formamide/hexane or formamide/hexane-benzene (3 : 2) system (p. 198). Detection is carried out by spraying with a diazonium salt.

Conversion of Sulfonic Acids to Acetohydroxamic Acids (25)

Sulfonic acids can be converted on reaction with thionyl chloride to corresponding chlorides, which on reaction with hydroxylamine change to sulfonyl-hydroxamic acids. These acids react in an alkaline medium with acetaldehyde with the formation of acetohydroxamic acids and sulfinic acids. Both products give colors with ferric chloride: the first red, the second red-orange.

$$R-SO_3H \xrightarrow[(-SO_2, -HCl)]{SOCl_2} R-SO_2Cl \xrightarrow[(-HCl)]{NH_2OH} R-SO_2-NHOH \xrightarrow{CH_3CHO}$$
$$\rightarrow CH_3-CO-NHOH + R-SO_2H$$

Reagents: thionyl chloride; alcoholic solution of hydroxylamine hydrochloride; 1−3% $FeCl_3$ solution; acetaldehyde; 5% NaOH; alcoholic HCl.

Procedure: A few drops of thionyl chloride are added to several milligrams of a sulfonic acid or its alkali salt placed in a small dish, and the mixture is evaporated to dryness on a water bath. After cooling, two drops of hydroxylamine solution are added, followed by a drop of acetaldehyde and several drops of NaOH solution until the solution is alkaline. After several minutes the mixture is acidified with alcoholic HCl and mixed with two drops of $FeCl_3$: if sulfonic acid was present in the sample, a red to violet-red color is obtained.

The detection of the sulfonic group can also be carried out by heating the tested substance with succinic acid at 200 °C, during which SO_2 escapes, which can be detected with iron (III) ferricyanide (26).

Among derivatives of sulfonic acid, the following are suitable for identification: heavy-metal salts, chlorides, and crystalline sulfonamides (formed from chlorides and organic bases). To identify anthraquinonesulfonic acids, use can be made of the exchange reaction in which the sulfo group is substituted by chlorine. Paper chromatography represents an important method for the identification of sulfonic acids.

Sulfonic Acid Salts

Metal salts are usually employed for special cases of identification. Cobalt and zinc salts have been proposed (27, 28), for example, for the identification of naphthalenesulfonic acids. Magnesium and barium salts are also utilized (29).

Among many salts of sulfonic acids with amines, the derivatives with p-toluidine, phenylhydrazine, and, principally, benzylisothiourea are most often used.

p-Toluidine salts (30−33) are prepared by mixing equal amounts (by weight) of the sodium salt of the sulfonic acid and p-toluidine hydrochloride in hot water. The amount of water is chosen so that the salt formed is dissolved while the solution is hot. It is advisable to add 2−3 drops of conc. HCl to the solution. After cooling, the precipitated salt is filtered and crystallized from water, alcohol, or a mixture thereof. The salt can often be prepared from 10−30 mg of sample and the method was even applied for the quantitative determination of sulfonic acids (34).

Among other bases, phenylhydrazine (35), guanidine derivatives (36), and pyridine (37) have also been employed.

It is advantageous to use S-benzylthiuronium chloride for the preparation of sulfonic acid salts, because this type of salt was prepared from numerous sulfonic acids (38−45).

General Method of Preparation of S-Benzylthiuronium Salts of Sulfonic Acids

Sodium sulfonate (2−3 mmole) is dissolved in a minimum amount of water [in case only free sulfonic acid is available, it is dissolved in 1.5 to 2 ml of 10% NaOH and neutralized (with respect to phenolphthalein) with HCl] and mixed with a solution of 250 mg of S-benzylthiuronium chloride in the required amount of water (in the case of disulfonic acid 500 mg of the reagent are used). The solutions are mixed in the cold by adding the sodium sulfonate solution dropwise to the reagent solution with continuous shaking. The separated salt is washed with water and crystallized by dissolving it in a minimum amount of boiling ethanol and by adding water dropwise to incipient turbidity.

In case an oily precipitate is formed on mixing the two solutions which will not crystallize even after deep cooling or scratching with a glass rod, the two solutions should be cooled to 0 °C before mixing and the sulfonic acid solution added dropwise and with continuous stirring to the reagent solution. Instead of an aqueous solution of S-benzylthiuronium chloride, a 15% ethanolic solution may be used (46). If neither of the given procedures leads to a solid derivative, then either the sulfonic acid sample is a mixture (paper chromatography may decide this question), or S-1-naphthylmethylthiuronium chloride should be used instead as reagent (47).

S-Benzylthiuronium Salt of 2-Naphthalenesulfonic Acid

Reagents: S-benzylthiuronium chloride, 2 N NaOH, 70% ethanol.

Procedure: Naphthalenesulfonic acid (100 mg) is dissolved in 2 ml of water and neutralized (with respect to phenolphthalein) with 2 N NaOH. S-Benzylthiuronium chloride (0.15 g) is dissolved in the required amount of hot ethanol, and this solution is added dropwise with continuous shaking to the sulfonic acid salt solution. After 10 min standing in the refrigerator the salt is filtered off under suction and washed with 1 ml of water. Yield, 0.23 g; mp, 187−192 °C; after crystallization from 3 ml of 70% ethanol, 0.18 g, mp 193 °C.

S-Benzylthiuronium Salt of Anthraquinone-2-sulfonic Acid

Reagents: S-benzylthiuronium chloride, 30% aqueous ethanol.

Procedure: Potassium anthraquinonesulfonate (0.3 g) is heated under continuous shaking with 5 ml of water to complete dissolution, and 0.3 g of S-benzylthiuronium chloride dissolved in 5 ml of water is added in several portions. The mixture is allowed to stand in a refrigerator for 1 hr and the product is filtered off under suction and washed with 5 ml of water. Yield, 0.4 g; mp, 208−210 °C. Crystallization of this product from 3 ml of 30% aqueous ethanol gives 0.3 g of product, mp 208−210 °C.

S-1-Naphthylmethylthiuronium Salt of *p*-Toluenesulfonic Acid

Reagent: S-1-naphthylmethylthiuronium chloride.

Procedure: 100 mg of the sodium salt of *p*-toluenesulfonic acid are dissolved in 1 ml of water and the solution is mixed with 0.2 g of S-1-naphthylmethylthiuronium chloride dissolved in 3.5 ml of hot water. The mixture is boiled while adding water until complete dissolution. After cooling in a refrigerator the separated salt is filtered off and washed twice with 1 ml of water. Yield, 0.25 g; mp, 156−157 °C; after crystallization from 5 ml of water the product weighed 80 mg, mp 162 °C.

Preparation of Sulfonic Acid Chlorides
and Their Conversion to Amides

This method is utilized only when none of the other procedures described leads to the preparation of a suitable sulfonic acid derivative. The prerequisite for this method is the availability of $1-2$ g of a dry sulfonic acid salt and the absence of other functional groups (except sulfonic) in the molecule which might react with PCl_5.

$$ArSO_3Na \xrightarrow[\text{(−POCl}_3)]{PCl_5} RSO_2Cl \xrightarrow[\text{(−HCl)}]{ArNH_2} RSO_2NHAr$$

General Method of Preparation of Sulfonyl Chlorides

Dry sulfonic acid (2 g) is rapidly triturated with $5-6$ g of PCl_5 and the mixture is transferred into a 20-ml flask where it is heated with occasional stirring for $4-6$ hr. This is carried out under reflux in an oil bath heated first to 110 °C, and then, after 2 hr, to 150 °C. After cooling, the reaction mixture is diluted with 10 ml of benzene, heated to incipient boiling, and after partial cooling, filtered through a small folded filter paper into a separatory funnel (50 ml volume). The residue in the flask is extracted once more with 10 ml of boiling benzene and the partly cooled benzene solution is filtered into the same separatory funnel. The combined benzene extracts are extracted twice with 10 ml of water, filtered, and worked up either directly to an N-substituted sulfonamide (procedure a), or the sulfonyl chloride might be isolated and purified by crystallization before it is used for identification (procedure b). It can also be further converted to sulfonamide (procedure c).

(a) Aniline (0.5 ml) is added to the benzene solution (see above) and the mixture is refluxed. After partial cooling the solution is extracted twice with 5 ml of 5% HCl and twice with water. The sulfonamide is isolated by evaporating benzene to dryness and dissolving the residue in 5 ml of ethanol, filtering with charcoal, and diluting the filtrate with water until a turbidity is formed. The turbid solution is then heated again and allowed to crystallize.

(b) In certain cases acid sulfonyl chlorides can be used for indentification (anthraquinonesulfonyl chlorides). After the evaporation of benzene (under reduced pressure, if necessary) the residue is dissolved in boiling light petroleum, or chloroform, or carbon tetrachloride.

(c) To the crude sulfonic acid chloride 10 ml of conc. aqueous ammonia are added followed by 2 g of powdered ammonium carbonate. After 5 min stirring with a rod the mixture is heated for 15 min in a water bath (80 °C). The hot solution is filtered (charcoal) and the sulfonamide separated on cooling is crystallized from hot water.

Anthraquinone-2-sulfonic Acid Chloride

Reagents: $POCl_3$, PCl_5, ether, ethanol, benzene.

Procedure: In a small dry flask provided with a reflux condenser (ground-glass joint) 1 g of potassium anthraquinone-2-sulfonate (well dried),

1.5 g of $POCl_3$, and 0.8 g of PCl_5 are gently refluxed for 3 hr. The reflux condenser is fitted with a calcium chloride tube. After cooling, the contents of the flask are added dropwise carefully and under continuous stirring into a mixture of 15 g of ice and 15 ml of water. The separated sulfonyl chloride is filtered off under suction on a filtration crucible and washed twice with 1 ml of ether. Yield, 0.41 g; mp, 170–171 °C. Crystallization from 16 ml of benzene yielded 0.14 g of product, mp, 197–198 °C.

Anilide of Anthraquinone-2-sulfonic Acid

Reagents: aniline, HCl 1 : 1, nitrobenzene, ether, ethanol.

Procedure: Into a 10-ml flask (ground-glass joint) provided with a reflux condenser 100 mg of anthraquinone-2-sulfonyl chloride and 0.5 ml of aniline are placed, and the mixture is refluxed on a boiling water bath for 1 hr. After cooling, 10 ml of hydrochloric acid 1 : 1 are added to the mixture, and the supension formed is thoroughly stirred. The anilide formed is filtered off under suction using a filtration crucible and washed twice with 5 ml of HCl 1 : 1, twice with 3 ml of water, 2 ml of ethanol, and 2 ml of ether. Yield, 65 mg; mp, 197–199 °C. After crystallization from 0.5 ml of nitrobenzene the product weighed 50 mg, mp 198–199 °C.

1-Chloroanthraquinone from Anthraquinone-1-sulfonic Acid

Reagents: conc. HCl, 10% aqueous $NaClO_3$ solution, ethanol.

Procedure: Into a 25-ml flask with a reflux condenser 0.5 g of potassium anthraquinone-1-sulfonate, 6 ml of water, and 2 ml of conc. HCl are placed. The mixture is boiled under reflux for 2 hr while adding 8 ml of chlorate solution dropwise through the condenser. The boiling is continued for an additional hour, and, after cooling, the mixture is filtered and the derivative on the filter is washed three times with 1 ml of warm water. Yield, 0.38 g; mp, 154–161 °C. Crystallization of this product from 30 ml of ethanol gave 0.20 g of product, mp 162 °C.

Paper Chromatography and Paper Electrophoresis of Sulfonic Acids

As sulfonic acids are strong acids, they present no difficulties during chromatography on paper, even when neutral solvents are used. Hence, they can be successfully chromatographed in basic (*n*-propanol-ammonia 2 : 1; *n*-butanol saturated with ammonia), acid (*n*-butanol-acetic acid-water 4 : 1 : 5), and neutral solvent systems (tert-butanol-ethanol-water 4 : 3 : 3).

In the systems mentioned the behavior of sulfonic acids is determined primarily by the number of sulfonic groups: as their number increases, the R_F values decrease. If the molecule of sulfonic acid contains additional OH or NH_2 groups, then in addition to the number of sulfonic groups, the mutual positions of these latter groups relative to the sulfonic groups has the main influence on the R_F value. Moreover, the alkalinity of the solvent system can be easily regulated by using papers impregnated with sodium hydrogen carbonate and n-propanol additioned by 5% $NaHCO_3$ as solvent (to 100 ml of n-propanol 50 ml of 5% aqueous $NaHCO_3$ are added with stirring; the separated part of $NaHCO_3$ is filtered off after 15 min); in this case the acidity of acidic hydroxy groups — for example, in naphtholsulfonic acids — becomes evident, i.e., they dissociate, which causes a diminution of their R_F values, in contrast to corresponding amino acids.

Generally, sulfonic acids (as well as alkyl sulfates) can be detected on chromatograms by spraying with a 0.05% aqueous pinacryptol yellow solution (48). When observed under UV light sulfonic acids appear as strongly yellow to orange fluorescing spots on a blue-green background. The sensitivity of this detection is in the range of micrograms and even lower.

Another, less-sensitive and unspecific method of detection consists in heating the chromatogram at 200 °C in a drying oven. On the areas containing sulfonic acids (but also Na^+) the paper carbonizes and black spots appear on a brown background. This detection is less sensitive; it requires several tens of micrograms of the substance. In addition to these detection methods, some acids can be revealed on the basis of their fluorescence (anthraquinonesulfonic acids, naphthol- and naphthylaminesulfonic acids) or by reactions specific for other functional groups present in the molecule, such as, for example, diazotization and coupling for the amino group, coupling with diazonium salts for substances with phenolic hydroxyls. More information can be found in the original literature (49 — 51).

For the separation of aliphatic and aromatic acids, especially from substances of nonionic character, paper electrophoresis can be applied successfully. With a potential drop of 10 V/cm and with 2 N ammonia or 1 N acetic acid as a buffer, the separation can be carried out in 2 hr. Mono-, di-, and polysulfonated derivatives are also separated, but not isomeric sulfonic acids. Detection is carried out as in paper chromatography.

5. Sulfonamides

Sulfonamides are weak acids and their lower-molecular members are soluble in 5% NaOH. Sulfonamides soluble in 5% NaOH can be easily hydrolyzed,

$$ArSO_2NHR \xrightarrow[H_2O]{HCl} RNH_2 + ArSO_3H + HCl$$

Sulfonamides are also hydrolyzed in 25% HCl at 160−200 °C (in a sealed ampoule) or by 30 min boiling with 80% H_2SO_4 or a mixture of 80% sulfuric acid and 80% phosphoric acid. The use of sulfuric acid may cause a partial degradation of the products of hydrolysis (especially higher-molecular amines). Concentrated hydrobromic acid (48%) in a mixture with phenol has been recommended (52) as an efficient and fast reagent for the hydrolysis of sulfonamides.

Although hydrolysis with hydrochloric acid requires a sealed tube, it is a safe method, because the hydrolysis products are obtained in a pure state, and can be isolated simply by evaporating the excess HCl on a water bath.

The products of cleavage are isolated according to their character. If the amine is volatile, it is steam distilled and identified in the distillate by a method similar to that described on p. 326, or, if not volatile, it is extracted with ether. Nonvolatile, ether-insoluble amines are isolated in the form of insoluble salts (picrates), or, if their salts are soluble (ethanolamines), ion exchangers are used for the liberation of bases (53).

General Method of Hydrolysis of Sulfonamides

A mixture (2 ml) of equal parts of 80% H_2SO_4 and 85% H_3PO_4 is heated in a test tube with 4−5 mmole of sulfonamide at a temperature of 150−160 °C by immersing the tube into an oil bath while stirring with a rod-like thermometer. As soon as the sulfonamide is dissolved (5−10 min) the reaction mixture is cooled and diluted carefully with 10 ml of water and then neutralized with cooling, and eventually slightly alkalized by adding 25% NaOH.

The alkaline solution contains (after the separation of the amine) the sodium salt of the sulfonic acid. It is neutralized, filtered with charcoal, and concentrated to 5−6 ml. This concentrate is used for the preparation of a derivative of the sulfonic acid (for the preparation of S-benzylthiuronium salt see p. 395).

Other Methods of Identification of Sulfonamides

To identify sulfonamides characterized by the presence of a $-SO_2NH_2$ group, the reaction with phthaloyl chloride and xanthydrol has been proposed (54). Thirteen N-xanthylsulfonamides have already been described. The preparation of N-xanthylsulfonamides is based on mixing 2 mmole of sulfonamide with 10 ml of a filtered solution of 200 mg of xanthydrol in acetic acid. The product is isolated after 30−60 min standing (occasional shaking) and crystallized from a dioxane-water mixture (3 : 1).

Cleavage of *p*-Toluenesulfanilide with Hydrobromic Acid
and Phenol

(Conversion of Aniline to Acetanilide)

Reagents: phenol, freshly distilled 58% HBr, 30% NaOH, acetic anhydride, potassium carbonate.

Procedure: In a 10-ml ground-joint flask a mixture of 0.1 g of *p*-toluene-
sulfanilide, 0.25 g of phenol, and 2 ml of hydrobromic acid is refluxed for
1 hr. After cooling, the reaction mixture is extracted in a separatory funnel
three times with 5 ml of ether and the acid solution is alkalized with 30%
NaOH (test with indicator paper). After cooling, it is extracted again three
times with 10 ml of ether and the combined extracts are washed with 5 ml of
30% NaOH and twice with 5 ml of water. The ethereal solution is dried over
potassium carbonate (2 g potassium carbonate are added to the solution and
the mixture is allowed to stand for 1 − 2 hr with occasional shaking), filtered
through a folded filter paper into a distillation flask, and mixed there with
0.15 ml of acetic anhydride. After 20 min standing the ether is distilled off.
Boiling water is added to the residue until dissolved (approx. 2.5 ml) and the
boiling solution is filtered with charcoal through a folded filter paper. The
separated derivative (sometimes prolonged standing in a refrigerator and
scratching is necessary) is filtered off under suction and washed with 1 ml
of water. Yield, 33 mg; mp, 112 − 114 °C.

Cleavage of Sulfapyridine with 25% HCl

(Identification of the Liberated 2-Aminopyridine in the Form of *p*-Toluene-
sulfonamide)

Reagents: 25% HCl, 10% NaOH, *p*-toluenesulfonyl chloride, benzene,
pyridine, 50% ethanol.

Procedure: Sulfapyridine (0.3 g) is refluxed in a 10-ml ground-glass-
joint flask with 4 ml of 25% HCl for 24 hr. The contents of the flask are
poured into an evaporation dish and evaporated on a water bath to dryness.
NaOH solution (10%; 5 ml) is added to the residue and the mixture is
extracted three times with 5 ml of benzene in a separatory funnel. The benzene extracts are dried thoroughly with solid sodium hydroxide and filtered
through cotton-wool into a 25-ml dry flask. *p*-Toluenesulfonyl chloride
(0.1 g) is added to the solution, followed by 1 ml pyridine, and the mixture is
refluxed for 30 min. Five milliliters of benzene are then distilled off and the
residue in the flask is filtered to get rid of the hydrochloride formed. After
evaporation of benzene from the filtrate, sulfonamide is obtained, which is
recrystallized from 6 ml of 50% ethanol. Yield, 22 mg; mp, 214 − 216 °C.

After a second crystallization from 2.2 ml of 50% ethanol the product weighed 18 mg, mp 216 °C.

Xanthyl Derivative of p-Toluenesulfonamide

Reagents: xanthydrol, acetic acid, aqueous dioxane (one part of water plus three parts of dioxane).

Procedure: Xanthydrol (0.25 g) is disssolved in a test tube in 10 ml of acetic acid, and 0.25 g of o-toluenesulfonamide is added to the solution. The mixture is allowed to stand at room temperature for 90 min and the separated derivative is filtered off on a filtration crucible and dried. Yield, 0.3 g; mp, 202 – 203 °C.

Paper and Thin-Layer Chromatography

Among the members of this class of substances, the derivatives of p-amino-benzenesulfonamide, used as drugs, have mainly been analyzed by paper chromatography. Available systems can be applied for their separation, such as, for example, n-butanol saturated with ammonia or n-butanol – acetic acid – water (4 : 1 : 5). However, these solvent systems sometimes do not give constant R_F values, and the shape of their spots is also pH-dependent, which is caused by the easy variation of the ammonia or acetic acid concentration in the developing solvent or in the atmosphere of the chromatographic chamber. It is therefore necessary in many instances to employ water-saturated n-butanol and papers impregnated with suitable buffers. At lower pH values the ionization of the amino group on the benzene ring can be achieved, while at higher pH values the sulfonamide group becomes ionized. This makes it possible to decrease the R_F values as necessary or also to make use of this effect for the separation of certain pairs. The use of solvent systems with formamide as the stationary phase is more suitable (55).

Thin-layer chromatography of sulfonamides can be carried out on silica gel G or on a nonadhering (loose) layer of silica gel. In the first case chloroform – methanol (80 : 15), cyclohexanc – aeetone – acetic acid (4 : 5 : 1), or acetone – methanol – diethylamine (9 : 1 : 1) mixtures are used (56); in the second case chloroform – methanol (95 : 5) or chloroform – methanol – acetic acid (94 : 5 : 1 or 90 : 5 : 5) are preferable (57).

Detection of p-aminobenzenesulfonamide derivatives is similar to that of aromatic amines, either with p-dimethylaminobenzaldehyde (p. 349), or by diazotization with vapors of nitrogen oxides and coupling (p. 349). Roux' reagent is also suitable: 10 g of sodium nitroprusside are dissolved in 100 ml of water, and the solution is mixed with 2 ml of 33% NaOH and 5 ml of 0.1 N $KMnO_4$, and is filtered; after spraying the paper, spots of various colors appear, which are observed in visible or ultraviolet light.

6. Sulfinic Acids

Sulfinic acids are generally unstable compounds (especially aliphatic ones), and they can be identified either as phenylhydrazine salts (58), or, better, by alkylation of their sodium salts with methyl iodide, during which sulfones are formed (59).

7. Isothiocyanates

The reactivity of the −NCS group is used for the identification of these compounds. Addition reactions take place in the cold (60, 61).

General Method of Preparation of Substituted Thioureas

Benzylamine (0.2 ml) dissolved in 3 ml of ethanol is mixed with 1.5 − 2 mmole of isothiocyanate and the mixture is gently boiled for 3 − 5 min. The product separated after cooling (or after addition of water) is isolated by filtration, and is crystallized from aqueous alcohol. If an oily product separates, the mixture should be undercooled and scratched intensely with a glass rode, or decanted, and a few drops of ethanol added to the remaining oil. If benzylamine is not available, or if the prepared derivative does not possess suitable properties, 1-naphthylamine or piperidine should be used.

8. S-Heterocycles

S-Heterocycles with a saturated carbon ring have similar properties to aliphatic sulfides, and for their identification the reactions described on p. 388 may be used (oxidation to sulfones, preparation of sulfilimines).

S-Heterocycles in which sulfur is bound in an aromatic system are oxidized with difficulty, or a deeper oxidative degradation may take place during which sulfur may be oxidized to sulfuric acid. For more information the original literature should be consulted (62, 63).

For the identification of sulfur heterocycles the measurement of mass spectra is employed (64, 65), and the quoted literature also contains information on the measurement of infrared spectra.

To identify thiophene and some of its substituted homologs, it is useful to apply the reaction of mercuric chloride. The general method is demonstrated by the example of the preparation of thiophene.

For the gas chromatography of S-heterocycles see (66).

Mercuration of Thiophene

Reagents: saturated aqueous mercuric chloride solution, 30% aqueous sodium acetate solution, ethanol.

Procedure: Thiophene (0.8 g), 80 ml of saturated mercuric chloride solution, 15 ml of 30% aqueous sodium acetate solution, and 10 ml of ethanol are mixed and allowed to stand at room temperature for 48 hr. The separated product is filtered off under suction and dissolved in 75 ml of boiling ethanol. The hot solution is filtered to eliminate the insoluble bis-derivative. The product which separates from the filtrate after cooling is filtered using a filtration crucible and dried. Yield, 0.4 g; mp, 183 – 184 °C.

Color Reactions of Thiophene

Reaction with Isatin

Thiophene and its derivatives, unless substituted at both α-positions, give a positive reaction with isatin in conc. sulfuric acid. Where both positions are free a blue color is formed; if one α-position is substituted, but the adjacent β-position is free, the color is green or violet. The structure of the colored substances formed is probably similar to that of the products described in the literature (67, 68).

This popular reaction is used for the detection of thiophene and its derivatives in aromatic hydrocarbons, in pyridine, etc., and also for colorimetry (69). However, it should be kept in mind that other sulfur compounds and five-membered heterocycles also give this reaction.

Test for Thiophene in Benzene

Reagents: 0.2 – 0.5% solution of isatin in conc. sulfuric acid.

Procedure: 1 ml of the reagent is added to 1 ml of the tested benzene; the mixture is shaken and observed. If the benzene contains thiophene, a blue color appears; if its concentration is very low, the color is green-blue.

Similar to isatin, other compounds containing a − CO − CO − grouping also react with thiophene, such as, for example, benzil, phenylglyoxylic acid, alloxan (70), and phenanthrenequinone (71, 72).

Among other color reactions of thiophene, the following may be mentioned: Liebermann reaction with nitrite and sulfuric acid (73); the very sensitive reaction with thaline (tetrahydro-*p*-hydroxyquinoline methyl ether) and dilute nitric acid (74); reaction with sulfuric acid, copper (II) acetate, and lactic acid (75); and, especially, the reaction with ceric ammonium nitrate (76). Of course, this reaction is not specific (see also p. 170), but it allows the differentiation of isomeric thiophene derivatives.

References

1. Böhme, H., and Stachel, H.-D.: Zeit. anal. Chem. **154**, 27 (1957).
2. Sporek, K. F., and Danyi, M. D.: Anal. Chem. **35**, 956 (1963).
3. Grote, J.: J. Biol. Chem. **93**, 25 (1931).
4. Havíř, J., Fidler, A., and Husak, R.: Acta Chim. Acad. Sci. Hung. **50**, 39 (1966).

5. Shinohara, K.: J. Biol. Chem. **109**, 665 (1933); **110**, 263 (1934); **112**, 671 (1936).
6. Schöberl, A., and Ludwig, E.: Ber. **70**, 1422 (1937).
7. Bost, R. W., Turner, J. O., and Norton, R. D.: J. Am. Chem. Soc. **54**, 1985 (1932).
8. Bost, R. W., Turner, J. O., and Conn, M. W.: J. Am. Chem. Soc. **55**, 4956 (1933).
9. Folkard, A. R., and Joyce, A. E.: J. Sci. Food Agric. **14**, 510 (1963).
10. Meyerson. S., Drews, H., and Fields, E. K.: Anal. Chem. **36**, 1294 (1964).
11. Jureček, M., Večeřa, M., and Gasparič, J.: Chem. Listy **47**, 1410 (1953); **48**, 542 (1954).
12. Gasparič, J., Večeřa, M., and Jureček, M.: Collection Czech. Chem. Commun. **23**, 97 (1958).
13. Veibel, S., and Nielsen, B. J.: Acta Chem. Scand. **10**, 1488 (1956).
14. Večeřa, M., Gasparič, J., Šnobl, D., and Jureček, M.: Collection Czech. Chem. Commun. **21**, 1284 (1956).
15. Večeřa, M., Gasparič, J., and Jureček, M.: Collection Czech. Chem. Commun. **24**, 640 (1959).
16. Večeřa, M., and Petránek, J.: Collection Czech. Chem. Commun. **21**, 912 (1956).
17. Petránek, J., Večeřa, M., and Jureček, M.: Collection Czech. Chem. Commun. **24**, 3637 (1959).
18. Petránek, J., and Večeřa, M.: Collection Czech. Chem. Commun. **24**, 2191 (1959).
19. Petránek, J., and Večeřa, M.: Collection Czech. Chem. Commun. **24**, 718 (1959).
20. Petránek, J.: J. Chromatog. **5**, 254 (1961).
21. Alikhanov, P. P., and Mashkova, L, Ya.: Gazov. Khromatogr. No. **3**, 156 (1965).
22. Klouček, B., Jehlička, V., and Gasparič, J.: Chem. Průmysl **10**, 624 (1960).
23. Feigl, F., and Lenzer, A.: Mikrochim. Acta **1**, 127 (1937).
24. Borecký, J.: Mikrochim. Acta **1962**, 1137.
25. Feigl, F., and Anger, V.: Mikrochemie **15**, 23 (1934).
26. Feigl, F., Hagenauer-Castro, D., and Jungreis, E.: Talanta **1**, 80 (1958).
27. Lynch, D. F. J.: J. Ind. Eng. Chem. **14**, 964 (1922).
28. Haller, H. L., and Lynch, D. F. J.: J. Ind. Eng. Chem. **16**, 273 (1924).
29. Mužík, F.: private communication.
30. Fieser, L. F.: Org. Syntheses Coll. Vol. 2, 482 (1936).
31. Barton, A. D., and Young, L.: J. Am. Chem. Soc. **65**, 294 (1943).
32. Fieser, L. F.: J. Am. Chem. Soc. **51**, 2463 (1929).
33. Dermer, W. H., and Dermer, O. C.: J. Org. Chem. **7**, 581 (1942).
34. Marron, T. U., and Schifferli, J.: Ind. Eng. Chem., Anal. Ed. **18**, 49 (1946).
35. Stempel, G. H., and Schaffel, G. S.: J. Am. Chem. Soc. **64**, 470 (1942).
36. Perrot, R., and Barghon, A.: Bull. Soc. Chim. France **1951**, 916.
37. Huntress, E. H., and Foote, G. L.: J. Am. Chem. Soc. **64**, 1017 (1942).
38. Chambers, R. F., and Scherer, P. C.: J. Ind. Eng. Chem. **16**, 1272 (1924).
39. Chambers, E., and Watt, G. W.: J. Org. Chem. **6**, 376 (1941).
40. Campaigne, E., and Suter, C. M.: J. Am. Chem. Soc. **64**, 3040 (1942).
41. Donleavy, J. J.: J. Am. Chem. Soc. **58**, 1004 (1936).
42. Hann, R. M.: J. Am. Chem. Soc. **57**, 2166 (1935).
43. Veibel, S.: J. Am. Chem. Soc. **67**, 1867 (1945).
44. Veibel, S., and Lillelund, H.: Bull. Soc. Chim. France (5) **5**, 1153 (1939).
45. Garner, W.: J. Soc. Dyers and Col. **52**, 302 (1936).
46. Večeřa, M., and Borecký, J.: Chem. Listy **51**, 974 (1957).
47. Bonner, W. A.: J. Am. Chem. Soc. **70**, 3508 (1948).

48. Borecký, J.: J. Chromatog. **2**, 612 (1959).
49. Latinák, J.: Collection Czech. Chem. Commun. **25**, 1649 (1960).
50. Spencer, G., and Nield, V.: Chem. and Ind. **1956**, 922.
51. Večeřa, M., Gasparič, J., and Borecký, J.: Chem. Listy **49**, 706 (1955); Collection Czech. Chem. Commun. **20**, 1380 (1955).
52. Snyder, H. R., and Heckert, R. E.: J. Am. Chem. Soc. **74**, 2006 (1952).
53. Večeřa, M., and Friedrich, K.: Chem. Listy **51**, 283 (1957).
54. Phillips, R. F., and Frank, V. S.: J. Org. Chem. **9**, 9 (1944).
55. Macek, K.: Českosl. Farm. **12**, 365 (1963).
56. Gänshirt, H., in: Stahl, E. (Editor): Dünnschicht-Chromatographie, p. 315. Springer Verlag, Berlin, 1962.
57. Gajdoš, M.: Československ. farm. **14**, 70 (1965).
58. Hälssig, A.: J. prakt. Chem. (2), **56**, 217 (1897).
59. Allen, P., Jr.: J. Org. Chem. **7**, 23 (1942).
60. Hofmann, A. W.: Ber. **1**, 27 (1869).
61. Billeter, O.: Ber. **26**, 1681 (1893).
62. Arndt, F., Nachtwey, P., Pusch, J.: Ber. **58**, 1633 (1925);
63. Arndt, F., and Bekir, N.: Ber. **63**, 2393 (1930).
64. Catalog of Mass Spectral Data by American Petroleum Institute, Research Project 44 at the Nat. Bur. Stand. Washington.
65. Hartough, H. D.: Thiophens. Interscience Publ., New York, 1952.
66. Berk, S.: J. Gas Chromatog. **4**, 386 (1966).
67. Steinkopf, W., and Roch, J.: Ann. **482**, 251 (1930).
68. Steinkopf, W., and Hempel, H.: Ann. **495**, 144 (1932).
69. Mckee, H. C., Herndon, L. K., and Withrow, J. R.: Anal. Chem. **20**, 301 (1948).
70. Ekkert, L.: Pharm. Zentralhalle **71**, 625 (1930).
71. Laubenheimer, A.: Ber. **8**, 224 (1875).
72. Odernheimer, E.: Ber. **17**, 1338 (1884).
73. Liebermann, C.: Ber. **20**, 3231 (1887).
74. Kreis, H.: Chem. Ztg. **26**, 523 (1902).
75. Christomanos, A. A.: Biochem. Z. **229**, 248 (1930).
76. Hartough, H. D.: Anal. Chem. **20**, 860 (1948); **23**, 1128 (1951).

CHAPTER XVIII

CARBONIC ACID DERIVATIVES

A series of important compounds can be obtained by the substitution of the oxygen atom or of the OH groups in carbonic acid $CO(OH)_2$ by groups containing sulfur, halogens, or nitrogen.

It is beyond the scope of this monograph to give an exhaustive review of all such compounds and to describe methods of their detection and identification, because in many instances the latter are not even elaborated. A general method of identification consists in their hydrolytic cleavage and the subsequent identification of the cleavage products. In some cases the methods described in other chapters can be employed.

We can give a review of recommended methods of detection and identification of the most important compounds of this class of substances, and demonstrate by practical examples a few typical procedures for their identification. If the substances of this class are poisonous, the reader should refer to specialized monographs on the detection and identification of such compounds (1).

1. Phosgene

A gas containing phosgene is introduced into a saturated aqueous solution of aniline (2):

$$COCl_2 + 2\,C_6H_5NH_2 \;\rightarrow\; CO \Big\langle {}^{NHC_6H_5}_{NHC_6H_5}$$

2. Cyanates

To carry out the detection of cyanates, the formation of a blue-violet precipitate formed with a solution of curpic salt in pyridine (3) or the specific reaction with semicarbazide hydrochloride giving rise to hydrazodicarbonamide $NH_2CONHNHCONH_2$ (4) may be employed.

On alkaline hydrolysis alkylisocyanates yield the corresponding amines,

$$RNCO + H_2O \;\rightarrow\; CO_2 + RNH_2$$

3. Cyanamide

A shining yellow precipitate of silver salt is formed with ammoniacal silver nitrate. With p-dimethylaminobenzaldehyde (p. 322) a yellow color is formed, while an orange to red color is created on heating with 1,2-naphthoquinone-4-sulfonate (p. 320).

Chromatography of cyanamide and related compounds can be effected in n-butanol−ethanol−water (4 : 1 : 1); detection is with ammoniacal silver nitrate (p. 176) or with p-dimethylaminobenzaldehyde (p. 349) (5).

4. Urea and Its Derivatives

Urea gives a yellow color with p-dimethylaminobenzaldehyde (6), a positive biuret reaction (p. 271), a color reaction with sodium nitroprusside in alkaline media in the presence of ammonium persulfate (red color) (7), a color reaction with phenol and sodium hypochlorite (8), a very sensitive reaction with phenylhydrazine and aniline (9), and a reaction with xanthydrol (10).

Urea and its derivatives can be well chromatographed in a series of neutral, acid, and ammonical solvent systems: for example, n-propanol−water (2 : 1), n-propanol−ammonia (2 : 1), or n-butanol−ethanol−water (4 : 1 : 1). Detection is carried out with p-dimethylaminobenzaldehyde.

N-N-Diphenylureas give a blue color with concentrated sulfuric acid in the presence of traces of an oxidant (KNO_3), similar to that given by diphenylamine (11).

For the identification of urea the preparation of xanthyl derivative can be employed.

For the identification of mono- and polymeric N-derivatives of carbonic acid (including the study of hydrolyzates) by thin-layer chromatography see (12).

Dixanthylurea

Reagents: 10% solution of xanthydrol in ethanol; glacial acetic acid.

Procedure: Urea (50 mg) is dissolved in 5 ml of glacial acetic acid, and 5 ml of the 10% ethanolic xanthydrol solution are added to this solution. After 10 min standing the separated crystals are filtered off using a filtration tube. Yield, 0.3 g; mp, 298−300 °C.

For the identification of substituted ureas acid hydrolysis is employed, and the formed amine or amines are identified by their conversion to a suitable derivative or by paper chromatography.

Cleavage of Symmetrical Diethyldiphenylurea

Reagents: conc. HCl, 5 N NaOH.

Procedure: Symmetrical diethyldiphenylurea (0.2 g) is heated carefully in a sealed tube for 6 hr at 200 °C with 2 ml of HCl. After opening the tube its contents are transferred into a separatory funnel and the tube is rinsed with 5 ml of water. The combined solutions are alkalized with 5 N NaOH and extracted three times with 10 ml of ether. The ethereal extracts are dried with anhydrous potassium carbonate and filtered into a distillation flask. Ether is evaporated and the residual oil (ethylaniline) is treated with 100 mg of l-naphthylisothiocyanate, applying the procedure described on p. 339. Yield, 30 mg; mp, 125 – 126 °C.

Guanidine

Guanidine gives a positive biuret reaction (p. 271), the reaction with sodium pentacyanoammoferroate (p. 361) (13) similar to urea, the reaction with thymol and hypochlorite (14), and the reaction with sodium 1,2-naphthoquinone-4-sulfonate (p. 320) (15). It is also precipitated with Nessler's reagent (p. 211). Guanidine gives a weakly soluble picrate, which can be prepared by precipitation of a neutral aqueous guanidine solution with an aqueous 1% ammonium picrate solution.

Thiourea

Similar to urea, thiourea also gives a reaction with *p*-dimethylaminobenzaldehyde (p. 322), it is precipitated with xanthydrol (p. 272), and is reduced with ammoniacal silver nitrate (p. 210). On reaction with monochloroacetic acid it is changed to rhodanin, which can be detected with sodium 1,2-naphthoquinone-4-sulfonate (16).

Reagents: 10% solution of monochloroacetic acid, 0.5% aqueous sodium 1,2-naphthoquinone-4-sulfonate, 2 N NaOH,

Procedure: One drop of the solution of monochloroacetic acid is added to the sample, the mixture is evaporated to dryness, and the residue is heated in a glycerol bath at 180 °C. After 1 – 3 min heating the mixture is isolated and additioned with one drop of the reagent and one drop of alkali; a violet color is formed.

Thiourea can be identified by alkylation with methyl iodide or ethyl bromide, and by the preparation of S-alkylthiuronium picrate or 3,5-dinitrobenzoate (see p. 139).

Thiuronium Salts

Thiuronium salts are identified by alkaline cleavage and conversion of the formed mercaptan to a 2,4-dinitrophenyl derivative with 2,4-dinitrochlorobenzene (see p. 386).

Cleavage of the S-Benzylthiuronium Salt

Reagents: N NaOH; 10% methanolic 2,4-dinitrochlorobenzene solution; methanol.

Procedure: S-Benzylthiuronium chloride (0.5 g) is refluxed in a 10-ml flask with 5 ml of N NaOH for 30 min. Then 10 ml of the 10% methanolic 2,4-dinitrochlorobenzene solution and 10 ml of methanol are added to the mixture and boiling is continued for another 30 min. The mixture is then filtered while hot through a folded filter, the filtrate is allowed to stand in a refrigerator, and the separated derivative is filtered off under suction using a filtration tube. Yield, 0.17 g; the melting point was unsharp. After recrystallization from 6.5 ml of ethanol the product weighed 0.1 g, mp 127 °C; after a second crystallization from 5 ml of ethanol the yield was 60 mg, mp 130 °C.

Carbonic Acid Esters

These are analyzed by identifying the alcohol set free on hydrolysis, either by converting it to a derivative or by gas chromatography.

Urethanes

After acid hydrolysis, best carried out with hydrochloric acid in a sealed tube, the alcohol and the amine set free (see the hydrolysis of acid amides on p. 272) are identified, if possible, after their separation.

Semicarbazides

$NH_2CONHNH_2$; for the preparation of semicarbazones see p. 228.

Isourea Derivatives

NH_2-C-OR; basic substances identified as picrates (p. 328).
$\|$
NH

Allophanic Acid Derivatives

$NH_2-C-NHCOOR$; they are identified after previous hydrolysis by
$\|$
O
identifying the alcoholic component; identification of O-alkyl is carried out by paper chromatography after the cleavage with hydriodic acid (p. 202).

Cyanuric Acid

$C_3N_3(OH)_3$; identification by preparation of p-nitrobenzyl esters (17). For chromatographic identification see (18).

Cyanuric Acid Trichloride
(2,4,6-Trichloro-1,3,5-triazine)

$C_3N_3Cl_3$; esterification with alcoholate (19, 20).

Melamine

$C_3H_6N_6$; for the preparation of picrate or other salts see (21) and (22), respectively. For paper and thin layer chromatography of melamine see (18).

References

1. Jacobson, J. B.: The Analytical Chemistry of Industrial Poisons, Hazards and Solvents. Interscience Publishers, New York, 1949.
2. Kling, A., and Schmutz, R.: Compt. Rend. **168**, 773, 891 (1919).
3. Werner, E. A.: J. Chem. Soc. **123**, 2577 (1923).
4. Leboucq, J.: J. Pharm. Chim. **5**, 531 (1927).
5. Milks, J. E., and Janes. R. H.: Anal. Chem. **28**, 846 (1956).
6. Cline, R. E., and Fink, R. M.: Anal. Chem. **28**, 47 (1956).
7. Pittarelli, E.: Arch. farmacol. sperim. **45**, 174 (1928); C. **1928** II, 2387.
8. Kirby-Berry, H.: Univ. Texas Publ. **5109**, 88 (1951).
9. Sanchez, A.: Ann. Chim. Anal. Appl. **18**, 65 (1936); C. 1906, I, 4771.
10. Fosse, R.: Compt. Rend. **158**, 1076 (1914); **159**, 253 (1915).
11. Gizycki, F. V., and Reppel, L.: Zeit, anal. Chem. **144**, 109 (1955).
12. Knappe, E., and Rohdewald, I.: Zeit. anal. Chem. **223**, 174 (1966).
13. Fearon, W. R.: Analyst **71**, 562 (1946).
14. Fearon, W. R.: Scient. Proc. Roy. Dublin Soc. **22**, 415 (1941); C. A. **35**, 7319 (1941).
15. Sullivan, M. X., and Hess, W. C.: J. Am. Chem. Soc. **58**, 47 (1936).
16. Feigl, F., and Gentil, V.: Anal. Chem. **29**, 1715 (1957).
17. Lyons, E., and Reid, E. E.: J. Am. Chem. Soc. **39**, 1733 (1917).
18. Cee A., and Gasparič J.: J. Chromatog. **56**, 342 (1971).
19. Diels, O., and Liebermann, M.: Ber. **36**, 3191 (1903).
20. Hofmann, A. W.: Ber. **19**, 2061 (1886).
21. Korinfskij, A. A.: Zavod. Lab. **11**, 816 (1945).
22. Ostrogovick, A.: Gazzetta Chim. Italiana **65**, 566 (1935).

INDEX

416